Engineering Mechanics: Statics

Engineering Mechanics: Statics

Contributors

Ramon F. Alvarez-Estrada et al.

AURIS
Reference

www.aurisreference.com

Engineering Mechanics: Statics

Contributors: Ramon F. Alvarez-Estrada et al.

Published by Auris Reference Limited
www.aurisreference.com

United Kingdom

Engineering Mechanics: Statics

ISBN: 978-1-78154-817-2

British Library Cataloguing in Publication Data
A CIP record for this book is available from the British Library

Printed in the United Kingdom

Exclusively distributed by CBS Publishers & Distributors Pvt. Ltd.

Sales & Distribution Rights only for India, Pakistan, Bangladesh, Sri Lanka, Nepal and Bhutan. This book is not to be sold outside these territories.

Contents

List of Abbreviations

ACF	autocorrelation functions
AFM	atomic force microscopy
ATP	adenosine triphosphate
DHM	digital holographic microscopy
DIC	differential interference contrast
DNA	deoxyribonucleic acid
ECM	extracellular matrix
EGFP	enhanced green fluorescent protein
FA	focal adhesion
FAK	focal adhesion kinase
FCS	fluorescence correlation spectroscopy
FDT	fluctuation dissipation theorem
FPM	Fourier phase microscopy
FRET	Forster resonance energy transfer
GFP	green fluorescent protein
GTPase	guanosine triphosphatase
HOPG	highly ordered pyrolytic graphite
MEF	mouse embryonic fibroblast
MLCK	myosin light chain kinase
MRTF	myocardin-related transcription factor
MTC	magnetic twisting cytometry
PAA	polyacrylamide
PC	phase contrast
PRIM	proximity imaging microscopy
PriSSM	PRI-based strain sensor modulus
QPM	quantitative phase microscopy
RICM	reflection interference contrast microscopy
RNA	ribonucleic acid
SGM	soft glassy material
SPRIE	surface plasmon resonance imaging ellipsometry
SRF	serum response factor
STM	scanning tunneling microscopy
TFM	traction force microscopy
TIRF	total internal reflection fluorescence
WT	wavelet transform

List of Contributors

Ramon F. Alvarez-Estrada
Departamento de Fisica Teorica I, Facultad de Ciencias Fisicas, Universidad Complutense, 28040 Madrid, Spain

F. Argoul
Laboratoire de Physique, CNRS, Ecole Normale Supérieure de Lyon, Lyon, France

B. Audit
Laboratoire de Physique, CNRS, Ecole Normale Supérieure de Lyon, Lyon, France

A. Arneodo
Laboratoire de Physique, CNRS, Ecole Normale Supérieure de Lyon, Lyon, France

Nikolai A. Magnitskii
LLC "New Inflow", Moscow, Russia

Nijalingappa Umakantha
Department of Physics, Karnatak University, Dharwad, India

Miriam Flores
Centro de Investigación en Ingeniería y Ciencias Aplicadas, CIICAp, Universidad Autónoma del Estado de Morelos, Cuernavaca, México

Gustavo Urquiza
Centro de Investigación en Ingeniería y Ciencias Aplicadas, CIICAp, Universidad Autónoma del Estado de Morelos, Cuernavaca, México

José María Rodríguez
Centro Nacional de Investigación y Desarrollo Tecnológico, Cuernavaca, México

Guang-jiong Ni
Department of Physics, Portland State University, Portland, USA
Department of Physics, Fudan University, Shanghai, China

Suqing Chen
Department of Physics, Fudan University, Shanghai, China

Jianjun Xu
Department of Physics, Fudan University, Shanghai, China

Messaoud Toumi
Department of Applied Sciences, University of Quebec at Chicoutimi, Saguenay, Canada

Mohamed Bouazara
Department of Applied Sciences, University of Quebec at Chicoutimi, Saguenay, Canada

Marc J. Richard
Department of Mechanical Engineering, Laval University, Quebec, Canada

Charles A. Osheku
Centre for Space Transport and Propulsion, National Space Research and Development Agency, Federal Ministry of Science and Technology, FCT, Abuja, PMB 437, Nigeria

J. AWREJCEWICZ
Department of Automatics and Biomechanics, Technical University of Lodz, 1/15 Stefanowskiego Street, 90924 Lodz, Poland

V. A. KRYSKO
Department of Mathematics, Saratov State University, 410054 Saratov, Russia

I. V. KRAVTSOVA
Department of Mathematics, Saratov State University, 410054 Saratov, Russia

Preface

Engineering mechanics is the application of mechanics to solve problems involving common engineering elements. Statics and dynamics are the basic subjects in the general field known as engineering mechanics. Statics is a branch of mechanics which studies the effects and distribution of forces of rigid bodies which are and remain at rest. The text *Engineering Mechanics: Statics* provides a clear and thorough presentation of the theory and application of engineering mechanics. Classical and quantum models in non-equilibrium statistical mechanics have been presented in first chapter. In second chapter, we review methods and techniques that have been implemented to decipher the cascade of temporal events which enable a cell to sense a mechanical stimulus and to elaborate a response to adapt or to counteract this perturbation. The theory of elementary particles based on Newtonian mechanics has been discussed in third chapter. A new approach to classical statistical mechanics has been focused in fourth chapter. In fifth chapter, the estimation of crack initiation life of a hydraulic Francis turbine runner has been presented. A complete analytical model of a modular tank vehicle has been developed in sixth chapter. Seventh chapter deals with discrete symmetry in relativistic quantum mechanics. Eighth chapter discusses on mechanics of static slip and energy dissipation in sandwich structures. In last chapter, we propose the method for the investigation of stochastic vibrations of deterministic mechanical systems represented by axially symmetric spherical shells.

Chapter 1

CLASSICAL AND QUANTUM MODELS IN NON-EQUILIBRIUM STATISTICAL MECHANICS: MOMENT METHODS AND LONG-TIME APPROXIMATIONS

Ramon F. Alvarez-Estrada

Departamento de Fisica Teorica I, Facultad de Ciencias Fisicas, Universidad Complutense, 28040 Madrid, Spain

ABSTRACT

We consider non-equilibrium open statistical systems, subject to potentials and to external "heat baths" (hb) at thermal equilibrium at temperature T (either with ab initio dissipation or without it). Boltzmann's classical equilibrium distributions generate, as Gaussian weight functions in momenta, orthogonal polynomials in momenta (the position-independent Hermite polynomials H_n's). The moments of non-equilibrium classical distributions, implied by the H_n's, fulfill a hierarchy: for long times, the lowest moment dominates the evolution towards thermal equilibrium, either with dissipation or without it (but under certain approximation). We revisit that hierarchy, whose solution depends on operator continued fractions. We review our generalization of that moment method to classical closed many-particle interacting systems with neither a hb nor ab initio dissipation: with initial states describing thermal equilibrium at T at large distances but non-equilibrium at finite distances, the moment method yields, approximately, irreversible thermalization of the whole system at T, for long times. Generalizations to non-equilibrium quantum interacting systems meet additional difficulties. Three of them are: (i) equilibrium distributions (represented through Wigner functions) are neither Gaussian in momenta nor known in closed form; (ii) they may depend on dissipation; and (iii) the orthogonal polynomials in momenta generated by them depend also on positions. We generalize the moment method, dealing with (i), (ii) and (iii), to some non-equilibrium one-particle quantum interacting systems. Open problems are discussed briefly.

INTRODUCTION

Introduction A quite wide and very interesting set of references, from different standpoints, related to and/or oriented towards the foundations of non-equilibrium Statistical Mechanics can be seen in [1]. Non-equilibrium statistical systems of classical particles are described by Liouville classical distribution z [2–5]. For non-equilibrium statistical systems of quantum particles, several distribution functions are available, Wigner functions (W) [3–7] being a suitable and simple possibility in a global sense (in spite of the fact that their positivity is not warranted). Open statistical systems (involving, typically, one or a few degrees of freedom) are those subject to external "heat baths" (hb) or reservoirs at thermal equilibrium at temperature T, with finite or small dissipation due to the hb. We shall regard vanishing dissipation as a mathematical idealization of the case with small friction. Closed statistical systems are those not subject to any external hb, and they are formed by an enormously large number of interacting particles. General equations describing, in differential form, the time evolutions of non-equilibrium closed interacting classical or quantum systems (with some initial off-equilibrium state) are well known [2–5]. On the other hand, approximate time evolution differential equations for the effective or relevant degrees of freedom for non-equilibrium open or closed (classical or quantum) interacting systems are known in certain cases [2–4,6,7]. Generically, those evolution equations turn out to be increasingly difficult to solve (even on a computer), as the number of particles in the system grows and/or one tries to describe the state of the system for increasingly long times (starting from some initial non-equilibrium state). On the other hand, there does exist certain global understanding of various classes of solutions of those evolution equations, at least in a number of cases. Anyway, to achieve wider and better knowledge of how open or closed (classical or quantum) statistical interacting systems evolve in time continues to play a key role in Statistical Physics. In particular, an issue of crucial importance is how irreversibility could arise in the long-time evolution of closed classical or quantum many-particle systems [2–5,8], or stated in other words, how an arrow of time could set in.

Non-equilibrium open one-particle statistical classical systems, subject to potentials and to dissipation due to the hb, lend themselves typically to perform certain constructions and approximations in a framework which is simpler than those met for other cases: see [9–12] and references therein. We shall remind them [9–12] firstly, in order to motivate our developments. In them, classical time-independent equilibrium distributions ($W_{c,eq}$), which are independent on dissipation, have been used as weight functions to generate families of orthogonal polynomials in momenta. The latter has been possible due to two distinguishing and simplifying features of $W_{c,eq}$: dependences on classical momenta and those on spatial coordinates factorize, and $W_{c,eq}$ is Gaussian in classical momenta. Then, the orthogonal polynomials generated by $W_{c,eq}$ turned out to be just the standard Hermite polynomials [13] and, moreover, they are independent on spatial coordinates (and on dissipation). Those orthogonal polynomials yield (by integrating over classical momenta) moments $W_{c,n}$ of the non-equilibrium classical distributions W_c. The non-equilibrium $\tilde{W}_{c,n}$'s depend on spatial coordinates and time (t) and on dissipation (but not on momenta), and fulfill an infinite linear hierarchy. A key issue was that, for long times, the lowest moment appears to dominate the evolution towards thermal equilibrium with the hb (and, hence, to become independent on the initial non-equilibrium state and on dissipation) [9,10].

Interesting open issues in [9–12], among others, include: (a) further study of the process leading to moments which depend only on spatial coordinates; (b) properties of the solutions of the non-equilibrium classical hierarchies for different potentials (vanishing or not at large distances) and their behaviour for long times; (c) generalizations to non-equilibrium closed classical interacting many-particle systems without external hb's (with potentials vanishing at large distances), with initial states at thermal equilibrium at temperature T at large distances but off-equilibrium at finite distances and with the possibility of obtaining, in a suitable long-time approximation, irreversibility; (d) generalizations of (c) when classical electromagnetic degrees of freedom are included; (e) extensions to open one-particle quantum systems by means of Wigner functions (either with or without dissipation); (f) closed quantum interacting many-particle systems, which clearly lead to far more difficult problems. The above non-equilibrium moment method relies, very crucially, on the previous knowledge and properties of equilibrium distributions. Being natural to work with quantum Wigner functions (although not mandatory), we anticipate certain genuine quantum difficulties: the equilibrium quantum Wigner functions are neither Gaussian in momenta nor known in closed form, and dependences on momenta and those on spatial coordinates do not factorize, in general. The issue of the possible dependence of the quantum equilibrium distributions on dissipation leads to additional complications. The use of other quantum distribution functions does not appear to yield improvements in a global sense.

We shall review in outline items (a)–(c), analyzed in our previous work [14–19]. We shall study further convergence aspects for issues (b) and (c), which, even if shortly, will provide some additional partial clarification, so far unpublished. We shall present a simple overview of [14] on long-time approximations and irreversibility in closed interacting many-particle classical systems, including some improvements from [16] and [18,19]. Item (d) has been dealt with in [15] in detail and we shall not add more here. And, finally, we shall concentrate on (e). The genuine difficulties of quantum cases anticipated above already show up in one-dimensional models. Some generalizations of the moment method to non-equilibrium open one-particle quantum interacting systems, without or with dissipation and to lowest order in Planck's constant, have already been studied [18,20]. Here, we shall concentrate on constructing the equilibrium Wigner functions, the families of orthogonal polynomials generated by the latter and non-equilibrium moments and equations at low order in the hierarchies, to all orders in Planck's constant, in one-dimensional models. Models in which the equilibrium Wigner functions either depend on dissipation or are independent on it will be analyzed. Item (f) above (closed many-particle interacting quantum systems and irreversibility issues) will not be addressed in this work. This paper is organized as follows. Section 2 reviews open one-dimensional classical systems with or without dissipation. Section 3 treats classical closed interacting many-particle systems. Section 4 deals with general aspects of quantum-mechanical one-dimensional models, in the idealized limit of vanishing dissipation. Section 5 is devoted to open quantum-mechanical one-dimensional models with quartic plus quadratic potential, also without dissipation, through an approach different from that in Section 4. Section 6 treats open quantum-mechanical one-dimensional models with quadratic plus quartic potential with dissipation. Various technical aspects in Sections 4, 5 and 6 are treated in Appendices A–D. Finally, Section 7 presents some conclusions and various discussions. A short account of the contents of Sections 5 and 6 has been presented orally in the 11th International Conference on Orthogonal Polynomials, Special Functions and Applications, held in Universidad Carlos III, Madrid, Spain (August 29 through September 2, 2011).

Open Classical One Particle Systems

One-Dimensional Case: Some General Aspects

Let a classical particle, with mass m, position x and momentum q, be subject to a real potential $V = V(x)$, in the presence of a hb at thermal equilibrium at temperature T. We shall employ the standard variable $\beta = (kBT)^{-1}$ (kB being Boltzmann's constant). To simplify matters, the potential will be supposed to be repulsive: $V(x) \geq 0$: either $V(x) \to 0$ as $|x| \mapsto +\infty$ (Subsection 2.3) or $V(x) \equiv 0$ (Subsection 2.4) or $V(x)$ will correspond to a harmonic oscillator (Subsections 2.5). The classical Hamiltonian of the particle is: $H_c = q^2/(2m)+V$. Let the classical particle be, at the initial time $t = 0$, out of thermal equilibrium with the hb, and have a probability distribution $W_{c,in} = W_{c,in}(x, q) (\geq 0)$ to be at the position x with momentum q. Then, the non-equilibrium particle could be, at time t($>$ 0), at the position x with momentum q, with probability distribution $W_c = W_c(x, q;t)(\geq 0)$. For instance, the hb could be air (at rest and at thermal equilibrium, at T ' 300 K) in a room, and the classical particle could be a virus or a grain of pollen, performing Brownian motion in air.

How does W_c evolve in time? A temporal evolution, including nonvanishing dissipation effects on the particle due to the hb (described, in turn, by the friction constant $\sigma > 0$), is provided by the irreversible Kramers equation [9–12]:

$$\frac{\partial W_c}{\partial t} + \frac{q}{m}\frac{\partial W_c}{\partial x} - \frac{\partial V}{\partial x}\frac{\partial W_c}{\partial q} = \frac{1}{\sigma}\frac{\partial[(q + (m/\beta))(\partial W_c/\partial q)]}{\partial q}$$

(1)

The equilibrium (or Boltzmann's) canonical distribution, the t-independent solution of Equation (1) describing thermal equilibrium of the particle with the hb, is: $W_c = W_{c,eq} = \exp[-\beta(q^2/(2m)+V)]$. The physics involved in the characterization of the equilibrium distribution (to be reached for very long t), which constitutes the core of Equilibrium Classical Statistical Mechanics, selects uniquely $W_{c,eq}$. Notice that $W_{c,eq}$ is σ-independent and, hence, it is not influenced by the dissipation mechanism embodied in the right-hand-side of Equation (1). The last property does not hold necessarily in the quantum case.

For another time evolution, let dissipation effects on the particle due to the h_b be supposed so small that they are discarded completely (vanishing friction). Then, the time evolution of W_c is given by the reversible Liouville equation:

$$\frac{\partial W_c}{\partial t} + \frac{q}{m}\frac{\partial W_c}{\partial x} - \frac{\partial V}{\partial x}\frac{\partial W_c}{\partial q} = 0$$

(2)

$W_{c,eq}$ is a t-independent solution of Equation (2). After some long-time approximation, $W_c \rightarrow W_{c,eq}$ in some sense: see Subsection 2.3.

Let $H_n(y)$ be the standard n-th Hermite polynomial [13]. We remind that the Hn(y)'s constitute an infinite family of orthogonal polynomials in y, with the weight $\exp[-y^2]$: $\int_{-\infty}^{+\infty} dy \exp[-y^2]H_n(y)H_{n'}(y) = \pi^{1/2}2^n n!\delta_{n,n'}$, $\delta_{n,n'}$ being the Kronecker delta symbol ($\delta_{n,n'} = 0, 1$,

If $n \neq n', n = n'$ respectively) [13]. For both Equations (1) and (2), we shall introduce the (normalized) non-equilibrium classical moments $W_{c,n} = W_{c,n}(x;t)$ ($n = 0, 1, 2, \dots$) of W_c

$$W_{c,n} = \int_{-\infty}^{+\infty} d(q/q_{eq})\frac{H_n(q/q_{eq})}{(\pi^{1/2}2^n n!)^{1/2}}W_c(x, q;t), \quad q_{eq} = (2m/\beta)^{1/2}$$

(3)

$W_{c,0}$ is the marginal probability distribution for x. If $W_c = W_{c,eq}$, then $W_{c,eq,0}$ is proportional to $\exp[-\beta V]$ and $W_{c,eq,n} = 0$ for n = 1, 2, ... Equation (3) can also be applied to the initial off-equilibrium distribution Wc,in and gives the initial moments, $W_{c,in,n}$. It will be very convenient to work also with the symmetrized moments $g_n = W_{c,eq,0}^{-1/2} W_{c,n}$ The time evolution of gn will be treated below.

Classical Non-Equilibrium and Formal Solution Using Operator Continued Fractions

Equations (1) and (3) yield the infinite irreversible three-term linear hierarchy for $g_{n's}$ (n = 0, 1, 2, . . . , g–1 = 0) [9–12,20]:

$$\frac{\partial g_n}{\partial t} = -M_{n,n+1}g_{n+1} - M_{n,n-1}g_{n-1} - \frac{n}{\sigma}g_n$$

(4)

$$M_{n,n\pm1}W_{c,s,n\pm1} \equiv [\frac{(n + (1/2)(1 \pm 1))K_BT}{m}]^{1/2}[\frac{\partial W_{c,s,n\pm1}}{\partial x} - \frac{(\pm1)W_{c,s,n\pm1}}{2K_BT}\frac{dV}{dx}]$$

(5)

On the other hand, Equations (2) and (3) yield the infinite reversible three-term linear hierarchy for gn's (n = 0, 1, 2, . . . , g_{-1} = 0) [14,16], which is the same as in Equation (4), with $\sigma^{-1} = 0.$. For the hierarchy Equation (4) and for the reversible one, with $\sigma^{-1} = 0.$, the initial condition is formed by the set of all $g_n = W_{c,eq,0}^{-1/2}W_{c,n}.$'. We shall treat both σ −1 > 0 and σ $^{-1}$ = 0 simultaneously, unless otherwise stated.

The $M_{n,n\pm1}$'s are linear first-order differential operators, none of which is Hermitian: except for the n-dependent factors, their structures are typical of those appearing in the study of the Smoluchowski equation [10]. $M_{n,n+1}$ is the adjoint of $-M_{n+1,n}$. Notice that, if $(dV/dx) \neq 0, M_{n,n-1}$ 1 does not commute with Mn,n+1. Let us consider the Laplace transforms $\tilde{W}_{c,n} = \tilde{W}_{c,n}(s) = \int_0^{+\infty} dt W_{c,n}\exp(-st)$. with inverse $W_{c,n} = \int_{c-i\infty}^{c+i\infty}(ds/2\pi i)\tilde{W}_{c,n}\exp(st)$ (c being real and such that $\tilde{w}_{c,n}(s)$ is analytic in the half-plane Res > c of the complex s-plane). It will also be useful to deal with the symmetrized Laplace transforms $\tilde{g}_n = W_{c,eq,0}^{-1/2}\tilde{W}_{c,n}$.This definition and Equation (4) yield the symmetrized three-term hierarchy for $\tilde{g}_n$.

$$(s + \frac{n}{\sigma})\tilde{g}_n = W_{c,eq,0}^{-1/2}W_{c,in,n} - M_{n,n+1}\tilde{g}_{n+1} - M_{n,n-1}\tilde{g}_{n-1}$$

(6)

for both σ $^{-1}$ > 0 (irreversible) and σ^{-1} = 0 (reversible).

The hierarchy Equation (6) can be solved formally by extending to it standard procedures for solving numerical three-term linear recurrence relations in terms of continued fractions (see, for instance, [10]). Thus, one neglects $\tilde{g}_{n'+1}(s)$ in Equation (6) for given n' , solves for $\tilde{g}_{n'}(s)$ in terms of $\tilde{g}_{n'-1}(s)$

proceeds to Equation (6) for n' − 1, solves for $\tilde{g}_{n'-1}(s)$ in terms of $\tilde{g}_{n'-2}(s)$ and so on. Then, one infers directly the general formal (continued-fraction) structure of the solution as $n' \to +\infty$ That formal procedure yields all $\tilde{g}_n(s)$, for any n = 1, . . . , in terms of sums of products of certain s-dependent linear operators $D[n'; s + \frac{n'}{\sigma}]$, $n' \geq n$, acting upon $\tilde{g}_{n-1}(s)$ and upon all $W_{c,eq,0}^{-1/2}W_{c,in,n'}$'s, n' ≥ n. The linear operators $D[n; s + \frac{n}{\sigma}]$'s are defined recurrently through:

$$D[n; s + \frac{n}{\sigma}] = [sI + \frac{n}{\sigma} - M_{n,n+1}D[n+1; s + \frac{n+1}{\sigma}]M_{n+1,n}]^{-1} \tag{7}$$

I is the unit operator. By iteration of Equation (7), D[n; s + n/σ] becomes a formal infinite continued fraction of products of the linear operators $M_{n,n+1}$ and $M_{n+1,n}$ (which do not commute if dV/dx 6= 0). For a simpler hierarchy and a clearer exposition, without loss of generality, let us assume that $W_{c,in,n'}$ = 0 for n' ≥ 1, with $W_{c,in,0}$ 6= 0 (and 6= $W_{c,eq,0}$). Also, let us choose some n_0(≥ 1). Then, Equation (6) yields:

$$(s + \frac{n}{\sigma})\tilde{g}_n = W_{c,eq,0}^{-1/2}W_{c,in,0}\delta_{n,0} - M_{n,n+1}\tilde{g}_{n+1} - M_{n,n-1}\tilde{g}_{n-1}, \quad n \leq n_0 - 1$$

$$\tilde{g}_n(s) = -D[n; s + \frac{n}{\sigma}]M_{n,n-1}\tilde{g}_{n-1}(s), \quad n \geq n_0$$

$\delta_{n,0}$ is the Kronecker delta symbol. If $\sigma^{-1} = 0$ and $n_0 = 1$, Equations (8) and (9) yield:

$$\tilde{g}_0(s) = D[0; s]W_{c,eq,0}^{-1/2}W_{c,in,0}$$

It should be clear that if $\sigma^{-1} = 0$ (absence of dissipation) the particle is not expected to reach irreversibly, for long time, thermal equilibrium with the hb. In such a case, Equation (10), in which no long-time approximation has been carried out so far, is as reversible as Equation (2).

Operator Continued Fractions and Long-Time Approximation

with initial condition $W_{c,eq,0}^{-1/2}W_{c,in,0}$. For the computations below, we shall interpret the right-hand-side of Equation (13) as $\int_{-\infty}^{+\infty} dx'(M_{0,1}D[1; \epsilon + \sigma^{-1}]M_{1,0})(x, x')g_0(x')$. Let: $(f_1, f_2) = \int_{-\infty}^{+\infty} dx f_1(x)^* f_2(x)$ for suitable functions f_1 and f_2. Due to the Hermiticity of $D[1; \epsilon + \sigma^{-1}]$: $(f_1, M_{0,1}D[1; \epsilon + \sigma^{-1}]M_{1,0}f_2) = (M_{0,1}D[1; \epsilon + \sigma^{-1}]M_{1,0}f_1, f_2)$, thereby checking that $M_{0,1}D[1; \epsilon + \sigma^{-1}]M_{1,0}$ is Hermitian. Moreover:

$(f_1, M_{0,1}D[1; \epsilon + \sigma^{-1}]M_{1,0}f_1) = -(M_{1,0}f_1, D[1; \epsilon + \sigma^{-1}]M_{1,0}f_1) \leq 0$ for arbitrary functions f_1, as all eigenvalues of $D[1; \epsilon + \sigma^{-1}]$ are ≥ 0. Let $f_\lambda(x)$ be an eigenfunction of the integral operator $M_{0,1}D[1; \epsilon + \sigma^{-1}]M_{1,0}$ with eigenvalue $\lambda(\leq 0)$. Then, $(M_{0,1}D[1; \epsilon + \sigma^{-1}]M_{1,0})(x, x') = \sum_\lambda \lambda f_\lambda(x)f_\lambda(x')^*$. $\sum_\lambda$ is a short-hand notation denoting integration over the whole spectrum of $M_{0,1}D[1; \epsilon + \sigma^{-1}]M_{1,0}$. By expanding $W_{c,eq,0}^{-1/2}W_{c,in,0} = \sum_\lambda g_{in,\lambda}f_\lambda(x)$, with x-independent $g_{in,\lambda}$, the solution of Equation (13) with the above initial condition is $g_0 = \sum_\lambda g_{in,\lambda}f_\lambda(x)\exp\lambda t$, which relaxes irreversibly as $t \to +\infty$ towards $g_{in,0}f_0(x)$, corresponding to $\lambda = 0$. At equilibrium, one has: $g_0 = W_{c,eq,0}^{1/2}$ (proportional to f_0), $M_{1,0}g_0 = 0$ and $g_n = 0$, $n = 1,2,3,\ldots$, consistently with [9,10]. Let us restrict the above to

$\sigma^{-1} = 0$. Then, Equation (13) with $\sigma^{-1} = 0$ is (at least, with $\epsilon > 0$) as irreversible as the standard heat equation: for long $t(> 0)$ the dominant moment is g_0, while any g_n with $n > 0$ is negligible, gn being the smaller, the larger n and $t(> 0)$, and so on for the Wc,n's. As stressed after Equation (10), irreversible thermalization does not occur if $\sigma^{-1} = 0$, in the absence of long-time approximations. Then, for $\sigma -1 = 0$, to have carried out mathematical approximations (say, D[n; s] $\simeq$ D[n; ϵ] for n $\geq$ n$_0$, but not for n $<$ n$_0$), which give rise to thermalization with the h$_b$ [like that based on Equation (13)], could be regarded as an alternative way of introducing irreversibility out of a reversible model [namely, Equation (2)] with vanishing dissipation ab initio. The latter interpretation with $\sigma^{-1} = 0$ is consistent for the one-dimensional case. We shall regard it as a mathematical introduction, which will be very helpful to deal later with non-equilibrium classical closed interacting many-particle systems without external hb's.

Convergence Properties for V $\equiv$ 0 and $\sigma^{-1} = 0$

We shall now undertake a more detailed analysis for $\sigma^{-1} = 0$ and V $\equiv$ 0, in order to understand the structure and convergence of the operator continued fractions D[n; s] generated by Equation (7). To fix the ideas, we shall take $\epsilon > 0$ first, so as to allow for $\to$ 0 later, thereby reviewing and extending [14]. Equation (13) becomes formally (using standard notations for continued fractions [21]):

$$\partial g_0/\partial t = \frac{k_B T}{m}\frac{\partial}{\partial x}D[1;\epsilon]\frac{\partial}{\partial x}g_0$$

(14)

$$D[1;\epsilon] = \frac{1}{\epsilon I+}\frac{2[(k_B T/m)(-\partial^2)/(\partial x^2)]}{\epsilon I+}\frac{3[(k_B T/m)(-\partial^2)/(\partial x^2)]}{\epsilon I+}\cdots$$

(15)

Let us perform a spatial Fourier transformation from configuration space (x) to wavevector space (k), by applying $(2\pi)^{-1/2}\int dx\exp(-ikx)$. Let $e(k) \equiv (2m)^{-1}k_B T k^2$. Then, the Fourier transforms of the operator continued fractions in Equation (7), for Res > 0 and n $\geq$ 0, and in Equation (14) are:

$$D_1[k;n;s] = [s + 2(n+1)e(k;N)D_1[k;n+1;s]^{-1}$$

(16)

$$\partial g_{0,1}/\partial t = i(k_B T/m)^{1/2}kD_1[k;1;\epsilon]i(k_B T/m)^{1/2}kg_{0,1}$$

(17)

$g_{0,1} = g_{0,1}(k, t)$ being the spatial Fourier transform of g_0. By iteration, $D_1[k; n; s]$ becomes the following ordinary continued fraction (in standard notations [21]):

$$D_1[k; n; s] = \frac{1}{s+} \frac{2e(k)(n+1)}{s+} \frac{2e(k)(n+2)}{s+} \cdots = \frac{1}{e(k)^{1/2}} \cdot [\frac{2^{-1}}{z+} \frac{2^{-1}(n+1)}{z+} \frac{2^{-1}(n+2)}{z+} \cdots \}$$ (18)

with $z = s/(2e(k)^{1/2})$. For real $s \geq 0$, $D_1[k; n; s]$ is real. On the other hand:

$$D_1[k; n; s = \epsilon = 0] = D_1[k; n; 0] = [2e(k)^{1/2}]^{-1} \frac{1}{0+} \frac{2^{-1}(n+1)}{0+} \frac{2^{-1}(n+2)}{0+} \cdots$$ (19)

with $i^n erfc(z) = \int_z^{+\infty} dz' i^{n-1} erfc(z')$, $i^0 erfc(z) = erfc(z)$ (the complementary error function) and $i^{-1} erfc(z) = (2/\pi^{1/2}) \exp(-z^2)$ [21]. One has: $D_1[k; n; \epsilon] \geq 0$. With $\epsilon > 0$ and by using [21], one gets: $D_1[k; 1; \epsilon] \to \epsilon^{-1}$ as $k \to 0$, while $D_1[k; 1; \epsilon] \to (\pi e(k))^{-1/2}$ as $| k | \to \infty$. For $s = 0$, the

behaviour of D_1 is different. Equation (18) gives:

$$D_1[k; n; s = \epsilon = 0] = D_1[k; n; 0] = [2e(k)^{1/2}]^{-1} \frac{1}{0+} \frac{2^{-1}(n+1)}{0+} \frac{2^{-1}(n+2)}{0+} \cdots$$ (20)

The continued fraction in Equation (20) can, in turn, be evaluated in terms of the standard Gamma function Γ [21]:

$$\frac{2^{-1}}{0+} \frac{2^{-1}(n+1)}{0+} \frac{2^{-1}(n+2)}{0+} \cdots = \frac{\Gamma((n/2) + 1/2)}{2\Gamma((n/2) + 1)} > 0, \ n \geq 1$$ (21)

For large n, the ratio in Equation (21) $\to n^{-1/2}$. Equation (19) behaves similarly. For $k \to 0$, $D_1[k; n; 0]$ diverges as k^{-1} (due to $e(k)^{-1/2}$). Then, $\int dk D_1[k; n; 0]$ also diverges near $k = 0$.

Classical Harmonic Oscillator: Operator Continued Fractions

For vanishing dissipation due to the hb, which is at thermal equilibrium at T, $D[n; s]$ has also been evaluated for a classical harmonic oscillator ($V = 2^{-1} m\omega^2 x^2$ with frequency $\omega > 0$) in one space dimension [16]. It will be methodologically adequate to outline those results here. The actual classical hierarchy turns out to be exactly solvable, by inspiring on known algebra for the quantum harmonic oscillator: compare with [10]. Here, we shall work with the dimensionless position and momentum variables $y \equiv [(2k_B T)^{-1} m]^{1/2} \omega x, \pi \equiv [(2k_B T m)^{-1}]^{1/2} q$. Accordingly, we shall deal with $W_c(x.q;t) = f(y, \pi;t) = f$. The non-equilibrium moments for f are introduced through Equation (3), with the corresponding changes. For convenience, we introduce, in the actual classical context, "annihilation" (a) and "creation" (a^+) operators: $a \equiv 2^{-1/2}(d/dy + y)$, $a^+ \equiv 2^{-1/2}(-d/dy + y)$ $[a, a^+] = +1$ $([A, B] \equiv AB - BA)$ Notice that a^+ and a are proportional to

$M_{n,n+1}$ and $M_{n,n-1}$, respectively. We proceed like in Subsections 2.1 and 2.2. Then, the hierarchy for the symmetrized $\tilde{g}_n(s)(=\tilde{g}_n(y,s))$, which is the actual counterpart of Equation (6), becomes (with $\sigma^{-1} = 0$):

$$s\tilde{g}_n = W_{c,eq,0}^{-1/2}W_{c,in,n} + \omega[(n+1)^{1/2}.a^+\tilde{g}_{n+1} - n^{1/2}.a\tilde{g}_{n-1}]$$

(22)

with the initial condition $W_{c,eq,0}^{-1/2}W_{c,in,n}$.. In the new variables, $W_{c,eq,0}^{1/2}$ is proportional to $\exp(-y^2/2)$ The actual counterpart of the operator D[n; s] in Subsection 2.2 is: D[n; s] = [sI + (n + 1)ω ^{2}a +D[n + 1; s]a] $^{-1}$. This D[n; s] can be evaluated in closed form [16], as we shall now outline. By using $a^+[sI + n\omega a^+a]^{-1} = [sI + n\omega(a^+a - 1)]^{-1}a^+$, and after some iterative algebra, one finds that the dependence of D[n; s] on a and on a $^+$ occurs only through the product a $^+$a and that, by rewriting $D[n;s] \equiv D[n;s;a^+a]$, one gets:

$$D[n;s;a^+a] = [sI + (n+1)\omega^2 a^+ a D[n+1;s;a^+a - 1]]^{-1}$$

(23)

We $f_{n'} = H_{n'}(y)\exp[-2^{-1}y^2]$, $n' = 0,1,2,\ldots$. Then,   the eigenfunctions of a $^+$a and of D[n; s; a $^+$a] are fn0. The eigenvalues of a $^+$a are n' which, through iteration of Equation (23), yield directly those of D[n; s; a $^+$a] as finite fractions, precisely due to the structure a $^+$a $^{-1}$. In other words, D[n; s] (which, in principle, is an infinite continued fraction of operators) becomes, by iterating Equation (23), a finite fraction in a +a. A posteriori, one confirms for the actual harmonic oscillator that the resulting $D[n;s = \epsilon]$ $(\epsilon > 0)$ is Hermitian and has nonnegative eigenvalues. See [16] for further details.

CLOSED CLASSICAL MANY-PARTICLE SYSTEMS: LONG-TIME APPROXIMATION AND ARROW OF TIME

We shall present an outline of the main developments, omitting lengthy arguments and equations, which can be seen in [14]. We treat a closed large system of many (N >> 1) classical nonrelativistic particles, in d spatial dimensions (d = 1, 2, 3), with spatial coordinates $x_1,\ldots,x_N$ ($\equiv$ [x]) and momenta $q_1,\ldots,q_N$ ($\equiv$ [q]). All particles, which are identical, have mass m. Let $x_{i,\alpha}$ and $q_{i,\alpha}$ be the Cartesian components of x_i and q_i, respectively (i = 1, . . . , N, α = 1, . . . d). Neither a hb nor external friction mechanisms nor external forces are assumed. The interaction potential is: $V = \Sigma_{i,j=1,i<j}^{N}V_{i,j}(|x_i - x_j|)$ and we suppose that all $V_{i,j}(|x_i - x_j|)$ are repulsive (≥ 0) and tend quickly to zero for large $|x_i - x_j|$. The physical idea is that the very large number of degrees of freedom (which be at thermal equilibrium with one another) in the system

play the role of a hb. The non-equilibrium classical distribution function is: $W_c = W_c([x], [q]; t)$. Boltzmann's equilibrium (canonical) distribution at absolute temperature T is : $W_{c,eq} = \exp[-\beta((2m)^{-1}\Sigma_{i=1}^{N}\Sigma_{\alpha=1}^{d}q_{i,\alpha}^2 + V)]$. The initial non-equilibrium distribution $W_{c,in}$, to be regarded as known, is quite arbitrary in practice. Instead of considering the most general initial state, we shall choose a class of $W_{c,in's}$, which will be: (i) qualitatively consistent with the idea [8,22], typical of Information Theory, that one should employ only distribution functions compatible with the limited information available which, in turn, refers to expectation values not of all possible dynamical variables but only of an observable subset of such variables (these ideas being imposed for t = 0 only, but not for t > 0); (ii) also consistent with standard variables employed in Equilibrium Statistical Mechanics and Fluid Dynamics [2–5]. Then, we shall treat a class of an explicit ansatze for " Wc,in which will depend on a finite number of functions (actually, 2 + d) of one single position vector, x: $\lambda_k = \lambda_k(\mathbf{x})$, $k = 0, 2$, and $\lambda_{1.\alpha} = \lambda_{1.\alpha}(\mathbf{x})$, $\alpha = 1, \dots, d$ d (all of them being independent on time and on momenta). The expression for $W_{c'in}$ in terms of λ_k and $\lambda_{1,a}$ has appeared previously [3,5] (and is related to the Massieu-Planck function [5]). In short, $W_{c,in}$ at t = 0 will be chosen to describe thermal equilibrium with homogeneous temperature T for large distances but non-equilibrium for intermediate and short distances (with spatial inhomogeneities). The ansatz is:

$$W_{c,in} = (N!)^{-1}\exp[-\int d^d\mathbf{x}(\lambda_0(\mathbf{x})\sum_{i=1}^{N}\delta^{(d)}(\mathbf{x}_i - \mathbf{x}) + \sum_{\alpha=1}^{d}\lambda_{1,\alpha}(\mathbf{x})\sum_{i=1}^{N}\frac{q_{i,\alpha}}{m}$$

$$\times \delta^{(d)}(\mathbf{x}_i - \mathbf{x}) + \lambda_2(\mathbf{x})\sum_{i=1}^{N}(\frac{\mathbf{q}_i^2}{2m} + \frac{1}{2}\sum_{j=1,j\neq i}^{N}V_{i,j}(|\mathbf{q}_i - \mathbf{q}_j|))\delta^{(d)}(\mathbf{x}_i - \mathbf{x})]$$

(24)

Consistently with (i)–(ii) above, the λ's will be uniquely determined (through standard recipes in Information Theory) in terms of 2 + d x-dependent observables (also independent on time and on momenta) typically employed in Fluid Dynamics, which, by assumption, are known at t = 0: mass density, fluid velocity and some suitable energy density [5,14]. For simplicity, no observables associated to angular momentum are included. How does the equilibrium temperature T appear in this closed system, without hb? We assume that $W_{c,in}$ describes thermal equilibrium at temperature T at large x but off-equilibrium at finite x. We accept that λ2(x) approaches quickly a non-vanishing constant, $\lambda_2(\infty)$, as | x | tends to ∞ along any direction and that the same holds for λ_0(x). A similar statement holds for $\lambda_{1,a}$(x), the corresponding (large-| x |) limiting value being zero. At finite x, the off-equilibrium λ_2(x), λ_0(x) and $\lambda_{1,\alpha}$(x) do depend on x and, so, differ from their respective constant (large-| x |) limiting values, which describe equilibrium. Then, consistency is achieved (T being thereby introduced) if, in the thermodynamical limit, $\lambda_2(\infty)$ tends to $(kBT)^{-1}$ (plus corrections which approach zero in that limit). The dominant contributions to various statistical averages at $t = 0$ come from large x, up to corrections which tend to vanish as N increases. The very large number of degrees of freedom involved in the largest part of the system (for large $|x|$) are at equilibrium at T. For consistency, the $2 + d$ x-dependent dynamical variables known at $t = 0$ (mass density, fluid velocity and some suitable energy density) have to fulfill the corresponding behaviour as $|x|$ tends to ∞ along any direction. For details, see [14]. The reversible Liouville equation reads:

$$\frac{\partial W_c}{\partial t} = \Sigma_{i=1}^{N}\Sigma_{\alpha=1}^{d}[\frac{\partial V}{\partial x_{i,\alpha}}\frac{\partial W_c}{\partial q_{i,\alpha}} - \frac{q_{i,\alpha}}{m}\frac{\partial W_c}{\partial x_{i,\alpha}}]$$

(25)

Let [n] denote a set of nonnegative integers $(n(i = 1, \alpha = 1), \ldots, n(i = N, \alpha = d))$ and let $n = \Sigma_{l=1}^{N}\Sigma_{\alpha=1}^{d}n(l, \alpha)$. Let $[dq] = \prod_{i=1}^{N}\prod_{\alpha=1}^{d}dq_{i,\alpha}$. We introduce non-equilibrium moments $W_{[n]}$ of W (using products of Hermite polynomials, by generalizing Equation (3) with all integrations in $(-\infty, +\infty)$):

$$\int [dq] \prod_{i=1}^{N} \prod_{\alpha=1}^{d} \frac{H_{n(i,\alpha)}(q_{i,\alpha}/(2mk_BT)^{1/2})}{(\pi^{1/2}2^{n(i,\alpha)}n(i,\alpha)!)^{1/2}} W_c([x], [q], t) \equiv W_c(x; [n]; t) = W_c([n])$$

(26)

If $W_c = W_{c,eq}$, then $W_{c,eq}([0])$ ($[0] = (0, 0, \ldots, 0) = (n(i = 1, \alpha = 1) = 0, \ldots, n(i = N, \alpha = d) = 0) = [n = 0]$) is proportional to $\exp[-\beta V]$ and $W_{c,eq}([n]) = 0$, $[n] \neq [0]$ (say, $n \neq 0$). Equation (26) can

also be applied to $W_{c,in}$ and gives the corresponding initial moments, $W_{c,in}([n])$. We shall work with the symmetrized moments $g([n]) = W_{c,eq}([0])^{-1/2}W_c([n])$. One gets an infinite reversible three-term linear recurrence for $g([n])$'s, generalizing Equations (4) and (5). It reads:

$$\frac{\partial g(n(1,1), \ldots, n(j, \beta), \ldots, n(N, d))}{\partial t}$$

$$= -\Sigma_{l=1}^{N}\Sigma_{\alpha=1}^{d}[M_{l,\alpha;n(l,\alpha);+}g(n(1,1), \ldots, n(l, \alpha) + 1, \ldots, n(N, d))$$

$$+ M_{l,\alpha;n(l,\alpha);-}g(n(1,1), \ldots, n(l, \alpha) - 1, \ldots, n(N, d))]$$

(27)

$$M_{l,\alpha;n(l,\alpha);+} = [\frac{(n(l,\alpha) + 1)k_BT}{m}]^{1/2}[\frac{\partial}{\partial x_{l,\alpha}} - \frac{1}{2k_BT}\frac{\partial V}{\partial x_{l,\alpha}}]$$

(28)

$$M_{l,\alpha;n(l,\alpha);-} = [\frac{n(l,\alpha)k_BT}{m}]^{1/2}[\frac{\partial}{\partial x_{l,\alpha}} + \frac{1}{2k_BT}\frac{\partial V}{\partial x_{l,\alpha}}]$$

(29)

The Laplace transform of the hierarchy Equation (27) is the actual many-particle counterpart of Equation (6) with $\sigma^{-1} = 0$. Such a Laplace transform (with $N \gg 1$) can be formally solved in terms of linear operators $D[[n]; s]$, which generalize the previous $D[n; s]$. For details, see [14]. All $D[[n]; s]$ are square matrices, due to the indices $i = 1, \ldots, N$ and $\alpha = 1, \ldots, d$. In turn, each matrix element in those square matrices is an ordinary linear integral operator, arising from the partial differential operators $M_{l,\alpha;n(l,\alpha);+}$ and $M_{l,\alpha;n(l,\alpha);-}$, as l, α and $n(l, \alpha)$ and n(l, α) vary. The $D[[n]; s]$ fulfill the following formal hierarchy [which generalizes Equation (7)]:

$$D[[n]; s] = [sI - M_{+,[n+1]}D[[n+1]; s]M_{-,[n]}]^{-1}$$

(30)

The linear operators $M_{\pm,[n]}$ are rectangular matrices, the elements of which are formed out of the partial differential operators $M_{l,\alpha;n(l,\alpha);+}$ and $M_{l,\alpha;n(l,\alpha);-}$. $M_{+,[n+1]}$ can be shown to be the adjoint of $-M_{-,[n]}$. By iterating Equation (30) indefinitely, one can express formally the linear operator $D[[n]; s]$ as an operator continued fraction, which depends on all partial differential operators $M_{l,\alpha n(l,\alpha);+}$ and $M_{l,\alpha;n(l,\alpha);-}$ and generalizes the operator continued fraction for $D[n; s]$. See [14]

By generalizing the iterative arguments in Subsection 2.3, it follows that, for both $V \ne 0$ and $V = 0$, $D[[n_0]; \epsilon]$ for > 0 is a Hermitian operator with non-negative eigenvalues for $n \ge n0 \ge 1$ [14].

A few remarks for the case $V \equiv 0$ may be clarifying. In the Laplace transform of Equation (27), let us perform a spatial Fourier transformation from configuration space $(x_1, \ldots, x_N)$ to wavevector space $(k_1, \ldots, k_N)$ $\equiv [k]$. Let $e(k; N) \equiv k_B T \sum_{j=1}^{N}(2m)^{-1}k_j^2$. Then, the Fourier transform $D_1[[k]; [n]; s]$ of $D[[n]; s]$ for Res > 0 is an ordinary continued fraction, given in Equations (16)–(18) with e(k) replaced by e(k; N). Then, with such a replacement, the properties of $D_1(k; n; s)$ given in Subsection 2.4 also hold for $D_1[[k]; [n]; s]$. Notice that $D_1[[k]; [n]; s = 0]$ diverges as $e(k; N)^{-1/2}$ if $e(k; N) \to 0$. On the other hand, and contrary to what happened for one particle in one spatial dimension [recall the comment after Equation (21)], $\int [dk] D_1[[k]; [n]; s = 0]$ converges near e(k; N) = 0. This would suggest that the actual counterpart of Equation (9) (containing D[[n];] acting on various $M_{l,\alpha;n(l,\alpha);-}g(n(1, 1), \ldots, n(l, \alpha) - 1, \ldots, n(N, d))$'s), when integrated over [k]'s to come back to [x]-space, would converge at small [k]'s as $\epsilon \to 0$, for the actual $V \ne 0$. Whether such a property holds is an open question.

In spite of the very involved structure of Equation (27) (and of its Laplace transform), we argue that a simple long-time approximation can be performed in it, for $V_{i,j} \geq 0$ (and vanishing quickly at large distances) and very large N (in the thermodynamical limit), which generalizes that in Subsection 2.3.

This approximation consists in fixing $s = \epsilon > 0$ (ϵ being small) in the whole hierarchy of operators D[[n]; s], for any $n \geq n_0 (> 0)$ which, then, become Hermitian operators $D[[n]; \epsilon]$ with no negative eigenvalues. It is crucial that s-dependences be kept in D[[n]; s], for $n < n_0$. Notice that the non-vanishing factors $n(l, \alpha)^{1/2}$ and $((n(l, \alpha) + 1))^{1/2}$ in $M_{l,\alpha;n(l,\alpha);-}$ and $M_{l,\alpha;n(l,\alpha);+}$, respectively, tend to reduce, as the n(l, α)'s increase, the relative importance of having fixed $s = \epsilon$ and the contribution of the latter in the $D[[n]; s = \epsilon]$'s with $n \geq n_0$. This is a genuine feature of the D[[n]; $s = \epsilon$]'s. See [14], where it was seen that, by imposing $n_0 > 2$, the long-time approximation is exactly consistent with all hydrodynamical balance equations. For simplicity, we discard all the initial moments $W_{c,in([n])}$ for $n \geq n_0$. We regard D[[n_0]; ϵ] as a fixed (s-independent) operator, yielding all g([n_0]) in terms of all g([n_0^{-1}]). The choice $\epsilon = 0$ was made in [14] By using [16] and [18,19], the arguments in [14] are easily seen to hold for $\epsilon > 0$. All that leads to a closed approximate hierarchy for g([n])'s (with initial moments $W_{c,in([n]))}$, with n < n0, which appears to yield an approximate irreversible evolution towards thermal equilibrium at T. g([n])'s and, then, Wc([n]) relax the quicker the larger n, provided that $n < n_0$. $W_c([0])$ would dominate the approach towards equilibrium for t → +∞. See [14]. All that appears to work based on the general properties of D[[n_0]; ϵ]: for quantitative studies, some ansatz or approximation should be provided directly for it. An arrow of time would follow approximately, in the present case.

As an extreme example, let $n_0 = 1$ (which is strictly consistent only with the hydrodynamical balance equation for mass): see [14] for $n_0 = 2, 3$. By making the above long-time approximation and taking inverse Laplace transforms, one finds directly the irreversible Smoluchowski-like equation for the [n = 0] moment, which generalizes Equation (13) ([n = 0] meaning n(1, 1) = 0, . . . , n(j, β) = 0, . . . , n(N, d) = 0):

$$\frac{\partial g([n = 0])}{\partial t} = \Sigma_{l=1}^{N} \Sigma_{\alpha=1}^{d} M_{l,\alpha;n(l,\alpha)=0;+} \times$$
$$(\Sigma_{l'=1}^{N} \Sigma_{\alpha'=1}^{d} [D[[n = 1]; \epsilon]]_{l,\alpha;l',\alpha'} M_{l',\alpha';n(l',\alpha')=1;-}) g([n = 0]) \qquad (31)$$

The operator $D[[n = 1]; \epsilon]$ (Hermitian, with non-negative eigenvalues) has, as a square matrix, the matrix elements $[D[[n = 1]; \epsilon]]_{l,\alpha;l',\alpha'}$. The initial condition is $W_{c,eq}([0])^{-1/2} W_{c,in}([0])$. Compare with [14].

OPEN QUANTUM-MECHANICAL ONE-DIMENSIONAL SYSTEM WITHOUT DISSIPATION: GENERAL ASPECTS

We shall consider a quantum Brownian particle qBp of mass m (> 0) and momentum operator $-i h^- (\partial/\partial x)$, in one spatial dimension x, with (Hermitian) quantum Hamiltonian:

$$H = -\frac{\hbar^2}{2m}\frac{\partial^2}{\partial x^2} + V$$

(32)

with a real potential V = V (x) ≥ 0 vanishing quickly for | x |→ +∞. h⁻ is Planck's constant. All eigenvalues E_j of H sweep the continuous positive real half-line: $0 \le E_j < +\infty$. Such a continuous spectrum has, typically, a double degeneracy, associated to different asympotic conditions at x → ±∞, with the same energy. The continuous variable j (−∞ < j < +∞) labels all states and distinguishes degenerate states. Thus, if φ_j (x) is an eigenfunction of H: $H \varphi_j$ (x) = $E_j \varphi j$ (x)

The example considered in Subsection 2.1 (namely, a Brownian particle moving in air at room temperature) could be invoked here as well: the particle should now be described by Quantum Mechanics. This Section presents a purely formal generalization of Subsections 2.1 and 2.2 for the non-equilibrium statistical evolution of a qBp subject to V (x), and in the presence of a hb at thermal equilibrium at T, in the idealized case of vanishing dissipation. The time evolution for t > 0 of the qBp is given by the density operator ρ = ρ(t) (a statistical mixture of quantum states), with the initial condition ρ(t = 0) = ρ_{in}. For t ≥ 0, ρ(t) is a Hermitian and positive-definite linear operator acting in a Hilbert space. Unless otherwise stated, we shall not impose that ρ(t) be normalized. The time evolution of the qBp in differential form is described by the operator equation ([H, ρ] = Hρ − ρH):

$$\frac{\partial \rho}{\partial t} = \frac{1}{i\hbar}[H, \rho]$$

(33)

We consider the matrix element

$\langle x - y | \rho(t) | x + y \rangle$ of $\rho(t)$ in generic eigenstates, $|x - y\rangle, |x + y\rangle$, of the quantum position operator. The quantum Wigner function W = W(x, q;t), determined by ρ, is [6,7]:

$$W(x, q; t) = \frac{1}{\pi \hbar} \int_{-\infty}^{+\infty} dy \exp[\frac{i2qy}{\hbar}]\langle x - y | \rho(t) | x + y \rangle$$

(34)

The initial non-equilibrium Wigner function at t = 0 is Win, given by Equation (34) if ρ = ρ_{in}. For t > 0, the exact dissipationless quantum master equation (QME) for W [6,7] is:

$$\frac{\partial W(x,q;t)}{\partial t} = -\frac{q}{m}\frac{\partial W(x,q;t)}{\partial x} + M_Q W \tag{36}$$

$$M_Q W = \int_{-\infty}^{+\infty} dq' W(x, q + q'; t) \int_{-\infty}^{+\infty} \frac{idy}{\pi\hbar^2}[V(x+y) - V(x-y)]$$

$$\times \exp[-\frac{i2q'y}{\hbar}] = \frac{dV}{dx}\frac{\partial W}{\partial q} - \frac{\hbar^2}{3!2^2}\frac{d^3V}{dx^3}\frac{\partial^3 W}{\partial q^3} + \cdots \tag{37}$$

As $h^- \to 0$, Equation (35) becomes formally, by dropping all h^--dependent terms (containing $\partial^n W/\partial p^n$ in Equation (36), n = 3, 5, . . .), the classical Liouville equation Equation (2), with $W \to Wc$ [6,7]. We shall assume that, as $|q| \mapsto +\infty$, $W(x, q; t) \to 0$ .quickly, for fixed x and long t. Then, $\int_{-\infty}^{+\infty} dq W(x, q; t) q^n$ converges, for any integer n ≥ 0. Equation (35) readily implies that $(\partial/\partial t) \int_{-\infty}^{+\infty} dx \int_{-\infty}^{+\infty} dq W(x, q; t) = 0$.

A stationary Wigner function is any t-independent solution of Equation (35). We are not interested on arbitrary stationary Wigner functions, but only on a very specific one (denoted by Weq), namely, that which accounts for the thermal equilibrium state of the qBp, at temperature T = $(k_B\beta)^{-1}$, with the hb. Like in the classical case, the solutions of Equations (35) and (36) are not expected to approach Weq exactly, unless some approximation be made. Weq arises from the canonical (t-independent) density operator ρ_{eq} = exp[−βH], with matrix elements:

$$\langle x - y|\rho_{eq}|x + y\rangle = \sum_j \exp[-\beta E_j]\varphi_j(x - y)\varphi_j(x + y)^* \tag{37}$$

Σ_j is a short-hand notation denoting integration over the whole continuous spectrum of j $(\int_{-\infty}^{+\infty} dj)$. In turn, ρ_{eq} determines $W_{eq}(x, p)$, through Equation (34):

$$W_{eq}(x, q) = \frac{1}{\pi\hbar}\int_{-\infty}^{+\infty} dy \exp[\frac{i2qy}{\hbar}] \sum_j \exp[-\beta E_j]\varphi_j(x - y)\varphi_j(x + y)^* \tag{38}$$

$$-\frac{q}{m}\frac{\partial W_{eq}}{\partial x} + M_Q W_{eq} = 0, \quad \frac{\partial W_{eq}}{\partial t} = 0 \tag{39}$$

We remark that if one has constructed some general stationary Wigner function, by solving Equation (39), further additional information should be fed in, so as to select uniquely that solution of Equation (39) which yields precisely W_{eq}. Such a situation will be met in Section 5. Some known genuine difficulties of the quantum case are that $W_{eq}(x, p)$ is neither Gaussian in q nor known in closed form, for a general V [6,7]. In Subsection 4.1 we shall start the generalizations of Subsections 2.1 and 2.2 to the actual quantum situation, so as to search for ways for solving or, at least, bypassing the genuine difficulties, mentioned above, of the quantum case. An alternative representation for W_{eq} has been given by Wigner [6]: we shall review it shortly in Subsection 4.2 and use it in Section 5.

Orthogonal Polynomials $H'_{Q,n}$ Generated by W_{eq} in Equation (38), Moments and Hierarchy We shall introduce the denumerably infinite family of all (unnormalized) polynomials in q $H'_{Q,n} = H'_{Q,n}(q)(n = 0, 1, 2, 3, \ldots)$, orthogonalized in q (for fixed x) by using the equilibrium Wigner function W_{eq} Equation (38) as weight function. Let q0 be some fixed (scaling) momentum: q_0 could equal q_{eq} [Equation (3)], but not necessarily. By choosing $H'_{Q,0}(q) = 1$, for $n \neq n'$ and any x (left unintegrated), we impose:

$$\int_{-\infty}^{+\infty} \frac{dq}{q_0} W_{eq}(x, q) H'_{Q,n}(q) H_{Q,n'}(q) = 0$$

(40)

The $H'_{Q,n}(q)$'s, depending parametrically on x for $n \geq 1$, generalize the standard Hermite polynomials, and will be used for the time evolution. We shall look for the $H'_{Q,n}(q)$'s as ($y_0 = q/q_0$):

$$H'_{Q,n}(q) = y_0^n + \sum_{j=1}^{n} \epsilon_{n,n-j} y_0^{n-j}$$

(41)

$\epsilon_{n,n-j}$ being q-independent (but x-dependent, in general): see Appendix A. The orthonormalized polynomials are $H'_{Q,n}(q)/(h'_n)^{1/2}$, with the (x-dependent) normalization factor:

$$h'_n \equiv \int_{-\infty}^{+\infty} \frac{dq}{q_0} W_{eq}(x, q) H'_{Q,n}(q)^2$$

(42)

The lowest normalization factors h'_0 and h'_1 are given in Appendix A.

We shall treat general situations in which the qBp could be out of thermal equilibrium with the hb at t = 0 (and, hence, for t > 0). We shall analyze them by using Equation (35), for the non-equilibrium W. The study will also enable to discuss, as a consistency check, the case in which the qBp be at thermal equilibrium with the hb at any t. The $H'_{Q,n}(q)$'s suggest the following new moments Wn (n = 0, 1, 2, . . .):

$$W_n = W_n(x;t) = \int_{-\infty}^{+\infty} \frac{dq}{q_0} H'_{Q,n}(q) W(x,q;t)$$

(43)

The initial condition $W_{in,n}$ for W_n is obtained by replacing W by the initial non-equilibrium Wigner function Win in Equation (43). One has the following (formal) expansion for W

$$W = W_{eq}(x,q) \sum_{n=0}^{+\infty} W_n(x;t) \frac{H'_{Q,n}(q)}{h'_n}$$

(44)

For $W = W_{eq}(x,q)$, Equation (43) yields $W_{eq,n} = 0$ if $n > 0$, and $W_{eq,0} = h_0$.

For the actual dissipationless case, the role of $H'_{Q,n}(q)$, of the W_n's and of Equation (44) can be appreciated through the following formal argument, which extends to the quantum case behaviours met in the classical case (Subsections 2.1–2.3). One could argue that, at least for some initial non-equilibrium conditions and after some approximations, $W_n(x;t)$ would approach or be related to $W_{eq,n}$. Then, for long-time and approximately, the dominant moment would be $W_0(x;t)$, while any $W_n(x;t)$ with $n > 0$ would be negligible, W_n being the smaller, the larger n and $t(> 0)$. One has: $\int_{-\infty}^{+\infty} dx \int_{-\infty}^{+\infty} d(q/q_0) W(x,q;t) = \int_{-\infty}^{+\infty} dx W_0(x;t)$. The above argument, even if it would not seem essentially wrong, is recognizedly loose for vanishing dissipation, unless some approximation be made. It will be revisited in Section 6, with dissipation.

The transformation of Equations (35) and (36) into a linear hierarchy for the new moments W_n is outlined in Appendix A. The hierarchy in Equations (74), (75) and so on is exact and very general, but it requires to know the eigenfunctions $\varphi_j(x)$ and the eigenvalues E_j. Our limited aim here was to show that orthogonal polynomials and a non-equilibrium hierarchy exist formally for Equation (35). Then, we shall neither delve further into this nor treat the issue of long-time approximations for Equations (74), (75) and so on. In Sections 5 and 6, we shall treat other alternatives, with repulsive quadratic plus quartic potentials.

The Wigner Representation for W_{eq} for High Temperature, Near the Classical Limit

We remind the following representation for W_{eq} [given in Equation (38)] as a series in the standard Hermite polynomials $H_n(q/qeq)$ [6]:

$$W_{eq} = W_{c,eq} \sum_{n=0}^{+\infty} a_{0,n} H_n\left(\frac{q}{q_{eq}}\right)$$

(45)

Wc,eq is Boltzmann's classical distribution function, given in Subsection 2.1, and qeq was given in Equation (3). The $a_{0,n} = a_{0,n}(x)$'s are coefficients, which depend on $\hbar$, V (x) and β, with $a_{0,n} \equiv 0$ for n = 1, 3, 5, ... The expansion Equation (45) holds for high temperature and near the classical limit (small β and $\hbar$). The $a_{0,n}$'s, n = 0, 2, 4, 6, 8, . . . , are non-vanishing in general: those for n = 0, 2, 4 have been given in [6]. Here, it will suffice to quote only $a_{0,0}$, through the first terms in a series expansion into powers of $\hbar$):

$$a_{0,0}(x) = 1 + \frac{\hbar^2}{8}\left[\frac{\beta^3}{3m}\left(\frac{dV}{dx}\right)^2 - \frac{2\beta^2}{3m}\frac{d^2V}{dx^2}\right] + \frac{\hbar^4\beta^2}{64m^2}\left[-\frac{4\beta}{15}\frac{d^4V}{dx^4} + \frac{\beta^4}{18}\left(\frac{dV}{dx}\right)^4\right.$$
$$\left. - \frac{22\beta^3}{45}\left(\frac{dV}{dx}\right)^2\frac{d^2V}{dx^2} + \frac{8\beta^2}{15}\frac{dV}{dx}\frac{d^3V}{dx^3} + \frac{2\beta^2}{5}\left(\frac{d^2V}{dx^2}\right)^2\right] + \cdots$$

(48)

All terms in Equation (46) contribute if $\hbar \neq 0$ and $\beta \neq 0$, even if $V = m\omega^2x^2/2$ ($a_{0,0}(x) \neq 1$, then). The key information, which does select uniquely the equilibrium Wigner function W_{eq}, out of the set of all stationary Wigner functions, is encoded in the factor $W_{c,eq}$ and in the fact that $W_{eq} \to W_{c,eq}$ for any V , if $\hbar \to 0$ and small β. This is, of course, consistent with Equation (46) and with $a_{0,0}(x) \to 1$ in the classical high-temperature limit for any V . As shown in [6], all the remaining non-vanishing $a_{0,n}(x)$'s, n = 2, 4, 6, 8, . . . , are determined recursively in terms of $a_{0,0}(x)$

QUANTUM-MECHANICAL ONE-DIMENSIONAL MODEL WITH QUADRATIC PLUS QUARTIC V : NO DISSIPATION

In this Section, we shall restrict to a quadratic plus quartic V ($\geq$ 0):

$$V = V(x) = \frac{m\omega^2x^2}{2} + V_1, \quad V_1 = V_1(x) = \frac{gx^4}{4!}$$

(47)

that is, we deal with a quantum anharmonic oscillator, with harmonic frequency ω(> 0). g, real and $\geq$ 0, is an anharmonicity parameter. Then, H is defined in a denumerably infinite Hilbert space. H has a denumerably infinite discrete spectrum of real eigenvalues E_j ($\geq$ 0), j = 1, . . . and there is no continuous spectrum. For the actual quadratic plus quartic V (x), the (formal) series expansion of the integral operator MQ into powers of $\hbar$ 2 reduces exactly to the first two terms shown in Equation (36). In this Section, we shall also bypass the genuine difficulties of the quantum case: the representation of Weq and the construction of the H_0 Q,n's will now differ considerably from those in Section 4.

Alternative Series for $W_{eq}(x, q)$, Orthogonal Polynomials Generated by It and Hierarchy

We shall deal with $W_{eq}(x, q)$ (as given in Equation (38) in general, and in Equation (45) for small h^- and β) for the special case in Equation (47). We shall recast $W_{eq}(x, q)$ into a new series in the standard Hermite polynomials, which should be not restricted to the regime of small h^- and β [as is Equation (45)]. For that purpose, we shall introduce:

$$\alpha = \frac{\hbar\omega[1 + \cosh(\beta\hbar\omega)]}{2\sinh(\beta\hbar\omega)}, \quad q_{Q,eq} = +[2m\alpha]^{1/2} \tag{48}$$

$$W_{eq}^{(0)} = W_{eq}^{(0)}(x, q) = \exp[-\frac{1}{\alpha}(\frac{q^2}{2m} + \frac{m\omega^2 x^2}{2})] \tag{49}$$

Notice that $q_{Q,eq} \neq q_{eq}$ (although $q_{Q,eq} \to q_{eq}$ for $\hbar\omega \to 0$). One looks for $W_{eq}(x, q)$ as the new formal series in the standard Hermite polynomials $H_n(q/q_{Q,eq})$:

$$W_{eq} = W_{eq}^{(0)} \sum_{n=0}^{+\infty} a_{eq,n} H_n(\frac{q}{q_{Q,eq}}) \tag{50}$$

with new coefficients $a_{eq,n} = a_{eq,n}(x)$ depending on h^-, m, ω, β and x. We emphasize that Equations (45) and (50) are different expansions. Based upon the well known ρ_{eq} for $g = 0$ [7,23–25], a direct computation shows that $W_{eq}^{(0)}$ is the exact Wigner function for the quantum harmonic oscillator for any m, ω, β and h^-. Then, if $g = 0$, it follows that $a_{eq,n} = 0$ for n = 1, 2, 3, 4, . . . and that $a_{eq,0}$ is x-independent. This justifies the interest of Equation (50): it can be expected to converge, at least, for small positive g for finite β.

t us characterize $a_{eq,0}$ first. As $\int_{-\infty}^{+\infty} (dq/q_{Q,eq}) W_{eq} = (q_{eq}/q_{Q,eq}) \int_{-\infty}^{+\infty} (dq/q_{eq}) W_{eq}$ and by using Equations (45) and (50), one gets:

$$a_{eq,0} = \frac{q_{eq}}{q_{Q,eq}} a_{0,0}(x) \exp[+\frac{1}{\alpha}\frac{m\omega^2 x^2}{2}] \exp[-\beta V] \tag{51}$$

If $g = 0$ (for finite β and h^-), the x-dependences of $a_{0,0}(x)$, $\exp[+(2\alpha)^{-1} m\omega^2 x^2]$ and $\exp[-\beta V]$ out in Equation (51) exactly, so that $a_{eq,0}$ is x-independent. For the new series Equation (50), the additional information which does select uniquely W_{∞}, out of the set of all stationary Wigner functions, is encoded in the factor $W_{eq}^{(0)}$ and in $a_{eq,0}(x)$ [determined, through Equation (51), in terms of $a_{0,0}$ given, in turn, in Equation (46)]. Equations (39), (50), (47) and (49) yield a three-term linear recurrence relation for the aeq,n's: see Appendix B.

By generalizing Equation (50), one could expand the non-equilibrium Wigner function in terms of the $H_n(q/q_{Q,eq})$'s, in terms of non-equilibrium moments a_n's. For large t(> 0) and, at least, for some suitable class of initial non-equilibrium conditions, one could argue whether the a_n's would approach approximately or be related, in some sense, to the $a_{eq,n}$'s which, as shown above, are non-vanishing. Then, for long-time, all an, n = 2, 4, 6, . . . would be non-vanishing as well and should be taken into account, which would make their analysis rather difficult. The above argument, which is rather loose in the absence of dissipation, will be revisited in Section 6 with friction included. Anyway, the above argument will provide motivation for the following developments. We shall come back to the denumerably infinite family of polynomials $H'_{Q,n}(q)$ $(n = 0, 1, 2, 3, \ldots)$, orthogonalized in q (for fixed x) by using Weq [Equation (38), in the form of Equation (50)] as weight function. Here, we shall use Equation (40), with q0 replaced by qQ,eq. For the actual V , given in Equation (47), we shall recast those $H'_{Q,n}(q) \equiv H_{Q,n}(y)$ $(y = q/q_{Q,eq})$ into another form. Strictly speaking, the $H'_{Q,n}(q)$'s considered in Equation (40) for a general V coincide with the $H_{Q,n}(y)$'s to be analyzed here only when Equation (47) holds. By assumption, we search for $H_{Q,n}(y)$'s to equal the standard Hermite polynomial $H_n(y)$ plus a remainder. The latter is another polynomial in y of degree smaller than n. We shall write:

$$H_{Q,n}(y) = H_n(y) + \sum_{j=1}^{n} \sigma_{n,n-j} H_{n-j}(y)$$

(52)

with n = 1, 2, 3, . . . and (y-independent) coefficients $\sigma_{n,n-j}$ (given in Appendix B), which depend on x and β (these two dependences not being explicited). The orthonormalized polynomials are $H_{Q,n}(q/q_{Q,eq})/h_{Q,n}^{1/2}$, with the (x-dependent) normalization factor:

$$h_{Q,n} \equiv \int_{-\infty}^{+\infty} \frac{dq}{q_{Q,eq}} W_{eq}(x, q) H_{Q,n}(q/q_{Q,eq})^2$$

(53)

See Appendix B. The alternative hierarchies off-equilibrium and at equilibrium following from $H'_{Q,n}(q) = H_{Q,n}(q/q_{Q,eq})$ and $h'_n = h_{Q,n}$ (with q_0 replaced by $q_{Q,eq}$) are outlined in Appendix B.

QUANTUM-MECHANICAL ONE-DIMENSIONAL MODEL WITH QUADRATIC PLUS QUARTIC V : DISSIPATION

First Model: Equilibrium Wigner Function Dependent on Dissipation

Here, we shall suppose that the evolution of the qBp, in the presence of the hb, is also subject to non-vanishing friction. We shall regard the qBp as an open quantum system, also described by Equations (32) and (47) in the same denumerably infinite Hilbert space. Recall that open quantum systems constitute a very active research field, which is common to several modern branches of quantum physics (quantum optics, laser theory, atoms and electromagnetic radiation in cavities, spin relaxation dynamics, decoherence, ion traps, quantum information, . . .) [23–31]. For t > 0, we shall assume that the QME for W, including dissipation, is:

$$\frac{\partial W(x,q;t)}{\partial t} = -\frac{q}{m}\frac{\partial W(x,q;t)}{\partial x} + M_Q W + M_D W$$

$$(54)$$

$$M_D W = [\gamma + \frac{\gamma}{2}(x\frac{\partial}{\partial x} + q\frac{\partial}{\partial q}) + \gamma\frac{\alpha}{2m\omega^2}(\frac{\partial^2}{\partial x^2} + m^2\omega^2\frac{\partial^2}{\partial q^2})]W$$

$$(55)$$

with the same M_Q given in Equation (36). The linear operator M_D, with the same α as in Equation (48), accounts for the dissipation on the qBp due to the hb. The key real parameter γ (≥ 0) accounts for friction effects. Equations (54) and (55) readily imply that $(\partial/\partial t)\int_{-\infty}^{+\infty} dx \int_{-\infty}^{+\infty} dq W(x,q;t) = 0.$ Equations (54) and (55) are: (a) the standard equations describing a single mode of the electromagnetic field inside a laser cavity [23,24], if g = 0; (b) a toy model for the inclusion of weak nonlinearities in (a) (say, due to a nonlinear medium within the cavity), along the lines pursued in [25], if $g \neq 0$.

First, we shall look for the physically acceptable equilibrium solution, denoted as $W_{eq,\gamma}(x,q)$, of Equation (54) $(\partial W_{eq,\gamma}(x,q)/\partial t = 0)$. An important property is that, provided that $g = 0$, $W_{eq}^{(0)}$ given in Equation (49) [with the same $q_{Q,eq}$ as in Equation (48)], is the equilibrium solution of Equation (54) (with $\gamma \neq 0$) as a direct computation shows. Then, if $g = 0$, $W_{eq,\gamma}(x,q) = W_{eq}^{(0)}$ which is γ-independent. However, for $g \neq 0$, a qualitatively new feature is that $W_{eq,\gamma}(x, q)$ is indeed γ-dependent and, hence, it is not given by the right-hand-sides of either Equation (34) (with $\rho = \exp(-\beta H)$) or Equation (50). This dependence of quantum equilibrium distributions in certain models, like the one in Equations (54) and (55), on the dissipation mechanism is one of the additional difficulties met in the quantum case, to be treated in this Subsection. We remind that, in order to avoid physically that difficulty, an interesting generalization of Equation (1) to the quantum case (to order $\hbar^2$), in the regime of small $\hbar$ and β, has been carried out in [32].

Then, by reminding Equation (50) for the frictionless case, we shall look for $W_{eq,\gamma}(x,q)$ if $\gamma \neq 0$ and $g \neq 0$, as the new series using the standard Hermite polynomials:

$$W_{eq,\gamma} = W_{eq}^{(0)} \sum_{n=0}^{+\infty} a_{eq,\gamma,n} H_n\left(\frac{q}{q_{Q,eq}}\right)$$

(56)

with new (γ-dependent) coefficients $a_{eq,\gamma,n}$ ($\neq a_{eq,n}$) see Appendix C. In order to select uniquely Weq,γ in the actual case with dissipation, we need a choice for $a_{eq,\gamma,0}$. We shall choose $a_{eq,\gamma,0} = a_{eq,0}$.as given in Equation (51) (γ-independent). Such a choice for $a_{eq,\gamma,0}$ seems a natural one (not leading to inconsistencies) although, recognizedly, it also seems less compelling than that for aeq,0 in Equation (51) (which was necessary, for consistency). Then, all $a_{eq,\gamma,n}(x)$'s, $n = 1, 2, 3, 4, \cdots$, are determined recurrently in terms of $a_{eq,\gamma,0}(x)$, through Equation (89), through the obvious generalizations of Equations (77) and (78). Notice that

$a_{eq,\gamma,n}(x) \neq 0$ for odd n' ($n' = 1, 3, 5, \ldots$). For $g = 0$,   with nonvanishing β, $\hbar$ and γ, $a_{eq,\gamma,n}(x) = 0$ for $n = 1, 2, 3, 4, \ldots$ while $a_{eq,\gamma,0}(x)$ is x-independent, consistently.

Like in the dissipationless case, we shall introduce the new denumerably infinite family of all polynomials in $y = q/q_{Q,eq}$, $H_{Q,\gamma,n} = H_{Q,\gamma,n}(y)(n = 0, 1, 2, 3, \ldots)$, orthogonalized in q by using the equilibrium Wigner function $W_{eq,\gamma}$ given in Equation (56) as weight function. Then:

$$\int_{-\infty}^{+\infty} \frac{dq}{q_{Q,eq}} W_{eq,\gamma}(x,q) H_{Q,\gamma,n}(y) H_{Q,\gamma,n'}(y) = 0, \, n \neq n'$$

(57)

We impose that $H_{Q,0}(y) = 1$ and that for $n \geq 1$, $H_{Q,\gamma,n}(y)$ equals the standard Hermite polynomial Hn(y) plus another polynomial in y of degree smaller than n. Then:

$$H_{Q,\gamma,n}(y) = H_n(y) + \sum_{j=1}^{n} \sigma_{\gamma,n,n-j} H_{n-j}(y)$$

(58)

with n = 1, 2, 3, ... The coefficients $\sigma_{\gamma,n,n-j}$ depend on x, β and γ: they are determined recurrently through Equation (57). See Appendix C. We omit the normalizing factors of $H_{Q,\gamma,n}(y)$.

Using the $H_{Q,\gamma,n(y)'s}$, we introduce the new non-equilibrium moments $W_{\gamma,n}$ (n = 0, 1, 2, . . .) of W:

$$W_{\gamma,n} = W_{\gamma,n}(x;t) = \int_{-\infty}^{+\infty} \frac{dq}{q_{Q,eq}} H_{Q,\gamma,,n}(y) W(x,q;t)$$

(59)

Notice that $\int_{-\infty}^{+\infty} dx \int_{-\infty}^{+\infty} d(q/q_{Q,eq}) W(x,q;t) = \int_{-\infty}^{+\infty} dx W_{\gamma,0}(x;t)$. At equilibrium, the moments of $W_{eq,\gamma}$ determined by Equation (59) are: $W_{eq,\gamma,n} = 0$ for $n = 1,2,3,4,...$ while $W_{eq,\gamma,0} = \pi^{1/2} \exp[-\frac{1}{\alpha} \frac{m\omega^2 x^2}{2}] a_{eq,,\gamma,0}$. Then, one can proceed to the non-equilibrium recurrence relation for the Wγ,n. It will suffice to give the lowest two equations in the recurrence in Appendix C.

Second Model: Equilibrium Wigner Function Independent on Dissipation (Lindblad's Theory)

A characteristic feature of Equations (54) and (55) was that the corresponding equilibrium Wigner function did depend on dissipation. Here, we shall treat another model for the evolution of the qBp, in the presence of the hb, also with the choice Equation (47) for V (x) and subject to non-vanishing dissipation, based upon [25,29,33,34] and to all orders in h⁻: then, the equilibrium Wigner function will be independent on dissipation. For an economical description, it will be convenient not to limit to the Wigner function W, but to employ also an equivalent operator formulation in terms of the density operator $\rho(t)$, which is uniquely determined by W. In fact, any matrix element of $\rho(t)$ is given, by performing an inverse Fourier transform of Equation (34), by:

$$\langle x - y | \rho(t) | x + y \rangle = \int_{-\infty}^{+\infty} dq \exp[\frac{-i2qy}{\hbar}] W(x,q;t)$$

(60)

Let a, a⁺ be the standard destruction and annihilation operators for the quantum harmonic oscillator:

$a = 2^{-1/2}[\alpha_0 x + \alpha_0^{-1}(\partial/\partial x)]$ and $a^+ = 2^{-1/2}[\alpha_0 x - \alpha_0^{-1}(\partial/\partial x)]$ ($\alpha_0 = (m\omega/\hbar)^{1/2}$).+ will always denote the adjoint. Then, with Equation (47), one has: H = H$_0$ + V$_1$, with $H_0 = \hbar\omega_0(a^+a + 1/2)$ and $V_1 = g_1(a + a^+)^4$ ($g_1 = g/[2^2 4! \alpha_0^4]$).

We shall suppose that the actual time evolution of the qBp, including dissipation, is a quantum Markov process, described by the following QME:

$$\frac{\partial\rho}{\partial t} = \frac{1}{i\hbar}[H,\rho] + D\rho$$

(61)

D ("the dissipator"), accounting for the dissipation on the qBp due to the hb, is a linear operator. In order to characterize D, we shall introduce first the following linear (t-independent) operators A(ω') and A⁺(ω') as:

$$[H, A(\omega')] = -\hbar\omega' A(\omega') \tag{62}$$

$$[H, A^+(\omega')] = \hbar\omega' A^+(\omega') \tag{63}$$

$$A^+(\omega') = A(-\omega') \tag{64}$$

$D, A(\omega')$ and $A^+(\omega')$ act on the actual denumerably infinite Hilbert space. Ω' (with dimension (time) $^{-1}$) is a real number, as shown in general [29]. $A(\omega')$ and $A^+(\omega')$ are named the eigen-operators of H [33,34] and also Lindblad's operators. They are supposed to be dimensionless. $A^+(\omega')$ is a traceless operator [29]. Equations (62) and (63) imply: $[H, A^+(\omega')A(\omega')] = 0.$Equation (64) implies that A(ω') is Hermitian only if ω' = 0: then, [A(ω' = 0), H] = 0. For the time being, we shall not deal with A(ω'), A+(ω 0) and the associated ω 0 for g1 6= 0: we shall turn to that pending task in Appendix D, for the sake of completeness.

We shall make more precise the model in Equation (61) by assuming that Dρ has the structure [33,34]:

$$D\rho = \sum_{\omega''} \gamma(\omega'')[[A(\omega'')\rho, A^+(\omega'')] + [A(\omega''), \rho A^+(\omega'')]] \tag{65}$$

The summation in Equation (65) is performed only over two real values: $\omega'' = \omega'$ and $\omega'' = -\omega'$. $\gamma(\omega'')$ are real and positive numbers: they have dimension (time) $^{-1}$ and, physically, they play the role of relaxation rates. Technically, Equations (61)–(65) (known as Lindblad's formulation for Markovian open quantum systems) give the most general Markovian and time-homogeneous master equation describing a non-unitary evolution of ρ that is trace preserving and completely positive for any ρ_{in}. See [25,29,31] for presentations of several alternative arguments leading to Equation (65). For g 6= 0 (say, g1 6= 0), the representation of Equation (61) [with Dρ given in Equation (65)] by means of the Wigner function is different from Equation (54) [with M_D given in Equation (55)]. It is crucial to notice that the canonical (t-independent) density operator ρeq(= exp(−βH)) with vanishing friction as considered in Subsection 4.1 (with [ρ_{eq}, H] = 0) fulfills:

$$\rho_{eq}A(\omega') = \exp[\beta\hbar\omega']A(\omega')\rho_{eq} \tag{66}$$

$$\rho_{eq}A^+(\omega') = \exp[-\beta\hbar\omega']A^+(\omega')\rho_{eq} \tag{67}$$

We shall not treat the class of models in Equations (61) and (65) in its full generality. Rather, by inspiring ourselves on [25,29], we shall add the assumption:

$$\gamma(-\omega') = \exp[-\beta\hbar\omega']\gamma(\omega') \tag{68}$$

Equations (61), (65) and (68) fully characterize the quantum model with dissipation on which we shall concentrate in this Subsection. Then, Equations (65), (66), (67) and (68) imply the crucial property:

$$D\rho_{eq} = 0 \tag{69}$$

See [29] for the proof of Equation (69). As $\partial\rho_{eq}/\partial t = 0$, ρ_{eq} is, indeed, an equilibrium state of the QME Equation (61) describing the qBp in the presence of the hb, with dissipation included through Equations (65) and (68). Then, the equilibrium Wigner function determined by ρeq coincides exactly with Weq in Equation (50). Thus, the orthogonal polynomials for the actual Lindblad's theory coincide with the $H_{Q,n}(q/qQ_{,eq})$'s studied in Subsection 5.1. The non-equilibrium moments WQ,n for the actual Lindblad's theory are also given in Equations (43) and (44), with $H'_{Q,n}(q) = H_{Q,n}(q/q_{Q,eq})$, $h'_n = h_{Q,n}$ and with q_0 replaced by $q_{Q,eq}$.

A very important example of A(ω') and the above formulation is provided by the QME for a quantized radiation mode with frequency ω in a laser cavity, with $V_1 = 0$, when dissipation is included [25,28–30].

Then, with g = 0 ($g_1 = 0$), the representation of Equation (61) and of Dρ [through Equations (65) and (68)] by means of the Wigner function can be shown to coincide with Equations (54) and (55).

At this stage, we shall focus shortly on our main target, namely, the interest of the WQ,n's (and, hence, of the $H_{Q,n's}$) for the time evolution, in the present framework with friction. Like in Subsection 5.1, we multiply Equation (61), with Dρ given in Equations (65) and (68), by $H_{Q,n}(q/q_{Q,eq})/q_{Q,eq}$ and integrate over q. That yields: $\partial W_{Q,n}(x,t)/\partial t = \Lambda_{n,1}(x,t) + \Lambda_{n,2}(x,t)$, in which we manipulate, so as to express the resulting equation solely in terms of the $W_{Q,n'}$'s. $\Lambda_{n,1}(x,t)$ is the contribution from 1 $\frac{1}{i\hbar}[H,\rho]$, which has been treated in Appendix B. $\Lambda_{n,2}(x,t)$ is the contribution from Dρ, given in Equation (65), which gives rise to new features. In order to deal with $\Lambda_{n,2}(x,t)$, we use, successively, the formal closure relation $I = \int dz|z\rangle\langle z|$. Equations (60) and (44), with $H'_{Q,n}(q)$, h'_n and q_0 replaced by $H_{Q,n}(q/q_{Q,eq})$, $h_{Q,n}$ and $q_{Q,eq}$, .respectively. As expected, the net result is: $\Lambda_{n,2}(x,t) = \sum_{n'=0}^{+\infty}\int dx'\lambda_{n,2;n'}(x,x')W_{Q,n'}(x';t)$, with some kernel $\lambda_{n,2;n'}(x,x')$, determined by A(ω'), $A^+(\omega')$ and the $H_{Q,n}(q/q_{Q,eq})$'s and $H_{Q,n'}(q/q_{Q,eq})$'s. We shall omit the complicated $\lambda_{n,2;n'}(x,x')$, because it will be

unnecessary for the discussion that follows.

First, let us consider the case in which the qBp is at thermal equilibrium, so that $W = W_{eq}$ in Equation (50). All equilibrium moments Weq,n_0 = 0 for n_0 = 1, 2, 3, 4, . . . and Weq,n_0=0 = $h_{Q,0}$ fulfill, for any n = 0, 1, 2, 3, . . . , not only $\Lambda n,1$ = 0 but, by virtue of Equation (69), $\Lambda_{n,2}$ = 0 as well.

Based upon the above equilibrium case, we consider next the time evolution for large t. It may be adequate to address here the discussion in Subsection 4.1, after Equation (44). One would argue that $W(x,q;t) - W_{eq}(x,q)$ should be the smaller (and eventually tend to zero), the larger t is (at least, for a suitable, possibly restricted, class of initial non-equilibrium conditions). Then, to argue that, for long-time, the dominant moment would be $W_{Q,0}(x;t)$, while any $W_{Q,n}(x;t)$ with $n > 0$ would be negligible ($W_{Q,n}$ being the smaller, the larger n and t(> 0)) would not seem essentially incorrect. This (even in the lack of a rigorous control) would suggest the following approximation scheme, for suitably large t and some restricted set of initial non-equilibrium conditions. In the hierarchy $\partial W_{Q,n}(x,t)/\partial t = \Lambda_{n,1}(x,t) + \Lambda_{n,2}(x,t)$, let us restrict to the equations for n = 0 and n = 1 for suitably large t, and we neglect in them all $W_{Q,n'}(x,t)$'s with $n' \geq 2$.

In the resulting equation for n = 1, we also neglect $\partial W_{Q,1}(x, t)/\partial t$, obtain formally $W_{Q,1}$ in terms of WQ,0 and reshuffle the last expression into the equation for $\partial W_{Q,0}(x,t)/\partial t = \Lambda_{0,1}(x,t) + \Lambda_{0,2}(x,t)$. This would provide an approximate evolution equation for $W_{Q,0}$(x, t) alone, for large t, describing the approach to thermal equilibrium of the qBp, with dissipation.

CONCLUSIONS, DISCUSSIONS AND OPEN PROBLEMS

This work contains two parts, very closely related to each other, summarized below with more detail. The first part (a review of previous work, including certain improved analysis) has dealt with classical systems: Sections 2 and 3. It is not warranted that the main issues of the latter (the long-time approximation and so on) could be extended to general quantum systems. As an attempt towards that, the second part has been devoted to one-dimensional quantum systems: Sections 4, 5 and 6, with new material dealing with more difficult problems and, hence, having a formal and limited scope. (1) We have extended previous work by other authors on open classical one-dimensional systems off-equilibrium, with dissipation. In the present work, we have considered those systems, both with or without dissipation, as well as closed classical many-particle ones. In all cases, equilibrium (Boltzmann's) classical canonical distribution functions, which are Gaussian in momenta, have been used as weight functions to generate families of orthogonal polynomials in momenta (namely, the standard Hermite polynomials), for fixed spatial coordinates. Distinguishing and simplifying features of all classical systems considered here are that dependences on classical momenta and those on spatial coordinates factorize (as no magnetic fields are considered), so that those orthogonal polynomials in momenta are independent on positions. Three-term linear hierarchies for the non-equilibrium classical moments (which depend only on the spatial coordinates) have been studied and solved formally, in terms of certain operator continued fractions. For an open classical one-dimensional system with dissipation, the equilibrium distribution is independent on the latter: the analysis (with repulsive potentials which vanish at large distance) shows that, for long times, the lowest moment dominates the evolution towards thermal equilibrium with the hb, while higher order moments are subdominant. The latter study has been extended, under certain long-time approximation, to the case without dissipation, which does provide valuable hints for generalizations to non-equilibrium closed classical interacting nonrelativistic many-particle systems, with repulsive potentials vanishing at large distance. Then, the very large number of degrees of freedom of the system plays the role of a hb, and the initial states are assumed to correspond to thermal equilibrium at large distances (thereby introducing the equilibrium temperature, T), but to non-equilibrium situations at finite distances. The canonical equilibrium distribution leads to orthogonal polynomials which are suitable products of Hermite polynomials and leads to generalize the non-equilibrium moment method. We emphasize that for closed classical three-dimensional many-particle interacting systems, the non-equilibrium hierarchy based upon moments treated here is radically different from the well known one due to Born, Bogoliubov, Green, Kirkwood and Yvon (BBGKY) [2,3] for classical distribution functions, and that the operator continued fractions appear to yield, in a long-time approximation, irreversibility and thermalization of the whole system at T, for long times (and, hence, an approximate arrow of time): then, the lowest moment also dominates the evolution towards thermal equilibrium. The conclusion also appears to hold for a non-equilibrium classical closed plasma, with the degrees of the classical electromagnetic field included [15]. The classical open one-dimensional case corresponding to a harmonic oscillator (with a repulsive potential which grows at large distance), with vanishing dissipation, has been discussed very shortly, for completeness. The analysis of non-equilibrium for classical systems, considered here, has been based on continued fractions of certain linear operators. Some crucial properties of those operator continued fractions (Hermiticity, absence of negative eigenvalues) have been inferred through the analysis of examples and iterative arguments. Moreover, those operators have been constructed in compact forms, in outline, for free particles and for a classical harmonic oscillator. We emphasize that such operators appear to be very interesting objects, which would deserve much more mathematical attention than the limited one devoted to them here.

(2) Non-equilibrium quantum interacting systems present various conceptually new difficulties. We have concentrated on descriptions through Wigner functions. We quote some quantum difficulties of the latter, on which we have focused: canonical equilibrium Wigner distributions W_{eq} are not Gaussian in momenta, their dependences on the latter and on spatial coordinates do not factorize and they may depend on the dissipation mechanism (for various open systems). Due to such difficulties, we have limited ourselves to open quantum one-dimensional interacting models, so as to investigate various procedures o construct the W_{eq}s, and the very construction of families of orthogonal polynomials in momenta which have those W_{eq}s as weight functions, to all orders in Planck's constant. We have considered, first, a general repulsive potential V (x) (vanishing suitably for $| x | \mapsto +\infty$) without dissipation due to the hb first and, later, we have turned to a repulsive quadratic plus quartic potential (either without or with dissipation). We have studied the $W_{eq's}$, in all those cases. For a repulsive quadratic plus quartic potential with dissipation, we have considered two models, inspired on laser theory: (i) one, in which W_{eq} does depend on dissipation; (ii) another one, using Lindblad's theory, in which W_{eq} is independent on dissipation. In all those cases, we have outlined the construction of the new families of orthogonal polynomials in momenta, which have the corresponding W_{eq} as weight function. We have employed those new families of orthogonal polynomials (depending parametrically also on spatial coordinates), in order to construct new moments of the non-equilibrium Wigner function. Hierarchies for the non-equilibrium moments (which depend only on spatial coordinates) have been treated in outline. Then, it seems possible to generalize the developments of the open classical case for the open quantum one, at least formally. The construction of stationary or equilibrium solutions of Equation (61) in a general case (in which Equations (68) and (69) do NOT hold) would proceed, if the eigen-operators are known, through a procedure similar to that in Subsection 6.1. See [31] for theorems and constructive methods to find stationary solutions of Equation (61) in general cases.

The issues of carrying through some limited control of convergence and of long-time approximations in the open quantum one-dimensional case (generalizing the ones for the classical cases in Sections 2 and 3) are more difficult and have not been addressed here. We have also omitted generalizations to quantum closed interacting many-particle systems (and, hence, issues related to quantum indistinguishability). Then, many questions remain open in the approach to non-equilibrium quantum systems through orthogonal polynomials, moment methods and long-time approximations, compared to classical cases.

ACKNOWLEDGEMENTS

The author is grateful to Craig Callender for inviting him to contribute to the Special Issue Arrow of Time of Entropy. The author acknowledges the financial support of Project FIS2008-01323, Ministerio de Ciencia e Innovacion, Spain. He is an associate member of BIFI (Instituto de Biocomputacion y Fisica de los Sistemas Complejos), Universidad de Zaragoza, Zaragoza, Spain. He thanks A. Rivas for discussions and facilities.

REFERENCES

1. Wallace, D. Reading list for the philosophy of statistical mechanics. Available online: http:// users.ox.ac.uk/ mert0130/papers/smreading.doc (accessed on 13 February 2012).

2. Kreuzer, H.J. Nonequilibrium Thermodynamics and Its Statistical Foundations; Clarendon Press: Oxford, UK, 1981.

3. Balescu, R. Equilibrium and Nonequilibrium Statistical Mechanics; John Wiley and Sons: New York, NY, USA, 1975.

4. Liboff, R.L. Kinetic Theory, 2nd ed.; John Wiley (Interscience): New York, NY, USA, 1998.

5. Zubarev, D.; Morozov, V.G.; Ropke, G. ¨ Statistical Mechanics of Nonequilibrium Processes; Akademie Verlag: Berlin, Germany, 1996; Volume I.

6. Wigner, E.P. On the quantum correction for thermodynamic equilibvrium. Phys. Rev. 1932, 40, 749–759.

7. Hillery, M.; O'Connell, R.F.; Scully, M.O.; Wigner, E.P. Distribution functions in physics: Fundamentals. Phys. Rep. 1984, 106, 121–167.

8. Penrose, O. Foundations of statistical mechanics. Rep. Prog. Phys. 1979, 42, 1937–2006.

9. Brinkman, H.C. Brownian motion in a field of force and the diffusion theory of chemical reactions. Physica 1956, 22, 29–34.

10. Risken, H. The Fokker-Planck Equation, 2nd ed.; Springer: Berlin, Heidelberg, Germany, 1989.

11. Coffey, W.T.; Kalmykov, Yu. P. ; Waldron, J.T. The Langevin Equation, 2nd ed.; World Scientific: Singapore, 2004.

12. Coffey, W.T.; Kalmykov, Yu. P.; Titov, S.V.; Mulligan, B.P. Wigner

function approach to the quantum Bronian motion of a particle in a potential. Phys. Chem. Chem. Phys. 2007, 9, 3361–3382.

13. Hochstrasser, U.W. Orthogonal polynomials. In Handbook of Mathematical Functions; Abramowitz, M., Stegun, I.A., Eds.; Dover: New York, NY, USA, 1965.

14. Alvarez-Estrada, R.F. New hierarchy for the Liouville equation, irreversibility and Fokker-Planck-like structures. Ann. Phys. (Leipzig) 2002, 11, 357–385.

15. Alvarez-Estrada, R.F. Liouville and Fokker-Planck dynamics for classical plasmas and radiation. Ann. Phys. (Leipzig) 2006, 15, 379–415.

16. Alvarez-Estrada, R.F. Nonequilibrium quasi-classical effective meson gas: Thermalization. Eur. Phys. J. A 2007, 31, 761–765.

17. Alvarez-Estrada, R.F. Nonequilibrium quantum anharmonic oscillator and scalar field: High temperature approximations. Ann. Phys. (Berlin) 2009, 18, 391–409.

18. Alvarez-Estrada, R.F. Brownian motion, quantum corrections and a generalization of the Hermite polynomials. J. Comput. Appl. Math. 2010, 233, 1453–1461.

19. Alvarez-Estrada, R.F. Classical systems: Moments, continued fractions, long-time approximations and irreversibility. AIP Conf. Proc. 2011, 1332, 261–262.

20. Alvarez-Estrada, R.F. Quantum Brownian motion and generalizations of the Hermite polynomials. J. Comput. Appl. Math. 2011, 236, 7–18.

21. Gautschi, W. Error functions and Fresnel integrals. In Handbook of Mathematical Functions; Abramowitz, M., Stegun, I.A., Eds.; Dover, New York, NY, USA, 1965.

22. Penrose, O.; Coveney, P.V. Is there a "canonical" non-equilibrium ensemble? Proc. R. Soc. Lond. 1994, A447, 631–646.

23. Louisell, W.H. Quantum Statistical Properties of Radiation; John Wiley and Sons: New York, NY, USA, 1973.

24. Haken, H. Laser Theory; Encyclopedia of Physics, Volume XXV/2c, Light and Matter Ic; Springer: Berlin, Heidelberg, Germany, 1970.

25. Gardiner, C.W.; Zoller, P. Quantum Noise, 3rd ed.; Springer: Berlin, Heidelberg, Germany, 2004.

26. Weiss, U. Quantum Dissipative Systems, 3rd ed.; World Scientific: Singapore, 2008.

27. Joos, E.; Zeh, H.D.; Kiefer, C.; Giulini, D.; Kupsch, J.; Stamatescu, I.-O. Decoherence and the Appearance of a Classical World in Quantum

Theory, 2nd ed.; Springer: Berlin, Heidelberg, Germany, 2003.

28. van Kampen, N.G. Stochastic Processes in Physics and Chemistry; Elsevier: Amsterdam, The Netherlands, 2001.

29. Breuer, H.-P.; Petruccione, F. The Theory of Open Quantum Systems; Oxford University Press: Oxford, UK, 2006.

30. Haroche, S.; Raimond, J.-M. Exploring the Quantum; Oxford University Press: Oxford, UK, 2008. 31. Rivas, A.; Huelga, S.F. Open Quantum Systems. An Introduction; Springer: Berlin, Heidelberg, Germany, 2011.

31. Coffey, W.T.; Kalmykov, Yu.P.; Titov, S.V.; Mulligan, B.P. Semiclassical Klein-Kramers and Smoluchowski equations for the Brownian motion of a particle in an external potential. J. Phys. A Math. Theor. 2007, 40, F91–F98.

32. Lindblad, G. On the generators of quantum dynamical semigroups Commun. Math. Phys. 1976, 48, 119–130.

33. Gorini, V.; Kossakowski, A.; Sudarshan, E.C.G. Completely positive semigroups of N-level systems. J. Math. Phys. 1976, 17, 821–825.

34. Garcia-Palacios, J.L.; Zueco, D. The Caldeira-Leggett quantum master equation in Wigner phase space: Continued-fraction solutions and applications to Brownian motion in periodic potentials. J. Phys. A Math. Gen. 2004, 37, 10735–10770.

Chapter 2

MECHANICAL SENSING OF LIVING SYSTEMS — FROM STATICS TO DYNAMICS

F. Argoul, B. Audit and A. Arneodo

Laboratoire de Physique, CNRS, Ecole Normale Supérieure de Lyon, Lyon, France

ABSTRACT

Living systems are fascinating sensing machines that outmatch all artificial machines. Our aim is to put a focus on the dynamics of mechanosensing in cellular systems through concepts and experimental approaches that have been developed during the past decades. By recognizing that a cellular system is not simply the intricate assembly of active and passive macromolecular actors but that it can also manifest scale-invariant and/or highly nonlinear global dynamics, biophysicists have opened a new domain of investigation of living systems. In this chapter, we review methods and techniques that have been implemented to decipher the cascade of temporal events which enable a cell to sense a mechanical stimulus and to elaborate a response to adapt or to counteract this perturbation. We mainly describe intrusive (mechanical probes) and nonintrusive (optical devices) experimental methods that have proved to be efficient for real-time characterization of stationary and nonstationary cellular dynamics. Finally, we discuss whether thermal fluctuations, which are inherent to living systems, are a source of coordination (e.g., synchronization) or randomization of the global dynamics of a cell.

INTRODUCTION

The concept of mechanical sensing dates back to the 19th century with the emergent theory of tone sensing proposed by H. Helmholtz and J. Muller [1, 2]. All living organisms have the ability to sense mechanical stress and/or hydrostatic pressure, either locally or globally. One of the most studied example is the touch perception of metazoa [3-6]. The first model of mechanosensing proposed by physiologists was constructed on the concept of mechanical receptors [3, 4, 7], *i.e.*, mechanical machineries that are able to transform a mechanical information in another signal that will afterward be interpreted by the cell and potentially transformed into an adaptative response [8]. This concept of receptor or transducer is inspired from mechanical engineering methods. The term mechanotransduction [5] has been more recently introduced to explain how a single cell transforms a mechanical stress through signaling pathways down to nuclear molecular processes.

Mechanosensing [9-11] is a complex mechanism that involves not only a whole range of molecular actors with nanometer-scale sensibility but also a dynamical integration and regulation of these molecular actors that allow a much larger scale ($\mu\mu$m to mm) response in amplitude, with strength and duration adapted to the perturbation. Our physical models of cellular nanomechanics still rarely consider active viscoelastic systems [12] and despite a recent increase of the rate of publications devoted to nanobiomechanics of cells, the concept of cellular biodynamics is still in its middle age. This relative slow progression comes from the necessity to introduce concepts of active matter [13-18] into biology. All cells interact physically with their surrounding tissue and they can establish their response on various timescales. To get a full understanding of cellular biodynamics, we actually need to master the most fundamental concepts of atomic and statistical physics, submicron-scale hydrodynamics and out-of-equilibrium nonlinear dynamics and to associate nanotechnologies with optogenetic, microfluidic tools and molecular and cellular biology methods to achieve a complete mechanogenetic characterization of living cells. Actually, a cell is able to combine biochemical submolecular and supramolecular active or passive interactions with micron-size mechanical and electromagnetic informations that we still have much difficulty to reproduce, with either our experimental devices or our huge computer machines.

There are three aspects that we would like to put forward in this chapter. The first one is the importance of molecular machines of living systems, also called molecular motors, which drive all the cell movements thanks to ATP consumption. These molecular machines are not fully deterministic motors but are in part driven by thermal fluctuations. The second aspect that seems important to us is the fact that fluctuations are predominant in biological systems, and that the fluctuation dissipation theorem [19] must not be taken as granted in all situations. This means that standard approaches in statistical physics and corresponding mathematical models must be used with caution. The third aspect is the importance of multiscale properties in mechanosensing: short-, middle-, and long-distance interactions contribute to the dynamics of cellular systems and more widely of living systems [20-29]. Because they are nonlinear and nonstationary, these interactions produce a global dynamics that each element could not achieve alone. However, the nature of these interactions is still the subject of current debate. We will illustrate different approaches that have been used so far to address this issue. The most striking example of the interplay of mechanics and cell dynamics is illustrated in cell migration and adhesion. Actually, the mobility of a cell is a subtle combination of two counteracting mechanisms: on the one hand, adhesion which tends to immobilize the cell and, on the other hand, protrusion/retraction mechanisms which modify the cell shape and assist its movement. These processes also underlie the ability of a cell to deform under a mechanical perturbation. The cytoskeletal dynamics and mechanics are univocally linked to cellular tension in cell adhesion [30-33]. Cellular movement by membrane protrusion and formation of new adhesions at the cell front cannot occur without a tight link of the cell cortex to the whole cytoskeleton (CSK), allowing the settling of traction forces that drive the cell forward in motion in synchrony with the disassembly of the rear fibers. External membrane protrusions are important components of the ability of a cell to migrate or interact with other neighboring cells [34-36]. These protrusions can be viewed as local instabilities of the cell cortex. They are not independent of the internal dynamics of the CSK: the microtubule plus ends associate with F-actin via plus-tip proteins and act as a scaffolding complex [37] that recruits further down other protein effectors involved in the actin network remodeling [38].

When cells are placed in adherent conditions, they rapidly develop integrin-mediated adhesion complexes that link the extracellular matrix (ECM) to the actin CSK. These transmembrane proteins are associated with a complex of proteins (vinculin, actin, paxillin, tensin, etc.) which allows very fast and reversible connectivity of the intracellular CSK to the outer membrane complexes [39, 40]. The integrin-based molecular complexes concentrate in small domains with different size and shapes that focalize the cell traction force on the ECM. In addition to their function as adhesion sites, matrix adhesion foci also participate in the adhesion-dependent signaling pathways via tyrosine kinases, tyrosine phosphatases, etc. Focal adhesion (FA) centers function as both adhesion and signal transduction hubs that communicate the external stresses of the ECM to the cell interior [39]. The maturation of these FA complexes cannot exist if the cell does not have a contractile machinery, *e.g.*, canceling the cell contractility by inhibiting the Rho GTPase or tyrosine kinase activity aborts this maturation. The higher the cell tension, the larger and more mature are the FA complexes. The FA complexes can therefore be considered as mechanosensing intermediates coupling the internal cell traction forces with the ECM ridigity [41-44].

The actin network can be viewed as a fluid-gel structure which plays both a passive (viscous) and an active (ATP-driven) role in the spatiotemporal dynamics of the cells. This network is dynamically intertwined with microtubules and interfilament networks in such a way that the leading edge of the cell undergoes retrograde flow away from this edge simultaneously to the cell migration. This retrograde flow occurs in two steps: (i) on short timescales, a fast flow in the most peripheral region of the cell, the lamellipodium (a 1-4 μ m width extension filled with a dense network of branching F-actin filaments); and (ii) on longer timescales, a slower centripetal flow over the broad (more than 10 μm width) lamellum [45]. The fast lamellipodium flow principally involves an F-actin network. The slower lamellum flow involves all the CSK filaments (actin stress fibers, microtubules, and intermediate filaments) and a relatively more sparse actin network. These two types of flow are each characterized by a specific organization of the cytoskeletal network and a different turnover rate. In addition, they are driven by distinct forces, namely actin assembly/ disassembly in lamellipodium and actomyosin contraction in lamellum [46]. This indicates that the whole dynamics of a cell during protrusion, traction, and migration is a highly correlated, multiscale (in time and in space) process that entails long-range and short-range mechanisms that can only be tackled using multiscale experimental concepts and methods.

PHYSICAL PROBES TO CAPTURE THE MECHANICAL RESPONSE OF LIVING CELLS

In this section, we concentrate on the nano and micromechanical tools which have been designed in the past two decades to record in real time the mechanics and rheology of a living cell, with the specific purpose to understand its mechanosensing properties. These methods can be classified into two groups: (i) the methods that introduce mechanical tracers inside the cell and follow their spatiotemporal dynamics and (ii) the methods remaining external to the cell and that bring a mechanical device (nano- or microscale in size) close to the cell to follow its response in real time. In each case, a few examples will be described as regard to their ability and efficiency in extracting characteristic temporal and/or spatial scales in the dynamics of cell adaptation to a mechanical stress.

Cellular Rheology From The Outer Membrane

Rheological properties of cells, their deformability under stress are key features of their ability to sense their environment. Recent studies of the microrheology of the intracellular medium have highlighted the fact that this viscoelastic medium is complex and cannot be modeled by the association of a finite set of elastic and viscous elements as usually done in mechanical engineering [25, 47]. Actually, the viscoelastic complex modulus of the cell medium exhibits a weak power-law behavior over a wide frequency range. Using magnetic twisting cytometry (MTC) coupled to an optical detection of the motion of a bead coupled to membrane RGD receptor, Fabry *et al.* [20, 21] succeeded in probing the cell surface dynamics in the frequency range from 0.01 Hz to 1 kHz. During the bead displacement on the cell surface (forced by a twisting magnetic field), the cell responds with an opposing torque that reflects the cell mechanical strength. The ratio of the complex torque T to the complex bead displacement $\tilde{d}$ in Fourier space is defined as the elastic modulus $\tilde{G}(f)$:

$$\tilde{G}(f) = f_g \frac{\tilde{T}(f)}{\tilde{d}(f)},$$

(1)

where the proportionality geometrical factor fgfg depends on the shape, thickness of the cell, and the degree of embedding of the bead in the cell cortex. $\tilde{G} = G' + iG''$, where G' is the shear modulus and G″ the loss modulus. The range of stress and deformation used in this study was limited to the linear response regime for the cell. These authors found for five types of adherent cell models that both G'G'and G″G″ increase with excitation frequency as a weak power-law over the whole frequency range. These power-law dependence of G' and G″ on frequency was also observed by other groups [22, 24, 48-50] and with other methods, such as atomic force microscopy [51, 52]. Except for a small additive viscous term that emerged only at high frequencies, mechanical responses collected from the cell surface did not appear to be tied to any specific frequency and in that respect was considered as (time) scale-invariant.

When a power-law behavior emerges in the rheological response of a cell, a wide range of frequencies is required to bring the experimental demonstration of the existence of scale invariance in the cell dynamical response to stress. Actually, a limited range of frequencies could still be parameterized by a combination of a small number of viscoelastic elements, as an exponential crossover between two regimes. The most impressive result of the above studies is the fact that all the curves captured from different cells of various types could be collapsed to single master curves typical of a soft glassy material (SGM) [25, 53], demonstrating the universality of this behavior [24, 54]. This universality law can be written as

$$\tilde{G}(f) = G'(f) + iG''(f) = G_0 \left(i\frac{f}{f_0} \right)^{x-1} + 2\pi i f \mu,$$

where x is a unifying parameter, G_0 and f_0 are cell-type-dependent scaling factors for stiffness and frequency, μ is an additive Newtonian viscosity term that is negligible for frequencies lower than 30 Hz. This equation tells us that (below 30 Hz) the phase angle φ of $\widetilde{G}$,

$$\phi = \tan^{-1}(\frac{G''}{G'}) = \frac{\pi}{2}(x-1),$$

(3)

is independent of the forcing frequency. This unifying parameter xx depends on the cell state; xxdecreases to 1 when the cell approaches an ideal elastic material (for instance, by increasing its contraction) whereas xx increases toward 2 (limit of a Newtonian viscous fluid) when the cell prestress is diminished (*e.g.*, by disrupting the actin CSK).

The common and generic features of SGMs are due to the fact that they are composed of numerous discrete elements which are interconnected in a random way via weak interactions. These materials are out-of-equilibrium metastable systems, very much like living cells. However, soft glassy dynamics as proposed by soft glass rheology theory [53] is not the only mechanism that can lead to scale-free mechanical behavior as expressed by power-law stress relaxation. Power-law behavior can also be produced by models containing a large number of viscoelastic compartments with a particular distribution of characteristic relaxation times $P(\tau) \sim \tau^{-x}$ that must be related to intracellular processes.

With the same MTC device, it was also possible to track the spontaneous motions (without magnetic twisting) of small beads linked to cell membrane integrin receptors on adherent cells. Bursac *et al.* [48] observed that these motions were intermittent with periods of confinement (stalling) punctuated by directed movement (hopping). Plotting the mean-square displacement of the beads versus time revealed that they were subdiffusing (stalling) at short times but superdiffusing at longer times (hoping) and that intermittent motions reflecting nanoscale CSK rearrangements depended on both the approach to kinetic arrest and energy release due to ATP hydrolysis. The percentage of hopping events in the bead motion was shown to be higher than in soft glassy systems where only thermal energy can push the system out of a microenergy well. Thanks to ATP-driven motors and polymerization/depolymerization cycles, the active properties of the cell CSK and cortex provide alternative ways to visit different microconfigurations and adjust to the ECM changes, with an effective local temperature which increases with the parameter x. Therefore, the exponent xx tells us the extent to which the cell behaves as a fluidic system (x~2) or as an elastic solid (x~1).

Cytoplasm Rheology

Microbeads engulfed inside living cells were used as tracers of the internal cellular activity [55-57]. Two different regimes of transport were observed: on the one hand, the passive fluctuations (local movements of the tracers) which characterize the local viscosity of the cytoplasm and, on the other hand, the active trajectories which are driven by molecular motors such as kinesins and dyneins along microtubules. The same type of experiments has also been performed more recently by nanoparticule tracing and manipulation inside A7 melanoma, MCF-10A and MCF-7 cells [58] with optical tweezers, leading again to the conclusion that the elastic modulus follows a power-law: $|G(f)| \sim f^\beta$, with $\beta = x - 1 = 0.15$, in agreement with previous measurements on the cell exterior [20, 21]. They also noted that the measured cytoplasmic modulus is approximately of a few Pa, much lower than previously estimated for its actin cortex [20]. Thus, these tweezers measurements confirmed the rubber-like elastic properties of the cytoplasm of these cells in two-dimensional (2D) adherent conditions.

Fluctuation Dissipation Theorem and Cell Rheology

The equilibrium fluctuation-dissipation theorem (FDT) [19] assumes that the response of a system to a small perturbation is hampered by spontaneous fluctuations at equilibrium (damping term). Understanding the nature and the amplitude of nonequilibrium forces driving the dynamics of cells out of equilibrium is a very important challenge for statistical physicists and biologists. There are many evidences that FDT fails in living cells [59-62] as well as in active gel systems [63], and it is therefore important to estimate the critical timescale at which active-force-driven fluctuations are predominant over thermal fluctuations. Using micron-size silica beads attached to the wall of a living adherent myoblast cell, Bohec et al. [64] have recently identified a crossover time ($\tau \sim 1s$) between thermally controlled fluctuations and active-force-driven fluctuations. This seems to corroborate that the short-time behavior has equilibrium-like properties, from which the subdiffusive nature of viscoelasticity emerges, while the long-time behavior is strongly governed by active nonequilibrium forces. Bohec et al. [64] also provided a quantitative estimation of the power dissipated by the active forces into the system, which turns out to be three orders of magnitude smaller than the chemical power injected into the underlying motors by the ATPase machinery.

Long-Range Cell Deformation Capture with An Atomic Force Microscope

As emerging in the late 1990s from scanning tunneling microscopy (STM) technologies, atomic force microscopy (AFM) was early recognized to provide a unique opportunity to investigate the structure, morphology, micromechanical properties, and biochemical signaling activity of cells under physiological environment, and this with high temporal and spatial resolutions [65, 66]. The principle of AFM is to bring directly in soft (or hard) contact a sharp-tip cantilever probe over a cell surface and to capture with piconewton sensitivity the interaction force of the tip with the cell surface. AFM is a very powerful technique that has been used to detect single biomolecules (receptors, lipids) on single cell surface without the need for fixation or staining. AFM has such sensitivity that it can be used to measure interaction between and within single biomolecules [67-69]. Beyond its preliminary application for imaging the topography of biological objects [70-72], AFM has become a multitask scanning probe versatile tool (antigen recognition, molecular and membrane flexibility, single molecule, gel, cell and tissue elasticity, electric current, conductance, near-field electromagnetic field) [73, 74]. AFM force spectroscopy can be applied to probe the elastic properties of a cell, either adherent or confined in a narrow chamber [75-78]. Its unique ability to detect and to map the cellular elasticity of living cells with a few tens of nanometers' resolution definitely outmatches the performance of other techniques such as magnetic or optical tweezers. However, it has as a main limitation that it cannot probe cell internal structure without crossing before the cell cortex. This difficulty has been partly overcome recently, thanks to a singular space-scale analysis of force-distance curves to disentangle the viscoelastic moduli of the cell cortex and of the underlying CSK [79]. Interestingly, 2D mapping of mean elastic modulus on a large variety of cells [75, 76, 80-83] was reconstructed, revealing for the first time intracellular interplay of mechanical forces in living cells.

The purpose of this paragraph is not to make a detailed review of AFM or to advertise its latest technological development which can be found in an increasing number of published reports [84-88], but rather to (i) pinpoint the few approaches which were focused on dynamical characteristics of living cells during large deformation and (ii) propose new research directions to perform real-time capture of the cell dynamics when the cell is not in a stationary phase. Unlike AFM-based microrheology measurements [51, 52] discussed in the previous paragraph which were limited to very small deformations, we consider now much larger deformations (more than 1/10 of the cell size) and their temporal and/or frequency decomposition [89]. To perform large deformation cell study, a new experimental strategy has been recently proposed that consists in exciting the cantilever and recording the cell response over a band of frequencies rather than at a single frequency [90].

Note that broadband excitation of the cantilever can also be achieved by thermal excitation [91]. When the probed object is in a stationary regime, power spectral analysis of cantilever fluctuations, based on Fourier analysis, is the best way to understand how the interaction of the cantilever tip is changed when coming in contact with the sample surface. When the sampled surface is not stationary, it is no longer possible to perform a simple spectral analysis which only displays an averaged decomposition of the signal in frequency domain. The lower frequency part of the power spectrum is biased by the cell dynamical adaptation to the cantilever stress. To circumvent this difficulty, time-frequency analysis based on the wavelet transform has recently been proposed [92-95]. The continuous wavelet transform (WT) performs the spectral analysis of the signal on a compact window (given by the wavelet) and allows, therefore, to follow how the cell mechanics changes during its strain-to-stress response. Such a study has been recently performed on HOPG surfaces [96, 97] and on living myoblasts [98].

Intracellular Stress Measurements

External forces that are transferred across integrins in FAs and channeled through the CSK can alter signaling activities deep inside the cell [99]. Another evidence for long-distance force transfer was provided by intracellular stress tomography measurements [89]. The long-distance force transmission mediated by intermediate filaments was observed in response to fluid flow-induced shear stress applied to the apical surface of endothelial cells [100]. Using the MTC technique, Laurent *et al.* [101] also found from alveolar epithelial cells, that the submembranous "cortical" CSK, which is mainly composed of actin, is less stiff and more responsive to external forces than the "deep" subcortical CSK, which also includes intermediate filaments and microtubules. Based on these findings, these authors concluded that "mechanical deformation is transmitted globally throughout the network, whereas the cell surface is able to 'sense' very local deformation forces." To analyze the distribution and dynamics of traction stress within individual FAs, Plotnikov *et al.* [102] applied high-resolution traction force microscopy (TFM) [103, 104] to mouse embryonic fibroblasts expressing enhanced green fluorescent protein (eGFP)-paxillin as FA marker. These fibroblasts were plated on fibronectin-coupled elastic polyacrylamide supports (PAA) of known rigidity embedded with a mixture of red and far-red fluorescent beads.

Cell-induced ECM deformation was visualized by spinning disk confocal microscopy, and traction fields were reconstructed at 0.7 μm resolution with Fourier transform traction cytometry [104]. FAs were found to exhibit tugging traction fluctuations on a wide range of ECM rigidities, but the choice of tugging versus stable traction states was shown to be regulated by both tension and a specific signaling pathway. These experiments suggest that strengthening the molecular clutch via the FAK/phosphopaxillin/vinculin pathway broadens the range of rigidities over which dynamic ECM rigidity sampling operates. The requirement for tugging focal adhesion traction in durotaxis suggests that tugging is a means of repeatedly sensing the local ECM rigidity landscape over time. Individual FAs within a single cell sense dynamically the sample rigidity by applying fluctuating pulling forces to the ECM and behave therefore as dynamical sensors to guide durotaxis.

BIOCHEMICAL SENSORS BASED ON FLUORESCENCE METHODS FOR CAPTURING CELL DYNAMICS: FROM THE NANO TO THE MICROSCALE

Fluorescence-Based Nanomechanical Sensors of Intra- And Inter-molecular Dynamics

In the late nineties, combined progress in the biology of fluorescent proteins, miniaturization of optical systems, and nanotechnologies have provided a tremendous asset throughout the investigation of the kinetic properties of macromolecules in living cells [105]. The chemical interaction between two molecular complexes of a metabolic pathway is conditioned by their ability to come in contact, which is often assisted by ATP driven molecular motors. Transport of protein actors in a randomly crowded space such as the cellular cytoplasm differs markedly from a batch reactor. A common form of biochemical regulation is allostery, where an effector molecule binds to a regulatory site and favors a global conformational change that alters further down the structure and function of the active site. Mechanical forces regulate receptor-ligand binding conformation through control of allosteric conformational changes [106]. This general idea of mechanical regulation of active site functions through allosteric-like regulation of a distal site is termed mechanochemistry and is well accepted for motor proteins [107, 108]. We focus here on cytoskeletal proteins, since they are directly involved in mechanosensing pathways; however, the approaches discussed below could be generalized to a wide variety of biochemical interactions.

To dissect how mechanical stress impacts the structure of cytoskeletal proteins, molecular labels have been designed by physicists to provide a fluorescence signal that could report on the molecular strain. Whereas many proteins have been shown *in vitro* or predicted by numerical simulations to undergo conformational changes in response to external mechanical stress, we had to wait until the 2000s to get the demonstration of these changes *in vivo*. Most of these molecular sensors use the Forster resonance energy transfer [109] (FRET). FRET is a technique that can measure the proximity or spatial distance between a donor and an acceptor molecule. It has been widely used to detect protein conformational changes, thanks to molecular constructions with various fluorescent proteins [110]. The principle of stress FRET sensors was elaborated by combining two mutants of a fluorescent protein [111] with either a stable $\alpha\alpha$ -helix linker [112, 113], a spectrin linker [114], or a spider silk domain as linker [115]. These force sensors share a common mechanism for interpreting force: tension in the host induces strain in the linker, leading to increased distance between the donor and acceptor. The dynamic range of these sensors is limited by the nearly linear relationship between FRET efficiency and strain [113].

Recently, using the high flexibility of the vinculin linker domain, a sensor based on force transmission through FAs was developed [115, 116]. When the head integrin domain Vh binds to talin, it recruits vinculin to FA, whereas on the other side the tail integrin domain Vt binds to F-actin and paxillin. This intermediate flexible vinculin linker plays an important role in the transmission of adhesion strength from the FAs to the actin CSK. This calibrated biosensor has piconewton (pN) sensitivity, and the tension across vinculin in stable FAs was estimated to 2.5 pN. It was also demonstrated that higher tension across vinculin favors adhesion assembly and enlargement, and conversely that low tension vinculin favors disassembly or sliding of FAs at the trailing edge of migrating cells. Finally, this study [115] revealed that FA stabilization under force requires both vinculin recruitment and force transmission, and surprisingly, that these processes can be controlled independently.

Another type of strain sensors was elaborated from proximity imaging microscopy (PRIM) combined with GFP dimers [117] and further called PRI-based strain sensor module (PriSSM). If two GFP molecules are brought into physical contact, changes in the ratio of fluorescence emitted when excited with 395 nm and 475 nm light occur. Proximity imaging exploits these changes to reveal homotypic protein-protein interactions *in vivo*. Unlike FRET, PRIM involves only 2 types of fluorescent excitation spectra corresponding to monomeric and dimeric GFP, so that an estimated excitation ratio should simply reflect a mixing ratio of the monomer and the dimer. By combining the GFP-based PRIM technique and myosin-actin as the model system, Iwai *et al.* [117] used this genetically encoded fluorescent sensor to visualize the interaction between myosin II and F-actin in *Dictyostelium* cells. Both spectroscopic and microscopic studies suggested that the fraction of PriSSM-myosin bound to F-actin is low in normal cells.

We have just given few examples of application of FRET to probe cellular internal molecular structures and their transformation under mechanical stress and association with molecular partners. If one can use this method to identify the mechanical organizing centers in a mechanotransduction pathway, the range of forces estimated *in vivo* can be biased by several limitations. FRET is sensitive to changes in distance between 0.5 and 2 times the Forster radius (*i.e.*, between 2 and 10 nm) and to the fluorescence lifetimes of the donor (free or engaged in FRET). Indeed, the dynamic range accessible to this technique is rather low because the interactions which are probed by FRET are typically in the range of a few ns (the fluorescence lifetime of the donor) and difficult to discriminate from background thermal or shot noise. The donor fluorescence lifetime decreases due to energy transfer in the excited state. In adhesion, molecules interact with their closer neighbors but also with other partners on much longer distances which can reach several tens of nm, and thus may be missed by FRET. Conversely, molecules that do not actually participate to a mechanotransduction complex can nevertheless be in juxtaposition and show FRET. FRET experiments suffer from additional artifacts: the method is aimed at probing the interaction of two partners but it is quite impossible to separate the fluorescence responses coming from multiple donor and acceptor interactions even if extensive controls for every FRET pair studied have been performed *a priori*. Finally, none of these methods can determine the number of molecules of a certain component and stoichiometry within an adhesion. A partial correction of these limitations has recently been proposed by coupling FRET biosensors with fluorescence lifetime imaging microscopy [118-120]. Note that this method also provides a millisecond temporal resolution.

Spanning Short- To Long-Range Interactions and Transport With Fluorescence Correlation Spectroscopy

Analyzing the fluorescence fluctuation signals offered a simple, high-resolution, quantitative method to probe the intracellular dynamics that other fluorescence imaging techniques could not afford. From a single fluctuation temporal signal it is possible to get several informations over a wide range of frequencies, such as molecular densities, interaction rate and stochiometry, intra- and extracellular transport (diffusion, advection, etc.). These fluctuation signals should therefore be a very good candidate to capture the multiscale properties of cells in space and in time. Fluorescence correlation spectroscopy (FCS) was originally developed [121] to measure diffusion coefficients and chemical rate constants of biomolecules in solution. It has also been applied successfully to characterize the nature of transport processes of colloidal particles in complex flows [122]. This method uses a focused laser beam to define a very small focal volume (~1 femtoliter) from which the fluctuations of fluorescence intensity are recorded. These fluctuations are analyzed in the nanosecond-to-hour temporal range, and can therefore give information about many different processes including transport, exchange and binding interactions, fluorescence bleaching or blinking. The characteristic times of these different processes are uncovered by computing autocorrelation functions (ACF). Modeling of these ACFs allows the estimation of diffusion, transport, and reaction rates, but it can also be generalized to cross-correlation analysis to quantify molecular interactions if two fluorophores are used simultaneously in the confocal volume [123-125]. In experiments where the fluorescence signal is too weak or lacks contrast, thereby preventing separation of features from background signals, the spatial and temporal fluorescence cross-correlation functions allow to recover enough contrast thanks to their temporal fluctuations. This method was recently applied by Chiu *et al.* [126] to capture actin flow as well as F-actin dynamics and location of F-actin bundles in human breast adenocarcinoma cells (MDA-MB-231) grown in three-dimensional collagen gels. By recording simultaneously the collagen signal using confocal reflection microscopy, they also showed that collagen fibers move in concert with the actin-bundle flow. This experimental study is an impressive demonstration of the impact of fluctuations on internal cell dynamics, and of the power of fluorescence cross-correlation methods to discriminate different molecular entities inside living cells to reach a quantitative model of the intracellular architecture that resembles a random obstacle network for diffusion proteins [125]. Correlation of fluorescence amplitude fluctuations with two colors was also applied to detect the presence of molecular complexes in FAs [127]. In addition to their participation in the structural linking of the ECM to actin filaments, these complexes also serve as signaling "hubs" that regulate many cellular processes, including their own assembly and turnover, migration, gene expression, apoptosis, and proliferation. Capturing the dynamics of assembly and disassembly of these complexes is therefore of major importance to understand how a cell senses its environment. Indeed, there is not a single irreversible event of adhesion but rather a fully orchestrated sequence of adhesion events, which may take from seconds to many minutes to be established. This is typically a multifrequency behavior and applying FCS was a definite step toward deciphering the intricate mechanisms of adhesion. Adhesion complexes and interacting protein actors comprise more than 100 different molecules; some are stably associated and others only transiently [128]. When staining FAK, paxillin (Pax) and vinculin (Vn), with enhanced green fluorescent protein (EGFP) and mCherry, respectively, Digman *et al.* [127] evidenced a heterogeneity in the dynamics and aggregation state of paxillin in different regions across the cell. Taking diffusion as the main mode of transport for adhesion molecules through the cytoplasm, they showed that exchange (binding-unbinding) kinetics with a broad range of rates (0.1-10^{-1} s) dominate in the vicinity of adhesion zones. They observed large clusters and complexes exchanging rather slowly in the vicinity of the disassembling adhesion regions, whereas small aggregates (largely monomers) were observed exchanging rapidly in assembling adhesion zones.

In a very recent paper, Baum *et al.* [125] used FCS to link protein mobility and cellular structure in single cells at high resolution. They mapped the mobility of inert monomers, trimers, and pentamers of the GFP domain on multiple length and timescales in the cytoplasm and nucleus via parallelized FCS measurements. From the perspective of these proteins that cover the range of size of most enzymes, they showed that the cellular interior appears as a porous medium made up by randomly distributed obstacles that reorganize in response to intra- and extracellular cues for small molecules and acts as a viscous medium on large polymeric molecules.

Confining Fluorescence Measurements with Near Field Optical Probes to Improve Sensitivity

Reflection interference contrast microscopy (RICM) [129, 130] has been used since the seventies for imaging the internal structure of cells adhering on solid surfaces. Due to a lack of quantitative interpretation of these images, this method was early abandoned. This technique relies on reflections from an incident beam passing through materials of different refractive indices. The interference of these reflected beams is either constructive or destructive, depending on the thickness and index of the layer of both the liquid medium and the cell in contact with the glass coverslip. More recently, thanks to fast progress in data acquisition and storage and improved modeling of the reflection signals, RICM was applied to a variety of biological situations, such as adhesion of vesicles and cells [131]. It has the practical advantage of not requiring any staining or labeling of the sample, and can be implemented with relative ease and very little investment on a standard inverted microscope. It can also be combined with several other microscopy techniques such as fluorescence or other scanning probe microscopies (AFM, optical or magnetic tweezers [130, 132]). Reflected light imaging has also been coupled to fluorescence excitation in total internal reflection fluorescence (TIRF) microscopy [133-135] to capture the cellular structures involved in FA complexes. As compared to transmission microscopy, this planar confinement (evanescent field) of light not only provides a higher signal-to-noise ratio but also minimizes photodamage to the cellular material [136]. Interestingly, the fact that RICM can be performed without staining the cellular sample was exploited to capture the spontaneous fluctuations (called Fluctuation Contrast RICM or Dynamical RICM) of a soft interface to identify the organization of specific ligand-receptor bonds in cellular adhesion [137, 138].

More recently, surface plasmon microscopy has been proposed for imaging internal structures of cells without staining [139 -144]. This microscopy offers also the possibility to recover both the amplitude and the phase of the reflected field and in some situations to retrieve the index of the layer in contact with gold without needing to know its thickness [145-148]. This microscopy combines total internal reflection of light with surface plasmon resonance excitation to achieve high contrast and high resolution images. Lately, surface plasmon resonance imaging ellipsometry (SPRIE) has been applied to capture cell-matrix adhesion dynamics and strength [149].

Beyond Fluorescence Methods: Quantitative Phase Microscopies for Living Cell Data Capture

In the fifties, phase contrast (PC) and differential interference contrast (DIC) microscopies [150] have revolutionized the biologist view of living systems, by inferring their morphometric features without the need for exogenous contrast agents [151]. However, both PC and DIC remain qualitative in terms of optical path-length measurement, since the relationship between the incoming light power and the optical phase of the image field is generally nonlinear. Quantifying the optical phase shifts associated with biological structures was expected to give access to important information about morphology and dynamics at the nanometer scale [152-154]. However, imaging large field of view samples required time-consuming raster scanning. Full-filed phase measurement techniques were also developed [155,156], providing simultaneous information from a large number of points on the sample. Fourier phase microscopy (FPM) [157], digital holographic microscopy (DHM) [158] and quantitative phase microscopy (QPM) [159-164] have recently been implemented to provide quantitative phase images of biological samples with remarkable sensitivity and stability over extended periods of time. Thanks to its sub-nanometer path-length stability over long periods and efficient algorithms to retrieve the phase maps from fringe patterns [161, 165], QPM is well suited for studying a wide range of temporal scales.

This technique has been applied to capture red blood cell fluctuations (spontaneous flickering), which manifest as submicron motions characterized by membrane displacements in the millisecond (or less) timescale. Amin *et al.* [166] showed that the frequency behavior of the complex modulus G(f) of healthy red blood cells is similar to that obtained in SGMs. Over the frequency range 5 - 50 Hz, the storage and dissipation moduli approach power-law behavior, $G'\propto f0.5\pm0.02 G'\propto f0.5\pm0.02$ and $G''\propto f0.7\pm0.05 G''\propto f0.7\pm0.05$, where the errors indicate cell-to-cell variations (N=13). As already discussed for MTC, the intermediate exponent of 0.7 tells us that normal (discoid) red blood cell membranes behave neither as purely elastic nor as purely viscous media, but as viscoelastic gels. For red blood cells switching to echinocyte and spherocyte shapes, the $G''G''$ exponent decreases consistently, indicating stronger confinement of the membrane viscous motions. Finally, above 35 Hz, the viscous modulus becomes dominant, *i.e.*, the cell transits toward a dissipation-dominated regime, which has been ascribed by the authors to the culture medium viscosity.

FROM FLUCTUATIONS TO DETERMINISTIC BEHAVIOR

Emergence of Coherent Dynamics In Cellular Systems

So far, most cellular models have been established at specific scales, those which focus on molecular mechanisms are not suited to pave macroscopic scales and inversely. Establishing a connection between the discrete stochastic microscopic and the continuous deterministic macroscopic descriptions of the same biological phenomenon is likely to give new clues toward the understanding of mechanotransduction and mechanosensing processes. The scale invariance properties of the cell rheology revealed by MTC [24, 54] suggest that for very small mechanical deformations, no characteristic timescale emerges. Even if fluctuations have been shown to play an essential role in many biological systems, *e.g.*, Brownian rachets [167-169], does that mean that the molecular motors dynamics is not cooperative or synchronized? Progressive molecular motors such as myosin, kinesin and dynein, RNA and DNA polymerases, and chaperonins are macromolecules which hydrolyze ATP while moving unidirectionally along a linear macromolecular track. These progressive transport processes cannot operate without the presence of thermal fluctuations, their directionality resulting from the rectification (asymmetrization) of the Brownian motion [170].

A typical mechanism that crawling cells use to probe their environments is called protrusion, which is a thin (sharp or flat) actin gel extension that the cells generate to move and invade their environment. These protrusions result from many dynamical multicale processes namely polymerization/depolymerization of cytoskeleton filaments (actin, microtubules, and inter-filaments), progressive molecular motors, and FA complexes recruitment [171]. These outer cellular extensions are called filopodia and lamellipodia depending on the shape and dynamics of the protrusion; they also vary with the presence of intra- and extracellular factors [172]. Protrusions grow and shrink in a random manner around the cell on a few minutes' timescales over micrometers. When protrusions are temporarily stabilized, adhesion mechanisms are triggered and the cell can develop traction forces on its ECM. If the cell is polarized, an imbalance between the protrusions at the cell ends may lead to a directional motion. Filopodia stochastic dynamics was shown to play a key role in turning the nerve growth cone to face the chemical signal of a specific partner cell [173-175]. In a recent experimental work, Caballero *et al.* [176] have illustrated the key role of fluctuating protrusions on ratchet-like structures in driving NIH3T3 cell migration. They have shown that stochasticity affects the short- and long-term cell trajectories. They confirmed with a theoretical model that an asymmetry in the protrusion fluctuations is sufficient for predicting the long-term motion, which can be described as a biased persistent random walk. Depending on the type of cells and their environment (ECM stiffness, culture medium), fairly nondeterministic ruffling- and bubbling-like shape dynamics [80, 177] or low-dimensional "periodic" and coherent dynamics [178] can be observed. When placed in conditions for adhesion and spreading on fibronectin-coated glass plates, mouse embryonic fibroblasts (MEFs) show two modes of spreading [178, 179]: on the one hand, anisotropic spreading extensions supported by randomly emerging filopodia [180] and, on the other hand, deterministic spreading extensions that otherwise are rather smooth and continuous. In the latter case, these smooth extensions were shown to be periodically interrupted with a period of about 24 s [178] and to depend on the rigidity of the ECM, integrin binding and myosin light chain kinase (MLCK) activation. Giannone *et al.* [178] suggested a local cytoskeletal signal transport via the actin cytoskeleton from the tips of the lamellipodia to the back where contraction can be activated to start a new cycle. However, in that situation the oscillating signal remained local and did not synchronize over the whole cell cortex. In different situations, global cortical oscillations in spreading cells were observed and attributed to a cyclic depolymerization of microtubules [181], to Arp2/3 complex [182], or to calcium oscillations [183, 184]. These studies question the nature of the transition from stochastic and dissipative [185, 186] local protrusions or membrane pearling [187, 188] processes to global periodic morphological protrusions. This accumulation of experimental evidences of the impact of fluctuations on cell dynamics together with our improved ability to quantify and to model them are pushing the whole cellular biology community to revisit our traditional models of cell shape and dynamics.

From Mechanotransduction to Mechanogenetics: Is There A Genomic Signature Of The Cell Dynamics?

We have seen above that the cell mechanosensing mechanisms involve many length and temporal scales; they are definitely out-of-equilibrium processes which can manifest as stochastic in some situations or low-dimensional periodic dynamics in other situations. Cellular systems have a unique property that no physical/chemical system can reproduce. Depending on the external perturbation, they have the ability to evolve as they synthesize some cytoskeletal elements and/or biochemical activators, which may drastically change gene expression and their mechanical phenotype. Cancer stem cells are vivid examples of very drastic transformations [189]. The mechanical environment of a cell has a direct impact on its genetic expression and reciprocally the interplay between the cell mechanics and its geometrical constraints is conditioned by the gene expression level of all the cytoskeletal and adhesion proteins. Large-scale cellular mechanosensing leads to an adaptive response of cell migration to stiffness gradients [11, 190]. This two-way communication initially termed as mechanotransduction could also be called mechanogenetics of a cell to enlighten the interplay of genomic and mechanical functions. Recent advances in cellular biology have put forward mechanical forces as major actuators in cell signaling in addition to biochemical pathways [191]. Within the cell, the cytoskeleton provides a physical continuity from the ECM down to the interior of the nucleus, enabling direct mechanical links between the cellular microenvironment and chromosome organization. Sensed mechanical signals influence information processing through complex cellular signaling and transcriptional networks that may or may not be specifically force dependent [192]. In many cases, these responses feedback to remodel the cytoskeleton and/or nuclear architecture and consequently modify also the mechanosensitive structures that were initially involved in the response. It has been shown that both integrin-mediated and cadherin-mediated adhesion foci enlarge and strengthen in response to tension in the range of a few tens of seconds [193]. On longer timescales, signaling pathways are activated over minutes (*e.g.* the small GTPase RhoA), which stimulates the formation of actin stress fibers [32], whereas gene expression pathways that operate over hours or days (*e.g.* the induction of vinculin through serum response factor

[194]) change the composition and structure of FAs and of the CSK. Although it seems reasonable to assume that cell mechanics and motility require coordinated protein biosynthesis, nowadays the links between cytoskeletal actin dynamics and correlated gene activities are still poorly understood. Olson *et al.* [194] recently filled this gap by discovering that globular G-actin polymerization can modulate myocardin-related transcription factor (MRTF) cofactors, thereby inducing the nuclear transcription serum response factor (SRF) and subsequently impacting the expression of genes encoding structural and regulatory effectors of actin dynamics. In cancer, the genome architecture is often impacted directly by mutations and/or translocations or chromatin rearrangements, but the influence of the cellular microenvironment may also change the spatiotemporal program of replication and gene expression [195-199]. Recent large-scale sequencing efforts have also helped scientists to delineate the enormous complexity of cancer and the degree to which signaling, drug resistance and genomic alterations vary from patient to patient and even within one patient [200].

ACKNOWLEDGEMENTS

This work was supported by the Agence National de la Recherche (ANR 10 BLANC 1615 and ANR-11 IDEX-0007-02 with the PRES-University of Lyon) and INSERM (Plan Cancer 2012 01-84862)

REFERENCES

1.   H. L. F. Helmholtz. *On the Sensation of Tone as a Physiological Basis for the Theory of Music* (Longmans, Green and Co, London, 1875).

2.   J. Müller. *Elements of Physiology* (Taylor and Walton, London, 1838).

3.   E. D. Adrian & K. Umrath. The impulse discharge from the pacinian corpuscle. *J Physiol* 68, 139-154 (1929).

4.   W. R. Loewenstein & M. Mendelson. Components of receptor adaptation in a pacinian corpuscle. *J Physiol* 177, 377-397 (1965).

5.   M. Chalfie. Neurosensory mechanotransduction. *Natur Rev Mol Cell Biol* 10, 44-52 (2009).

6.   E. A. Lumpkin, K. L. Marshall & A. M. Nelson. The cell biology of touch. *J Cell Biol* 191, 237-248 (2010).

7.   W. R. Loewenstein. Excitation and changes in adaptation by stretch of mechanoreceptors. *J Physiol* 133, 588-602 (1956).

8.   O. P. Hamill & B. Martinac. Molecular basis of mechanotransduction in

living cells. *Physiol Rev* 81, 685-740 (2001).

9. T. Luo, K. Mohan, P. A. Iglesias & D. N. Robinson. Molecular mechanisms of cellular mechanosensing. *Natur Mater* 12, 1064-1071 (2013).

10. H. B. Schiller & R. Fässler. Mechanosensitivity and compositional dynamics of cell-matrix adhesions. *EMBO Rep* 14, 509-519 (2013).

11. B. Ladoux & A. Nicolas. Physically based principles of cell adhesion mechanosensitivity in tissues. *Rep Progress Phys* 75, 116601 (2012).

12. A. C. Callan-Jones & F. Jülicher. Hydrodynamics of active permeating gels. *New J Phys* 13, 093027 (2011).

13. F. Julicher, K. Kruse, J. Prost & J. Joanny. Active behavior of the cytoskeleton. *Phys Rep* 449, 3-28 (2007).

14. G. H. Koenderink, Z. Dogic, F. Nakamura, P. M. Bendix, F. C. MacKintosh, J. H. Hartwig, T. P. Stossel & D. A. Weitz. An active biopolymer network controlled by molecular motors. *Proc Natl Acad Sci USA* 106, 15192-15197 (2009).

15. J.-F. Joanny & S. Ramaswamy. A drop of active matter. *J Fluid Mech* 705, 46-57 (2012).

16. M. C. Marchetti, J. F. Joanny, S. Ramaswamy, T. B. Liverpool, J. Prost, M. Rao & R. A. Simha. Hydrodynamics of soft active matter. *Rev Modern Phys* 85, 1143-1189 (2013).

17. P. Marcq. Spatio-temporal dynamics of an active, polar, viscoelastic ring. *Eur Physic J E* 37, 1-8 (2014).

18. H. Berthoumieux, J.-L. Maître, C.-P. Heisenberg, E. K. Paluch, F. Jülicher & G. Salbreux. Active elastic thin shell theory for cellular deformations. *New J Phys* 16, 065005 (2014).

19. J. Weber. Fluctuation dissipation theorem. *Phys Rev* 101, 1620-1626 (1956).

20. B. Fabry, G. Maksym, J. Butler, M. Glogauer, D. Navajas & J. Fredberg. Scaling the microrheology of living cells. *Phys Rev Lett* 87, 148102 (2001).

21. B. Fabry, G. N. Maksym, J. P. Butler, M. Glogauer, D. Navajas, N. A. Taback, E. J. Millet & J. J. Fredberg. Time scale and other invariants of integrative mechanical behavior in living cells. *Phys Rev E* 68, 041914 (2003).

22. G. Lenormand, E. Millet, B. Fabry, J. P. Butler & J. J. Fredberg. Linearity and time-scale invariance of the creep function in living cells. *J Royal Soc, Interface* 1, 91-97 (2004).

23. C. T. Lim, E. H. Zhou, A. Li, S. R. K. Vedula & H. X. Fu. Experimental techniques for single cell and single molecule biomechanics. *Mater Sci Engin: C* 26, 1278-1288 (2006).

24. X. Trepat, G. Lenormand & J. J. Fredberg. Universality in cell mechanics. *Soft Matter* 4, 1750-1759 (2008).

25. B. D. Hoffman & J. C. Crocker. Cell mechanics: dissecting the physical responses of cells to force. *Ann Rev Biomed Engin* 11, 259-288 (2009).

26. M. A. Stolarska, Y. Kim & H. G. Othmer. Multi-scale models of cell and tissue dynamics. *Philo Transact Royal Soc. Series A* 367, 3525-3553 (2009).

27. D. Mitrossilis, J. Fouchard, A. Guiroy, N. Desprat, N. Rodriguez, B. Fabry & A. Asnacios. Single-cell response to stiffness exhibits muscle-like behavior. *Proc Natl Acad Sci USA* 106, 18243-18248 (2009).

28. D. Mitrossilis, J. Fouchard, D. Pereira, F. Postic, A. Richert, M. Saint-Jean & A. Asnacios. Real-time single-cell response to stiffness. *Proc Natl Acad Sci USA* 107, 16518-16523 (2010).

29. A. K. Miri, H. K. Heris, L. Mongeau & F. Javid. Nanoscale viscoelasticity of extracellular matrix proteins in soft tissues: a multiscale approach. *J Mechan Beh Biomed Mater* 30, 196-204 (2014).

30. J. Small, T. Stradal, E. Vignal & K. Rottner. The lamellipodium: where motility begins. *Trends Cell Biol* 12, 112-120 (2002).

31. T. D. Pollard & G. G. Borisy. Cellular motility driven by assembly and disassembly of actin filaments. *Cell* 112, 453-465 (2003).

32. J. T. Parsons, A. R. Horwitz & M. A. Schwartz. Cell adhesion: integrating cytoskeletal dynamics and cellular tension. *Natur Rev Mol Cell Biol* 11, 633-643 (2010).

33. A. J. Ridley. Life at the leading edge. *Cell* 145, 1012-1022 (2011).

34. A. Pierres, A.-M. Benoliel, D. Touchard & P. Bongrand. How cells tiptoe on adhesive surfaces before sticking. *Biophys J* 94, 4114-4122 (2008).

35. E. S. Welf & J. M. Haugh. Stochastic dynamics of membrane protrusion mediated by the DOCK180/Rac pathway in migrating cells. *Cell Mol Bioengin* 3, 30-39 (2010).

36. K. Lam Hui, C. Wang, B. Grooman, J. Wayt & A. Upadhyaya. Membrane dynamics correlate with formation of signaling clusters during cell spreading. *Biophys J* 102, 1524-1533 (2012).

37. E. Sackmann & A.-S. Smith. Physics of cell adhesion: some lessons from cell-mimetic systems. *Soft Matter* 10, 1644-1659 (2014).

38. A. E. Carlsson. Actin dynamics: from nanoscale to microscale. *Ann Rev*

Biophy 39, 91-110 (2010).

39. S. K. Mitra, D. A. Hanson & D. D. Schlaepfer. Focal adhesion kinase: in command and control of cell motility. *Natur Rev Mol Cell Biol* 6, 56-68 (2005).

40. P. W. Oakes & M. L. Gardel. Stressing the limits of focal adhesion mechanosensitivity. *Curr Opin Cell Biol* 30, 68-73 (2014).

41. D. Riveline, E. Zamir, N. Q. Balaban, U. S. Schwarz, T. Ishizaki, S. Narumiya, Z. Kam, B. Geiger & A. D. Bershadsky. Focal contacts as mechanosensors: externally applied local mechanical force induces growth of focal contacts by an mDia1-dependent and ROCK-independent mechanism. *J Cell Biol* 153, 1175-1186 (2001).

42. A. Bershadsky, M. Kozlov & B. Geiger. Adhesion-mediated mechanosensitivity: a time to experiment, and a time to theorize. *Curr Opin Cell Biol* 18, 472-481 (2006).

43. A. D. Bershadsky, C. Ballestrem, L. Carramusa, Y. Zilberman, B. Gilquin, S. Khochbin, A. Y. Alexandrova, A. B. Verkhovsky, T. Shemesh & M. M. Kozlov. Assembly and mechanosensory function of focal adhesions: experiments and models. *Eur J Cell Biol* 85, 165-173 (2006).

44. B. Geiger, J. P. Spatz & A. D. Bershadsky. Environmental sensing through focal adhesions. *Natur Rev Mol Cell Biol* 10, 21-33 (2009).

45. P. Vallotton, G. Danuser, S. Bohnet, J.-J. Meister & A. B. Verkhovsky. Tracking retrograde flow in keratocytes: news from the front. *Mol Biol Cell* 16, 1223-1231 (2005).

46. A. Ponti, M. Machacek, S. L. Gupton, C. M. Waterman-Storer & G. Danuser. Two distinct actin networks drive the protrusion of migrating cells. *Science* 305, 1782-1787 (2004).

47. W. N. Findley, J. S. Lai & K. Onaran. *Creep and Relaxation of Nonlinear Viscoelastic Materials* (Dover Publications Inc., 1990).

48. P. Bursac, G. Lenormand, B. Fabry, M. Oliver, D. A. Weitz, V. Viasnoff, J. P. Butler & J. J. Fredberg. Cytoskeletal remodelling and slow dynamics in the living cell. *Natur Mater* 4, 557-561 (2005).

49. D. Icard-Arcizet, O. Cardoso, A. Richert & S. Hénon. Cell stiffening in response to external stress is correlated to actin recruitment. *Biophys J* 94, 2906-2913 (2008).

50. A. Asnacios, S. Hénon, J. Browaeys & F. Gallet. Microrheology of living cells at different times and length scales. In *Cell Mechanics: From Scale-Based Models to Multiscale Modeling*, 5-28. Chapman & Hall/CRC, Boca Raton (2010).

51. J. Alcaraz, L. Buscemi, M. Grabulosa, X. Trepat, B. Fabry, R. Farré & D. Navajas. Microrheology of human lung epithelial cells measured by atomic force microscopy. *Biophys J* 84, 2071-2079 (2003).

52. E. Moeendarbary, L. Valon, M. Fritzsche, A. R. Harris, D. A. Moulding, A. J. Thrasher, E. Stride, L. Mahadevan & G. T. Charras. The cytoplasm of living cells behaves as a poroelastic material. *Natur Mater* 12, 253-261 (2013).

53. P. Sollich. Rheological constitutive equation for model of soft glassy materials. *Physic Rev E* 58, 738-759 (1998).

54. E. H. Zhou, X. Trepat, C. Y. Park, G. Lenormand, M. N. Oliver, S. M. Mijailovich, C. Hardin, D. A. Weitz, J. P. Butler & J. J. Fredberg. Universal behavior of the osmotically compressed cell and its analogy to the colloidal glass transition. *Proc Natl Acad Sci USA* 106, 10632-10637 (2009).

55. D. Heinrich & E. Sackmann. Active mechanical stabilization of the viscoplastic intracellular space of Dictyostelia cells by microtubule-actin crosstalk. *Acta Biomaterialia* 2, 619-631 (2006).

56. D. Arcizet, B. Meier, E. Sackmann, J. O. Rädler & D. Heinrich. Temporal analysis of active and passive transport in living cells. *Physic Rev Lett* 101, 248103 (2008).

57. M. Otten, A. Nandi, D. Arcizet, M. Gorelashvili, B. Lindner & D. Heinrich. Local motion analysis reveals impact of the dynamic cytoskeleton on intracellular subdiffusion. *Biophysic J* 102, 758-767 (2012).

58. M. Guo, A. J. Ehrlicher, S. Mahammad, H. Fabich, M. H. Jensen, J. R. Moore, J. J. Fredberg, R. D. Goldman & D. A. Weitz. The role of vimentin intermediate filaments in cortical and cytoplasmic mechanics. *Biophysic Journal* 105, 1562-1568 (2013).

59. A. W. C. Lau, B. D. Hoffman, A. Davies, J. C. Crocker & T. C. Lubensky. Microrheology, stress fluctuations, and active behavior of living cells. *Physic Rev Lett* 91, 198101 (2003).

60. D. Mizuno, C. Tardin, C. F. Schmidt & F. C. Mackintosh. Nonequilibrium mechanics of active cytoskeletal networks. *Science (New York, N.Y.)* 315, 370-3 (2007).

61. C. Wilhelm. Out-of-equilibrium microrheology inside living cells. *Physic Rev Lett* 101, 028101 (2008).

62. R. H. Pritchard, Y. Y. S. Huang & E. M. Terentjev. Mechanics of biological networks: from the cell cytoskeleton to connective tissue. *Soft Matter* 10, 1864-1884 (2014).

63.   A. Basu, J. F. Joanny, F. Jülicher & J. Prost. Thermal and non-thermal fluctuations in active polar gels. *Eur Physic J E* 27, 149-60 (2008).

64.   P. Bohec, F. Gallet, C. Maes, S. Safaverdi, P. Visco & F. van Wijland. Probing active forces via a fluctuation-dissipation relation: application to living cells. *EPL (Europhys Lett)* 102, 50005 (2013).

65.   G. Binnig & C. F. Quate. Atomic force microscope. *Physic Rev Lett* 56, 930-933 (1986).

66.   D. J. Müller & Y. F. Dufrêne. Atomic force microscopy as a multifunctional molecular toolbox in nanobiotechnology. *Natur Nanotechnol* 3, 261-269 (2008).

67.   A. F. Oberhauser, P. K. Hansma, M. Carrion-Vazquez & J. M. Fernandez. Stepwise unfolding of titin under force-clamp atomic force microscopy. *Proc Natl Acad Sci USA* 98, 468-472 (2001).

68.   M. Rief, M. Gautel, F. Oesterhelt, J. M. Fernandez & H. E. Gaub. Reversible unfolding of individual titin immunoglobulin domains by AFM. *Science* 276, 1109-1112 (1997).

69.   A. E. X. Brown, R. I. Litvinov, D. E. Discher & J. W. Weisel. Forced unfolding of coiled-coils in fibrinogen by single-molecule AFM. *Biophysic J* 92, L39-L41 (2007).

70.   P. Milani, G. Chevereau, C. Vaillant, B. Audit, Z. Haftek-Terreau, M. Marilley, P. Bouvet, F. Argoul & A. Arneodo. Nucleosome positioning by genomic excluding-energy barriers. *Proc Natl Acad Sci USA* 106, 22257-22262 (2009).

71.   F. Montel, H. Menoni, M. Castelnovo, J. Bednar, S. Dimitrov, D. Angelov & C. Faivre-Moskalenko. The dynamics of individual nucleosomes controls the chromatin condensation pathway: direct atomic force microscopy visualization of variant chromatin. *Biophysic J* 97, 544-553 (2009).

72.   J. Moukhtar, C. Faivre-Moskalenko, P. Milani, B. Audit, C. Vaillant, E. Fontaine, F. Mongelard, G. Lavorel, P. St-Jean, P. Bouvet, F. Argoul & A. Arneodo. Effect of genomic long-range correlations on DNA persistence length: from theory to single molecule experiments. *J Physic Chem* B 114, 5125-5143 (2010).

73.   P. Hinterdorfer. Molecular recognition studies using the atomic force microscope. *Meth Cell Biol* 68, 115-139 (2002).

74.   Z. Deng, V. Lulevich, F. T. Liu & G. Y. Liu. Applications of atomic force microscopy in biophysical chemistry of cells. *J Physic Chem B* 114, 5971-5982 (2010).

75. M. Radmacher, M. Fritz, C. M. Kacher, J. P. Cleveland & P. K. Hansma. Measuring the viscoelastic properties of human platelets with the atomic force microscope. *Biophysic J* 70, 556-567 (1996).

76. M. Prass, K. Jacobson, A. Mogilner & M. Radmacher. Direct measurement of the lamellipodial protrusive force in a migrating cell. *J Cell Biol* 174, 767-772 (2006).

77. C. Rotsch, F. Braet, E. Wisse & M. Radmacher. AFM imaging and elasticity measurements on living rat liver macrophages. *Cell Biol Int* 21, 685-696 (1997).

78. M. J. Rosenbluth, W. A. Lam & D. A. Fletcher. Supplemental material for: force microscopy of non adherent cells. *Biophysic J* 90, 4-6 (2006).

79. S. Digiuni, A. Berne-Dedieu, C. Martinez-Torres, J. Szecsi, M. Bendahmane, A. Arneodo & F. Argoul. Single cell wall nonlinear mechanics revealed by a multi-scale analysis of AFM force-indentation curves. *Biophysic J.* 108, 2235-2248 (2015).

80. C. Rotsch, K. Jacobson & M. Radmacher. Dimensional and mechanical dynamics of active and stable edges in motile fibroblasts investigated by using atomic force microscopy. *Proc Natl Acad Sci USA* 96, 921-926 (1999).

81. R. Matzke, K. Jacobson & M. Radmacher. Direct, high-resolution measurement of furrow stiffening during division of adherent cells. *Natur Cell Biol* 3, 607-610 (2001).

82. S. E. Cross, Y.-S. Jin, J. Rao & J. K. Gimzewski. Nanomechanical analysis of cells from cancer patients. *Natur Nanotechnol* 2, 780-783 (2007).

83. P. Milani, M. Gholamirad, J. Traas, A. Arnéodo, A. Boudaoud, F. Argoul & O. Hamant. In vivo analysis of local wall stiffness at the shoot apical meristem in Arabidopsis using atomic force microscopy. *Plant J* 67, 1116-1123 (2011).

84. J. Helenius, C.-P. Heisenberg, H. E. Gaub & D. J. Muller. Single-cell force spectroscopy. *J Cell Sci* 121, 1785-1791 (2008).

85. B. N. Johnson & R. Mutharasan. Biosensing using dynamic-mode cantilever sensors: A review. *Biosens Bioelectron* 32, 1-18 (2012).

86. Y. Suzuki, N. Sakai, A. Yoshida, Y. Uekusa, A. Yagi, Y. Imaoka, S. Ito, K. Karaki & K. Takeyasu. High-speed atomic force microscopy combined with inverted optical microscopy for studying cellular events. *Scient Rep* 3, 2131 (2013).

87. A. Pietuch & A. Janshoff. Mechanics of spreading cells probed by atomic force microscopy. *Open Biol* 3, 130084 (2013).

88. A. M. Whited & P. S.-H. Park. Atomic force microscopy: a multifaceted tool to study membrane proteins and their interactions with ligands. *Biochim Biophys Acta* 1838, 56-68 (2014).

89. N. Wang & Z. Suo. Long-distance propagation of forces in a cell. *Biochem Biophysic Res Commun* 328, 1133-1138 (2005).

90. B. J. Rodriguez, C. Callahan, S. V. Kalinin & R. Proksch. Dual-frequency resonance-tracking atomic force microscopy. *Nanotechnology* 18, 475504 (2007).

91. D. O. Koralek, W. F. Heinz, M. D. Antonik, A. Baik & J. H. Hoh. Probing deep interaction potentials with white-noise-driven atomic force microscope cantilevers. *Appl Phys Lett* 76, 2952 (2000).

92. J. Morlet, G. Arenss, E. Fourgeau & D. Giard. Wave propagation and sampling theory-Part I: Complex signal and scattering in multilayered media.*Geophysics* 47, 203-221 (1982).

93. J. Morlet, G. Arensz, E. Fourgeau & D. Giard. Wave propagation and sampling theory-Part II: Sampling theory and complex waves. *Geophysics* 47, 222-236 (1982).

94. R. A. Carmona, W. L. Hwang & B. Torresani. *Practical Time-Frequency Analysis* (Academic Press, San Diego, 1998).

95. Y. Meyer (ed). *Wavelets and their Applications*, (Springer, Berlin, 1992).

96. G. Malegori & G. Ferrini. Wavelet transforms to probe long- and short-range forces by thermally excited dynamic force spectroscopy. *Nanotechnology* 22, 195702 (2011).

97. V. Pukhova, F. Banfi & G. Ferrini. Complex force dynamics in atomic force microscopy resolved by wavelet transforms. *Nanotechnology* 24, 505716 (2013).

98. L. Streppa, C. Martinez-Torres, L. Berguiga, F. Ratti, E. Goillot, A. Devin, L. Schaeffer, A. Arneodo & F. Argoul. Myoblasts are committed to maintaining a minimal acto-myosin driven cytoskeletal tension for differentiating into myotubes: an AFM study. *Phys Biol*, submitted (2015).

99. D. E. Ingber. Tensegrity: the architectural basis of cellular mechanotransduction. *Ann Rev Physiol* 59, 575-599 (1997).

100. S. R. Heidemann, S. Kaech, R. E. Buxbaum & A. Matus. Direct observations of the mechanical behaviors of the cytoskeleton in living fibroblasts. *J Cell Biol* 145, 109-122 (1999).

101. V. M. Laurent, R. Fodil, P. Cañadas, S. Féréol, B. Louis, E. Planus &

D. Isabey. Partitioning of cortical and deep cytoskeleton responses from transient magnetic bead twisting. *Anna Biomed Engin* 31, 1263-1278 (2003).

102. S. V. Plotnikov, A. M. Pasapera, B. Sabass & C. M. Waterman. Force fluctuations within focal adhesions mediate ECM-rigidity sensing to guide directed cell migration. *Cell* 151, 1513-1527 (2012).

103. M. L. Gardel, K. E. Kasza, C. P. Brangwynne, J. Liu & D. A. Weitz. Chapter 19: Mechanical response of cytoskeletal networks. *Meth Cell Biol* 89, 487-519 (2008).

104. B. Sabass, M. L. Gardel, C. M. Waterman & U. S. Schwarz. High resolution traction force microscopy based on experimental and computational advances. *Biophysic J* 94, 207-220 (2008).

105. J. Lippincott-Schwartz, E. Snapp & A. Kenworthy. Studying protein dynamics in living cells. *Natur Rev Mol Cell Biol* 2, 444-456 (2001).

106. W. E. Thomas. Mechanochemistry of receptor-ligand bonds. *Curr Opin Struct Biol* 19, 50-5 (2009).

107. R. D. Astumian & M. Bier. Mechanochemical coupling of the motion of molecular motors to ATP hydrolysis. *Biophysic J* 70, 637-653 (1996).

108. D. Keller & C. Bustamante. The mechanochemistry of molecular motors. *Biophysic J* 78, 541-556 (2000).

109. T. Förster. Expermentelle und theoretische untersuchung des zwischengmolekularen übergangs von elektronenanregungsenergie. *Naturforsch* 4A, 321-327 (1949).

110. I. T. Li, E. Pham & K. Truong. Protein biosensors based on the principle of fluorescence resonance energy transfer for monitoring cellular dynamics. *Biotechnol Lett* 28, 1971-1982 (2006).

111. A. Miyawaki & R. Y. Tsien. Monitoring protein conformations and interactions by fluorescence resonance energy transfer between mutants of green fluorescent protein. *Meth Enzymol* 327, 472-500 (2000).

112. G. Y. Chen, T. Thundat, E. A. Wachter & R. J. Warmack. Adsorption induced surface stress and its effects on resonance frequency of microcantilevers. *J Appl Phys* 77, 3618-3622 (1995).

113. F. Meng, T. M. Suchyna & F. Sachs. A fluorescence energy transfer-based mechanical stress sensor for specific proteins in situ. *FEBS J* 275, 3072-3087 (2008).

114. F. Meng, T. M. Suchyna, E. Lazakovitch, R. M. Gronostajski & F. Sachs. Real time FRET based detection of mechanical stress in cytoskeletal and extracellular matrix proteins. *Cell Mol Bioengin* 4, 148-159 (2011).

115. C. Grashoff, B. D. Hoffman, M. D. Brenner, R. Zhou, M. Parsons, M. T. Yang, M. A. McLean, S. G. Sligar, C. S. Chen, T. Ha & M. A. Schwartz. Measuring mechanical tension across vinculin reveals regulation of focal adhesion dynamics. *Nature* 466, 263-266 (2010).

116. H. Chen & D. C. Chan. Emerging functions of mammalian mitochondrial fusion and fission. *Human Mol Genet* 14, R283-R289 (2005).

117. S. Iwai & T. Q. P. Uyeda. Visualizing myosin-actin interaction with a genetically-encoded fluorescent strain sensor. *Proc Natl Acad Sci USA* 105, 16882-16887 (2008).

118. S. Padilla-Parra, N. Audugé, M. Coppey-Moisan & M. Tramier. Quantitative FRET analysis by fast acquisition time domain FLIM at high spatial resolution in living cells. *Biophysic J* 95, 2976-2988 (2008).

119. S. Padilla-Parra, N. Audugé, H. Lalucque, J. C. Mevel, M. Coppey-Moisan & M. Tramier. Quantitative comparison of different fluorescent protein couples for fast FRET-FLIM acquisition. *Biophysic J* 97, 2368-2376 (2009).

120. E. Hinde, M. A. Digman, C. Welch & K. M. Hahn. Biosensor FRET detection by the phasor approach to fluorescence lifetime imaging microscopy (FLIM). *Microsc Res Techniq* 75, 271-281 (2013).

121. D. Magde, E. L. Elson & W. W. Webb. Fluorescence correlation spectroscopy. II. An experimental realization. *Biopolymers* 13, 29-61 (1974).

122. C. Ybert, F. Nadal, R. Salomé, F. Argoul & L. Bourdieu. Electrically induced microflows probed by fluorescence correlation spectroscopy. *Eur Physic J E* 16, 259-266 (2005).

123. A. I. Bachir, K. E. Kubow & A. R. Horwitz. Fluorescence fluctuations approaches to the study of adhesion and signaling. *Meth Enzymol* 519, 167-201 (2013).

124. K. Bacia, S. A. Kim & P. Schwille. Fluorescence cross-correlation spectroscopy in living cells. *Natur Methods* 3, 83-89 (2006).

125. M. Baum, F. Erdel, M. Wachsmuth & K. Rippe. Retrieving the intracellular topology from multi-scale protein mobility mapping in living cells. *Natur Commun* 5, 4494 (2014).

126. C. L. Chiu, M. A. Digman & E. Gratton. Measuring actin flow in 3D cell protrusions. *Biophysic J* 105, 1746-1755 (2013).

127. M. A. Digman, P. W. Wiseman, C. Choi, A. R. Horwitz & E. Gratton. Stoichiometry of molecular complexes at adhesions in living cells. *Proc Natl Acad Sci USA* 106, 2170-2175 (2009).

128. R. Zaidel-Bar, S. Itzkovitz, A. Ma'ayan, R. Iyengar & B. Geiger. Functional atlas of the integrin adhesome. *Natur Cell Biol* 9, 858-867 (2009).

129. J. Bereiter-Hahn, C. H. Fox & B. Thorell. Quantitative reflection contrast microscopy of living cells. *J Cell Biol* 82, 767-779 (1979).

130. L. Limozin & K. Sengupta. Quantitative reflection interference contrast microscopy (RICM) in soft matter and cell adhesion. *Eur J Chemic Phys Physic Chem* 10, 2752-2768 (2009).

131. H. Verschueren. Interference reflection microscopy in cell biology: methodology and applications. *J Cell Sci* 75, 279-301 (1985).

132. A.-S. Smith, K. Sengupta, S. Goennenwein, U. Seifert & E. Sackmann. Force-induced growth of adhesion domains is controlled by receptor mobility. *Proc Natl Acad Sci USA* 105, 6906-6911 (2008).

133. B. D. Axelrod. Total internal reflection fluorescence microscopy in cell biology. *Methods Enzymol* 361, 1-33 (2003).

134. T. Bretschneider, S. Diez, K. Anderson, J. Heuser, M. Clarke, A. Müller-Taubenberger, J. Köhler & G. Gerisch. Dynamic actin patterns and Arp2/3 assembly at the substrate-attached surface of motile cells. *Curr Biol* 14, 1-10 (2004).

135. M. E. Berginski, E. A. Vitriol, K. M. Hahn & S. M. Gomez. High-resolution quantification of focal adhesion spatiotemporal dynamics in living cells. *PLoS ONE* 6 (2011).

136. M. Cebecauer, J. Humpolíčková & J. Rossy. Advanced imaging of cellular signaling events. *Methods Enzymol* 505, 273-289 (2012).

137. A.-S. Smith. Physics challenged by cells. *Natur Phys* 6, 726-729 (2010).

138. A. Zidovska & E. Sackmann. Brownian motion of nucleated cell envelopes impedes adhesion. *Physic Rev Lett* 96, 048103 (2006).

139. M. G. Somekh. Surface plasmon fluorescence microscopy: an analysis. *J Microsc* 206, 120-131 (2002).

140. L. Berguiga, S. Zhang, F. Argoul & J. Elezgaray. High-resolution surface-plasmon imaging in air and in water: V(z) curve and operating conditions. *Optics Lett* 32, 509-511 (2007).

141. M. Mahadi Abdul Jamil, M. Youseffi, P. C. Twigg, S. T. Britland, S. Liu, C. W. See, J. Zhang, M. G. Somekh & M. C. T. Denyer. High resolution imaging of bio-molecular binding studies using a widefield surface plasmon microscope. *Sensors Actuators B: Chemic* 129, 566-574 (2008).

142. M. G. Somekh, G. Stabler, S. Liu, J. Zhang & C. W. See. Wide field high resolution surface plasmon interference microscopy. *Optics Lett* 34,

3110-3112 (2009).

143. L. Berguiga, T. Roland, K. Monier, J. Elezgaray & F. Argoul. Amplitude and phase images of cellular structures with a scanning surface plasmon microscope. *Optics Expr* 19, 6571-6586 (2011).

144. E. Boyer-Provera, A. Rossi, L. Oriol, C. Dumontet, A. Plesa, L. Berguiga, J. Elezgaray, A. Arneodo & F. Argoul. Wavelet-based decomposition of high resolution surface plasmon microscopy V (Z) curves at visible and near infrared wavelengths. *Optics Expr* 21, 7456-7477 (2013).

145. J. Elezgaray, L. Berguiga & F. Argoul. Plasmon-based tomographic microscopy. *J Optic Soc Am A* 31, 155-161 (2014).

146. L. Berguiga, E. Boyer-Provera, C. Martinez-Torres, J. Elezgaray, A. Arneodo & F. Argoul. Guided wave microscopy: mastering the inverse problem.*Optics Lett* 38, 4269-4272 (2013).

147. V. Yashunsky, V. Lirtsman, M. Golosovsky, D. Davidov & B. Aroeti. Real-time monitoring of epithelial cell-cell and cell-substrate interactions by infrared surface plasmon spectroscopy. *Biophysic J* 99, 4028-36 (2010).

148. V. Yashunsky, T. Marciano, V. Lirtsman, M. Golosovsky, D. Davidov & B. Aroeti. Real-time sensing of cell morphology by infrared waveguide spectroscopy. *PLoS ONE* 7, e48454 (2012).

149. S.-H. Kim, W. Chegal, J. Doh, H. M. Cho & D. W. Moon. Study of cell-matrix adhesion dynamics using surface plasmon resonance imaging ellipsometry. *Biophysic J* 100, 1819-1828 (2011).

150. F. Zernike. How I discovered phase contrast. *Science* 121, 345-349 (1955).

151. D. J. Stephens & V. J. Allan. Light microscopy techniques for live cell imaging. *Science (New York, N.Y.)* 300, 82-86 (2003).

152. C. G. Rylander, D. P. Davé, T. Akkin, T. E. Milner, K. R. Diller & A. J. Welch. Quantitative phase-contrast imaging of cells with phase-sensitive optical coherence microscopy. *Optics Lett* 29, 1509-1511 (2004).

153. M. A. Choma, A. K. Ellerbee, C. Yang, T. L. Creazzo & J. A. Izatt. Spectral-domain phase microscopy. *Optics Lett* 30, 1162-1164 (2005).

154. C. Joo, T. Akkin, B. Cense, B. H. Park & J. F. de Boer. Spectral-domain optical coherence phase microscopy for quantitative phase-contrast imaging. *Optics Lett* 30, 2131-2133 (2005).

155. G. A. Dunn, D. Zicha & P. E. Fraylich. Rapid, microtubule-dependent fluctuations of the cell margin. *J Cell Sci* 110, 3091-3098 (1997).

156. C. J. Mann, L. Yu, C.-M. Lo & M. K. Kim. High-resolution quantitative phase-contrast microscopy by digital holography. *Optics Expr* 13, 8693-

8698 (2005).

157. G. Popescu, L. P. Deflores, J. C. Vaughan, K. Badizadegan, H. Iwai, R. R. Dasari & M. S. Feld. Fourier phase microscopy for investigation of biological structures and dynamics. *Optics Lett* 29, 2503-2505 (2004).

158. B. Rappaz, P. Marquet, E. Cuche, Y. Emery, C. Depeursinge & P. Magistretti. Measurement of the integral refractive index and dynamic cell morphometry of living cells with digital holographic microscopy. *Optics Expr* 13, 9361-9373 (2005).

159. V. P. Tychinskii. Coherent phase microscopy of intracellular processes. *Physics-Uspekhi* 44, 617-629 (2001).

160. V. P. Tychinskii. Coherent phase microscopy of intracellular processes. *UFN* 171, 649-662 (2001).

161. G. Popescu, T. Ikeda, R. R. Dasari & M. S. Feld. Diffraction phase microscopy for quantifying cell structure and dynamics. *Optics Lett* 31, 775-777 (2006).

162. P. Bon, G. Maucort & B. Wattellier. Quadriwave lateral shearing interferometry for quantitative phase microscopy of living cells. *Optics Expr* 17, 468-470 (2009).

163. G. Popescu. *Quantitative Phase Imaging of Cells and Tissues* (McGraw Hill, New York, 2011).

164. P. Bon, J. Savatier, M. Merlin, B. Wattellier & S. Monneret. Optical detection and measurement of living cell morphometric features with single-shot quantitative phase microscopy. *J Biomed Optics* 17, 076004 (2012).

165. C. Martinez-Torres, L. Berguiga, L. Streppa, E. Boyer-Provera, L. Schaeffer, J. Elezgaray, A. Arneodo & F. Argoul. Diffraction phase microscopy: retrieving phase contours on living cells with a wavelet-based space-scale analysis. *J Biomed Optics* 19, 036007 (2014).

166. M. S. Amin, Y. Park, N. Lue, R. R. Dasari, K. Badizadegan, M. S. Feld & G. Popescu. Microrheology of red blood cell membranes using dynamic scattering microscopy. *Optics Expr* 15, 17001-17009 (2007).

167. M. Bier. Brownian ratchets in physics and biology. *Contemp Phys* 38, 371-379 (1997).

168. D. Dan, A. M. Jayannavar & G. I. Menon. A biologically inspired ratchet model of two coupled Brownian motors. *Physica A* 318, 40-47 (2003).

169. K. Kruse & D. Riveline. Spontaneous mechanical oscillations. Implications for developing organisms. *Curr Topics Develop Biol* 95, 67-91 (2011).

170. M. O. Magnasco. Forced thermal ratchets. *Physic Rev Lett* 71, 1477-1481 (1993).

171. T. D. Pollard, L. Blanchoin & R. D. Mullins. Molecular mechanims controlling actin filament dynamics in nonmuscle cells. *Ann Rev Biophys Biomol Struct* 29, 545-576 (2000).

172. K. Keren, Z. Pincus, G. M. Allen, E. L. Barnhart, G. Marriott, A. Mogilner & J. A. Theriot. Mechanism of shape determination in motile cells. *Nature*453, 475-480 (2008).

173. J. Q. Zheng, J. J. Wan & M. M. Poo. Essential role of filopodia in chemotropic turning of nerve growth cone induced by a glutamate gradient. *J Neurosci* 16, 1140-1149 (1996).

174. D. J. Odde & H. M. Buettner. Autocorrelation function and power spectrum of two-state random processes used in neurite guidance. *Biophys J* 75, 1189-1196 (1998).

175. T. Betz, D. Koch, B. Stuhrmann, A. Ehrlicher & J. Käs. Statistical analysis of neuronal growth: edge dynamics and the effect of a focused laser on growth cone motility. *New J Phys* 9, 426 (2007).

176. D. Caballero, R. Voituriez & D. Riveline. Protrusion fluctuations direct cell motion. *Biophys J* 107, 34-42 (2014).

177. P. W. Oakes, D. C. Patel, N. A. Morin, D. P. Zitterbart, B. Fabry, J. S. Reichner & J. X. Tang. Neutrophil morphology and migration are affected by substrate elasticity. *Blood* 114, 1387-1396 (2009).

178. G. Giannone, B. J. Dubin-Thaler, H.-G. Döbereiner, N. Kieffer, A. R. Bresnick & M. P. Sheetz. Periodic lamellipodial contractions correlate with rearward actin waves. *Cell* 116, 431-443 (2004).

179. B. J. Dubin-Thaler, G. Giannone, H.-G. Döbereiner & M. P. Sheetz. Nanometer analysis of cell spreading on matrix-coated surfaces reveals two distinct cell states and STEPs. *Biophys J* 86, 1794-806 (2004).

180. G. L. Ryan, N. Watanabe & D. Vavylonis. A review of models of fluctuating protrusion and retraction patterns at the leading edge of motile cells. *Cytoskeleton* 69, 195-206 (2012).

181. O. J. Pletjushkina, Z. Rajfur, P. Pomorski, T. N. Oliver, J. M. Vasiliev & K. A. Jacobson. Induction of cortical oscillations in spreading cells by depolymerization of microtubules. *Cell Mot Cytoskeleton* 48, 235-244 (2001).

182. G. L. Ryan, H. M. Petroccia, N. Watanabe & D. Vavylonis. Excitable actin dynamics in lamellipodial protrusion and retraction. *Biophys J* 102, 1493-1502 (2012).

183. M. Kapustina, T. C. Elston & K. Jacobson. Compression and dilation of the membrane-cortex layer generates rapid changes in cell shape. *J Cell Biol*200, 95-108 (2013).

184. G. Salbreux, J. F. Joanny, J. Prost & P. Pullarkat. Shape oscillations of non-adhering fibroblast cells. *Phys Biol* 4, 268-84 (2007).

185. J. Étienne & A. Duperray. Initial dynamics of cell spreading are governed by dissipation in the actin cortex. *Biophys J* 101, 611-621 (2011).

186. E. Bernitt, C. G. Koh, N. Gov & H.-G. Döbereiner. Dynamics of actin waves on patterned substrates: a quantitative analysis of circular dorsal ruffles.*PloS One* 10, e0115857 (2015).

187. D. Heinrich, M. Ecke, M. Jasnin, U. Engel & G. Gerisch. Reversible membrane pearling in live cells upon destruction of the actin cortex. *Biophys J*106, 1079-1091 (2014).

188. N. Nisenholz, K. Rajendran, Q. Dang, H. Chen, R. Kemkemer, R. Krishnan & A. Zemel. Active mechanics and dynamics of cell spreading on elastic substrates. *Soft Matter* 10, 7234-7246 (2014).

189. K. Polyak & R. A. Weinberg. Transitions between epithelial and mesenchymal states: acquisition of malignant and stem cell traits. *Natur Rev Can*9, 265-273 (2009).

190. L. Trichet, J. Le Digabel, R. J. Hawkins, S. R. K. Vedula, M. Gupta, C. Ribrault, P. Hersen, R. Voituriez & B. Ladoux. Evidence of a large-scale mechanosensing mechanism for cellular adaptation to substrate stiffness. *Proc Natl Acad Sci USA* 109, 6933-6938 (2012).

191. R. P. Martins, J. D. Finan, F. Guilak & D. A. Lee. Mechanical regulation of nuclear structure and function. *Ann Rev Biomed Engin* 14, 431-455 (2012).

192. B. D. Hoffman, C. Grashoff & M. A. Schwartz. Dynamic molecular processes mediate cellular mechanotransduction. *Nature* 475, 316-323 (2011).

193. M. A. Schwartz & D. W. DeSimone. Cell adhesion receptors in mechanotransduction. *Curr Opin Cell Biol* 20, 551-556 (2008).

194. E. N. Olson & A. Nordheim. Linking actin dynamics and gene transcription to drive cellular motile functions. *Natur Rev Mol Cell Biol* 11, 353-365 (2010).

195. M. J. Bissell & W. C. Hines. Why don't we get more cancer? A proposed role of the microenvironment in restraining cancer progression. *Natur Med*17, 320-329 (2011).

196. F. Michor, J. Liphardt, M. Ferrari & J. Widom. What does physics have

to do with cancer? *Natur Rev Cancer* 11, 657-670 (2011).

197. B. Schuster-Böckler & B. Lehner. Chromatin organization is a major influence on regional mutation rates in human cancer cells. *Nature* 488, 504-508 (2012).

198. M. J. M. Magbanua & J. W. Park. Advances in genomic characterization of circulating tumor cells. *Can Metast Rev* 33, 757-69 (2014).

199. P. Polak, R. Karlić, A. Koren, R. Thurman, R. Sandstrom, M. S. Lawrence, A. Reynolds, E. Rynes, K. Vlahoviček, J. A. Stamatoyannopoulos & S. R. Sunyaev. Cell-of-origin chromatin organization shapes the mutational landscape of cancer. *Nature* 518, 360-364 (2015).

200. N. Gjorevski, E. Boghaert & C. M. Nelson. Regulation of epithelial-mesenhymal transition by transmission of mechanical stress through epithelial tissues. *Can Microenviron* 5, 29-38 (2012).

Chapter 3

THEORY OF ELEMENTARY PARTICLES BASED ON NEWTONIAN MECHANICS

Nikolai A. Magnitskii[1]

[1]LLC "New Inflow", Moscow, Russia

INTRODUCTION

The basis of modern conception of the world consists of two phenomenological theories (theory of quantum mechanics and theory of relativity), both largely inconsistent, but, in a number of cases, suitable for evaluation of experimental data. Both of these theories have one thing in common: their authors are convicted in limitations of laws and equations of classical mechanics, in absolute validity of Maxwell equations and in essential distinction of laws and mechanisms of the device of macrocosm and microcosm. Nevertheless, such assurance, being dominant in physics in the last hundred years, hasn't resulted in creation of unifying fundamental physical theory, nor in essential understanding of principal physical conceptions, such as: electric, magnetic and gravitational fields, matter and antimatter, velocity of light, electron, photon and other elementary particles, internal energy, mass, charge, spin, quantum properties, Planck constant, fine structure constant and many others. All laws and the equations of modern physics are attempts to approximate description of the results of natural experiments, rather than strict theoretical (mathematical) findings from the general and uniform laws and mechanisms of the device of the world surrounding us. Moreover, some conclusions from modern physics equations contradict experimental data such as infinite energy or mass of point charge.

In papers (Magnitskii, 2010a, 2011a) bases of the unifying fundamental physical theory which a single postulate is the postulate on existence of physical vacuum (ether) are briefly stated. It is shown, that all basic equations of classical electrodynamics, quantum mechanics and gravitation theory can be derived from two nonlinear equations, which define dynamics of physical vacuum in three-dimensional Euclidean space and, in turn, are derived from equations of Newtonian mechanics. Furthermore, clear and sane definitions are given to all principal physical conceptions from above through the parameters of physical vacuum, namely its density and propagation velocity of various density's perturbations. Thereby, it is shown that a set of generally unrelated geometric, algebraic and stochastic linear theories of modern physics, which are fudged to agree with experimental data and operating with concepts of multidimensional spaces and space-time continuums, can be replaced with one nonlinear theory of physical vacuum in ordinary three-dimensional Euclidean space, based exclusively on laws of classical mechanics.

In the present paper research of system of equations of physical vacuum is continued with the purpose of studying and the description of processes of a birth of elementary particles and their properties. A system of equations of electrodynamics of the physical vacuum, generalizing classical system of Maxwell›s equations and invariant under Galilean transformations is deduced. Definition of the photon is given and process of its curling and a birth from the curled photon of a pair of elementary particles possessing charge, mass and spin are described. The model of an elementary particle is constructed, definitions of its electric and gravitational fields are given and absence of a magnetic field is proved. Coulomb›s law and Schrodinger's and Dirac's equations for electric field and also the law of universal gravitation for gravitational field are deduced. Definitions of electron, positron, proton, antiproton and neutron are given, and absence of graviton is proved. The elementary model of atom of hydrogen is constructed.

Postulate. All fields and material objects in the Universe are various perturbations of physical vacuum, which is dense compressible inviscid medium in three-dimensional Euclidean space with coordinates

$$\vec{r} = (x, y, z)^T$$

having in every time station t density

$$\rho(\vec{r}, t)$$

and perturbation propagation velocity vector

$$\vec{u}(\vec{r}, t) = (u_1(\vec{r}, t), u_2(\vec{r}, t), u_3(\vec{r}, t))^T .$$

With such problem definition, it's natural to consider that no external forces apply any tension on elements of physical vacuum. Therefore, in compliance with Newtonian mechanics equations of physical vacuum dynamics in the neighborhood of homogeneous stationary state of its density

$$\rho_0$$

should be as follows:

$$\frac{\partial \rho}{\partial t} + div(\rho \vec{u}) = 0, \quad \frac{\partial(\rho \vec{u})}{\partial t} + (\vec{u} \cdot \nabla)(\rho \vec{u}) = 0, \tag{1}$$

where first equation is an equation of continuity, and second is the momentum equation. Let's notice, that the physical vacuum has no mass and in this connection dimension of its density does not coincide with dimension of substance (matter).

ELECTRODYNAMICS OF PHYSICAL VACUUM

Let's consider a case in which perturbation propagation velocity $\vec{u}$ has a certain direction in physical vacuum set by unit vector $\vec{n}$. Solutions of the system of equations (1) we shall search in the form of

$$\vec{u}(\xi, t) = v(\xi, t)\vec{n} + w(\xi, t)\vec{m}, \quad \xi = (\vec{r} \cdot \vec{n}), \quad (\vec{m} \cdot \vec{n}) = 0, \quad \rho = \rho(\xi, t). \tag{2}$$

Note that the vector of perturbation propagation velocity in physical vacuum can have both transverse and longitudinal components in relation to the direction of propagation of perturbations. Substituting expression for the vector $\vec{u}$ in equations (1) and taking into account, that

$$(w\vec{m} \cdot \nabla)(\rho\vec{u}(\xi,t)) = 0, \ (v\vec{n} \cdot \nabla)(\rho\vec{u}(\xi,t)) = v\frac{\partial(\rho\vec{u})}{\partial\xi}, \ div(\rho w\vec{m}) = 0, \ div(\rho v\vec{n}) = \frac{\partial(\rho v)}{\partial\xi}, \tag{3}$$

one can obtain a system of the equations for functions $\rho(\xi,t), v(\xi,t), w(\xi,t)$:

$$\frac{\partial \rho}{\partial t} + \frac{\partial(\rho v)}{\partial \xi} = 0, \quad \frac{\partial(\rho v)}{\partial t}\vec{n} + v\frac{\partial(\rho v)}{\partial \xi}\vec{n} = 0, \quad \frac{\partial(\rho w)}{\partial t}\vec{m} + v\frac{\partial(\rho w)}{\partial \xi}\vec{m} = 0, \tag{4}$$

which we call the system of the equations of electrodynamics of physical vacuum

Plane Electromagnetic Waves. Photon structure

In the particular case of transverse fluctuations of physical vacuum of constant density $(\rho(\xi,t) = \rho_0 = const)$ and distribution of these fluctuations in a longitudinal direction with constant velocity

$v(\xi,t) = c$ the system of equations (3) can be reduced to one equation in one complex variable $w(\xi,t)$:

$$\frac{\partial(\rho w)}{\partial t}\vec{m} + c\frac{\partial(\rho w)}{\partial \xi}\vec{m} = 0. \tag{5}$$

Let's introduce into consideration vectors of electric $\vec{E}$ and magnetic $\vec{H}$ fields intensities by the formulas:

$$\vec{H} = c\,rot(\rho\,\vec{u}), \quad \vec{E} = c(\vec{n}\cdot\nabla)(\rho\,\vec{u}).$$

In the general case of propagation of perturbations in compressible physical vacuum of variable density the vector of electric field intensity has both transverse and longitudinal components, and its divergence is not zero and can be interpreted as linear density of a charge (see item. 2.3). In the considered case of propagation of perturbations in physical vacuum of constant density with constant velocity only transverse component of a vector of electric field intensity is not zero, and its divergence is equal to zero. It is also clear that so defined vector of magnetic field intensity has only a transverse component, divergence of which also is equal to zero, and the vector

$c\rho\vec{u} = \vec{A}$ is the vector of potential in classical electrodynamics.

Applying to the equation (4) consistently the operators $c\,rot$ and $c(\vec{n}\cdot\nabla)$ and taking into account, that in the considered case

$$rot\,\vec{H} = rot(c\,rot(\rho\,w\,\vec{m})) = -c\nabla^2(\rho\,w\,\vec{m}) = -c\frac{\partial^2(\rho w)}{\partial\xi^2}\vec{m}, \quad \vec{E} = c(\vec{n}\cdot\nabla)(\rho\,w\,\vec{m}) = c\frac{\partial(\rho w)}{\partial\xi}\vec{m},$$

we shall obtain the classical system of Maxwell's equations describing the propagation of electromagnetic waves in the so-called empty space (vacuum):

$$\frac{\partial\vec{H}}{\partial t} + c\,rot\,\vec{E} = 0, \quad div\,\vec{H} = 0,$$

$$\frac{\partial\vec{E}}{\partial t} - c\,rot\,\vec{H} = 0, \quad div\,\vec{E} = 0. \tag{6}$$

The system of equations (6) has a solution in the form

$$\vec{E} = \vec{E}_0\,e^{i(\omega t - k\xi)}, \quad \vec{H} = \vec{H}_0\,e^{i(\omega t - k\xi)}, \quad \omega = kc. \tag{7}$$

It is considered to be, that the real parts of complex expressions (7) have physical sense. They determine an in-phase plane transverse electromagnetic wave, propagating with a speed of light c in any direction set by an unit vector $\vec{n}$. The unique characteristic of a classical plane electromagnetic wave is its frequency ω (or its wavelength $\lambda = 2\pi c/\omega$). Note, that in-phase vectors of electric and magnetic fields intensities periodically vanish simultaneously that contradicts the law of conservation of energy and raises doubts about validity of classical interpretation of an electromagnetic wave in which a change of the electric field causes a change in the magnetic field and vice versa. In turn, the equation (4) has as its solution a spiral wave of constant amplitude w_0

$$w(\xi,t)\vec{m} = (w^* + w_0\,e^{i(\omega t - k\xi)})\vec{m}, \quad \omega = kc, \tag{8}$$

propagating with velocity c in physical vacuum in a direction of a vector $\vec{n}$ with conservation of energy carried by the wave and having arbitrary constant shift w* in a direction of a vector $\vec{m}$. In such ormulation the speed of light cc in empty space has a clear physical sense - it is the propagation velocity of perturbations of physical vacuum of constant density in the absence of matter (the birth process of elementary particles of matter and antimatter as a result of perturbations of physical vacuum is described in Sec. 3).And since in this case the vectors $\vec{E}$ and $\vec{H}$ of a classical plane electromagnetic wave are a directional derivative and a rotor of a vector $c\rho_0 w(\xi,t)\vec{m}$, it is possible to conclude, that the classical electromagnetic wave (7) is an artificial form and is completely determined by the spiral wave (8) of perturbations propagation in physical vacuum, and

$$\vec{E}_0 = -ikc\rho_0 w_0 \vec{m}, \quad \vec{H}_0 = -ikc\rho_0 w_0 [\vec{m}\cdot\vec{n}].$$

Suppose, for example, the transverse wave is propagated in physical vacuum in the direction of the axis

y, so $w_0\vec{m} = (w_{0x},0,w_{0z})^T$. Then $\xi = y$. Then

$$\vec{E} = ck\rho_0(w_{0x},0,w_{0z})^T \sin(\omega t - ky) = (E_{0x},0,E_{0z})^T \sin(\omega t - ky) = \vec{E}_0 \sin(\omega t - ky),$$

$$\vec{H} = ck\rho_0(w_{0z},0,-w_{0x})^T \sin(\omega t - ky) = (E_{0z},0,-E_{0x})^T \sin(\omega t - ky) = \vec{H}_0 \sin(\omega t - ky).$$

That is, in full accordance with classical electrodynamics, vectors $\vec{E}_0$ and $\vec{H}_0$ r are perpendicular to the axis y and perpendicular to each other, and their moduli are equal (Fig. 1a). In Fig. 1b for comparison the propagation of the spiral wave (8) in the physical vacuum of constant density is represented.

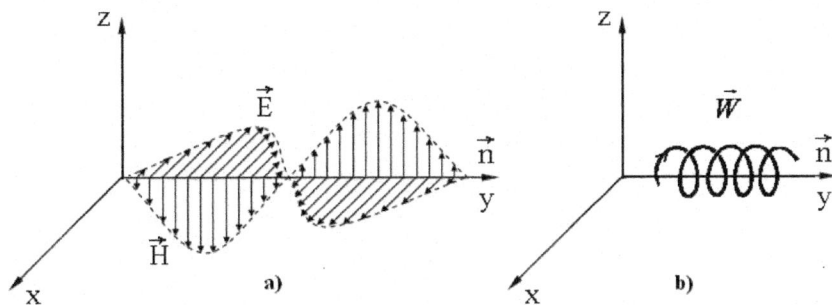

Figure 1: Propagation of a classical plane electromagnetic wave (a) and a spiral wave of physical vacuum (b).

Now we can compare the spiral wave in the physical vacuum, obtained as the solution of the equation (4), and the classical electromagnetic wave, obtained as the solution of system of Maxwell's equations (6). Both waves have an arbitrary frequencies and corresponding wavelengths, so the two solutions describe all plane transverse electromagnetic waves existing in nature. However, it is easy to see from the above analysis, that the vectors of classical electric and magnetic fields are artificial vectors, namely, the derivatives of the same true vector of the velocity perturbations propagation in the physical vacuum. Furthermore, a classical electromagnetic wave (Fig. 1a) does not allow to correctly define the concept of a quantum of electromagnetic waves (photon), because it except for wavelength λ needs also knowledge of the oscillation amplitude. The kind of a spiral wave of perturbations propagation in physical vacuum allows the unique determination of the photon - it›s a part of the cylindrical volume of the physical vacuum under a spiral of a wavelength λ and radius $r_0 = c/\omega = \lambda/2\pi$. Wave motion on a spiral inside the given volume occurs with a constant angular velocity ω, and linear velocity reaches its maximum value (the speed of light c) on the lateral surface of the cylinder. Exactly such photon colliding with an obstacle and being compressed is capable to generate elementary particles and antiparticles in the form of balls of radius r_0 (for more details about the birth of elementary particles, see Sec. 3). In addition, among the solutions of Maxwell's equations (6) in the form of classical electromagnetic waves, in principle, there are no solutions corresponding to the constant shift w* of transverse wave of physical vacuum (8). This, as it will be shown below, is the main reason that Maxwell's equations are not invariant under Galilean transformations, and, moreover, they cannot be modified so that they would satisfy these transformations.

Galileo Transformations of Electrodynamics Equations

Consider an inertial rest reference frame $O(x,y,z)$ and moving relative to it uniformly and rectilinearly with constant velocity $\vec{v}$ reference frame $O'(x',y',z')$. Without loss of generality, we assume that the respective axes are parallel to each other. Galilean transformations corresponding to common sense and centuries of experience are called transformations of coordinates and time in the transition from one inertial reference frame to another:

$$\vec{r}' = \vec{r} - \vec{v}t, \quad t' = t, \quad \vec{u}' = \vec{u} - \vec{v}.$$

Galilean transformation implies the same time in all frames of reference (absolute time). It is known also that all equations of classical mechanics are written the same in any inertial reference system, i.e. they are invariant under Galilean transformations. Let's show that any law, mathematical notation of which represents the full time derivative of any function $f(\vec{r},t)$ of coordinates and time is invariant under the Galilean transformations. Indeed, taking into account, that $t'=t$ and $\nabla'=\nabla$

we shall obtain

$$\frac{df(\vec{r},t)}{dt}=\frac{\partial f(\vec{r},t)}{\partial t}+(\vec{u}\cdot\nabla)(f(\vec{r},t))=\frac{\partial f'(\vec{r}',t')}{\partial t}+((\vec{u}'+\vec{v})\cdot\nabla)(f'(\vec{r}',t'))=\frac{\partial f'(\vec{r}',t')}{\partial t'}\frac{\partial t'}{\partial t}$$

$$+\frac{\partial f'(\vec{r}',t')}{\partial \vec{r}'}\frac{\partial \vec{r}'}{\partial t}+((\vec{u}'+\vec{v})\cdot\nabla)(f'(\vec{r}',t'))=\frac{\partial f'(\vec{r}',t')}{\partial t'}-(\vec{v}\cdot\nabla)(f'(\vec{r}',t'))+$$

$$((\vec{u}'+\vec{v})\cdot\nabla)(f'(\vec{r}',t'))=\frac{\partial f'(\vec{r}',t')}{\partial t'}+(\vec{u}'\cdot\nabla)(f'(\vec{r}',t'))=\frac{df'(\vec{r}',t')}{dt'}.$$

From this assertion follows immediately that the physical vacuum equations (1) are invariant under the Galilean transformations, since

$$\frac{\partial(\rho\vec{u})}{\partial t}+(\vec{u}\cdot\nabla)(\rho\vec{u})=\frac{d(\rho\vec{u})}{dt},\quad \frac{\partial\rho}{\partial t}+div(\rho\vec{u})=\frac{d\rho}{dt}+\rho(\nabla\cdot\vec{u}).$$

Also the system of equations of electrodynamics of physical vacuum (3) is invariant under the Galilean transformation that follows from the system of equations (1).Now consider in reference frames $O(x,y,z)$

a spiral wave of perturbations of physical vacuum of the form

$$\vec{u}(\xi,t)=c\vec{n}+w(\xi,t)\vec{m}=c\vec{n}+w_0e^{i(\omega t-k\xi)}\vec{m},\ \omega=kc,\ \xi=(\vec{r}\cdot\vec{n}),\ (\vec{m}\cdot\vec{n})=0. \tag{10}$$

As it shown above, to this solution of system of equations (1) with the function $w(\xi,t)$ satisfying theequation (4) there corresponds a classical electromagnetic wave, electric and magnetic fields intensities vectors of which are the directional derivative and the rotor of the vector $c\rho_0 w(\xi,t)\vec{m}$. In accordance with the Galilean transformations the considered solution has the form in the frame of reference

$O'(x',y',z')$

$$\vec{u}'(\xi',t)=c\vec{n}-\vec{v}+w_0e^{i(\omega't-k\xi')}\vec{m},\ \xi'=(\vec{r}'\cdot\vec{n})=\xi-(\vec{v}\cdot\vec{n})t,$$
$$\omega'=\omega-k(\vec{v}\cdot\vec{n})=k(c-(\vec{v}\cdot\vec{n}))=kc'.$$

Expanding now the vector $(\vec{n},\vec{m}):\vec{v}=(\vec{v}\cdot\vec{n})\vec{n}-w^*\vec{m}$, we obtain

$$\vec{u}'(\xi',t)=c'\vec{n}+w'\vec{m}=c'\vec{n}+(w^*+w_0e^{i(\omega't-k\xi')})\vec{m},\ \omega'=kc'. \tag{11}$$

Solution (11) is the solution of equations (1) and (3) in the reference frame. $O'(x',y',z')$. However, to obtain such solution from system of Maxwell's equations (6) is fundamentally impossible, even in case of failure of the postulate of the constancy of the speed of light with a replacement in (6) C on c'. The reason is that the differentiation of the solution (11) eliminates a constant shift w* of transverse component of velocity of perturbations propagation. Note also that the transition from the solution (10)to the solution (11) is accompanied by the Doppler effect, that is changing of the oscillation frequency $\omega' = \omega - k(\vec{v} \cdot \vec{n})$. When a radiation source located in a reference frame $O(x,y,z)$ moves in the direction of an observer which is in the reference frame $O'(x',y',z')$, the oscillation frequency increases $((\vec{v} \cdot \vec{n}) < 0)$, and at movement in an opposite direction - decreases $((\vec{v} \cdot \vec{n}) > 0)$

From the above it follows that, in contrast to the equations of a spiral wave (3) which are invariant under Galilean transformations, Maxwell's equations (6) describe the propagation of plane electromagnetic waves in moving inertial reference frames only approximately for small w*c. It is well known that the main cause of occurrence of the special theory of relativity in the early twentieth century were contradictions between electrodynamics, described by Maxwell›s equations and classical mechanics, governed by the equations and Newton›s laws. During the crisis of world science it was necessary to make a choice between two possibilities:

a. either to admit that Maxwell›s equations are not absolutely correct and are need to be changed so that they should satisfy the Galilean transformations;

b. or to recognize that equations of classical mechanics are not quite cor-rect and should be considered only as an approximation to the true equations, satisfying the Lorentz transformations.

Unfortunately, world science has chosen the second option, despite the reasoned objections of many outstanding scientists of the last century, among which the first is the name of Nikola Tesla (Tesla, 2003). The way chosen by world science has led to an absolutization of speed of light and Maxwell›s equations and has led to full termination of researches in the field of search more general equations of electrodynamics satisfying the principle of Galilean relativity. The present research proves that the correct way to exit from the crisis of science in early twentieth century was not in updating the equations of classical mechanics with the use of relativistic additives but, on the contrary, in finding the equations generalizing Maxwell›s equations and satisfying the Galilean transformations.

Longitudinal Electromagnetic Waves. Currents

Consider the general case of propagation of spiral waves (2) in physical vacuum of variable density. As shown in Sec. 2.1, these waves are solutions of the equations of electrodynamics of physical vacuum(3). Applying to the sum of the second and the third equations of system (3) consistently the operators crot and $c(\vec{n}\cdot\nabla)$ we obtain for the electric and magnetic fields intensities vectors defined by formulas(5), the system of equations

$$\frac{\partial \vec{H}}{\partial t} + v\,rot\,\vec{E} + \frac{\partial v}{\partial \xi}\vec{H} = 0, \quad div\,\vec{H} = 0,$$

$$\frac{\partial \vec{E}}{\partial t} - v\,rot\,\vec{H} + \frac{\partial v}{\partial \xi}\vec{E} + cv\frac{\partial^2(\rho v)}{\partial \xi^2}\vec{n} = 0, \quad div\,\vec{E} = c\frac{\partial^2(\rho v)}{\partial \xi^2}. \tag{12}$$

Note that in this case the electric field intensity vector $\vec{E}$ has a nonzero longitudinal component even at v= c =const This component is determined by small periodic compression-tension of density of physical vacuum in a longitudinal direction of propagation of electromagnetic wave. Let's introduce into consideration the linear charge density ρch and current density j r by the formulas

$$4\pi\rho_{ch} = div\,\vec{E} = div(c\frac{\partial(\rho v)}{\partial \xi}\vec{n}) = c\frac{\partial^2(\rho v)}{\partial \xi^2} = c\nabla^2(\rho v), \quad \vec{j} = \rho_{ch}v\vec{n}.$$

Then from (12) we shall obtain the system of equations

$$\frac{\partial \vec{H}}{\partial t} + v\,rot\,\vec{E} + \frac{\partial v}{\partial \xi}\vec{H} = 0, \quad div\,\vec{H} = 0,$$

$$\frac{\partial \vec{E}}{\partial t} - v\,rot\,\vec{H} + \frac{\partial v}{\partial \xi}\vec{E} + 4\pi\vec{j} = 0, \quad div\,\vec{E} = 4\pi\rho_{ch}.$$

The system of equations (13) at $v = c = const$ is a classical system of Maxwell's equations in the presence of charges and currents. It follows from here that charges and currents can exist in physical vacuum even at the absence of substance (matter) in it. Thus, a current in the sense of classical system of Maxwell's equations (13) at $v = c = const$ is not the motion of charges, but it is the second derivative (Laplacian) from propagating with the speed of light longitudinal wave of periodic compression - stretching of density of physical vacuum. Note that the substance (matter) is formed by elementary particles with the space charge and being waves of compression - stretching of density of physical vacuum, propagating along the parallels of spheres of radius $r \le r_0$ (see Sec. 3). Therefore, in substance the propagation of longitudinal waves (currents) also is possible.

As it is already mentioned above, the classical system of Maxwell›s equations describing propagation of electromagnetic waves in presence of charges and currents can be obtained from (13) at $v = c$. However, in general, the velocity of propagation of longitudinal waves in physical vacuum is not constant, but undergoes small periodic oscillations around the constant c. Therefore, the generalized system of equations of electrodynamics (13) has a much wider spectrum of solutions in comparison with the classical system of Maxwell's equations. In addition, the first two equations of system (13) at $v = c = const$ representing the Faraday's law of induction

$$\frac{\partial \vec{H}}{\partial t} + c\, rot\, \vec{E} = 0, \quad div\, \vec{H} = 0,$$

can be obtained by applying the operatorcrot directly to the linearized second equation of the physical vacuum equations (1). Therefore, these equations can be considered approximately always satisfied, but it is impossible to say about the second pair of equations of system (13), which are not always executed. Moreover, as follows from the analysis of item 2.2, the system of equations (13) and, consequently, the system of Maxwell's equations are not absolutely correct for the reason that they do not satisfy the Galilean transformations and describe the propagation of electromagnetic waves in moving inertial reference frames only approximately for small velocities of movement of such systems relatively to the speed of light. In all cases of the description of processes of propagation of both transverse and longitudinal waves in physical vacuum the system of equations (3) is correct. For the description of other more complex perturbations of physical vacuum connected, for example, with a birth of elementary particles and their electric and gravitational fields, it is necessary to use directly the equations of physical vacuum (1) (see Sec. 3).

ELEMENTARY PARTICLES OF A MATTER

We show in this section that processes of a birth of elementary particles of matter and antimatter from the physical vacuum (ether), as well as all basic quantum-mechanical properties of elementary particles can be obtained from the system of equations (1) written in spherical system of coordinates:

$$\frac{\partial \rho}{\partial t} + \frac{1}{r^2}\frac{\partial(r^2\rho V)}{\partial r} + \frac{1}{r\sin\theta}\frac{\partial(\rho\Omega\sin\theta)}{\partial\theta} + \frac{1}{r\sin\theta}\frac{\partial(\rho W)}{\partial\varphi} = 0,$$

$$\frac{\partial(\rho V)}{\partial t} + V\frac{\partial(\rho V)}{\partial r} + \frac{\Omega}{r}\frac{\partial(\rho V)}{\partial\theta} + \frac{W}{r\sin\theta}\frac{\partial(\rho V)}{\partial\varphi} = 0, (\vec{r})$$

$$\frac{\partial(\rho\Omega)}{\partial t} + V\frac{\partial(\rho\Omega)}{\partial r} + \frac{\Omega}{r}\frac{\partial(\rho\Omega)}{\partial\theta} + \frac{W}{r\sin\theta}\frac{\partial(\rho\Omega)}{\partial\varphi} = 0, (\vec{\theta})$$

$$\frac{\partial(\rho W)}{\partial t} + V\frac{\partial(\rho W)}{\partial r} + \frac{\Omega}{r}\frac{\partial(\rho W)}{\partial\theta} + \frac{W}{r\sin\theta}\frac{\partial(\rho W)}{\partial\varphi} = 0, (\vec{\varphi})$$

$$(14)$$

where

$$\vec{u} = (V_r, V_\theta, V_\varphi)^T, \quad V_r = V, \quad V_\theta = \Omega, \quad V_\varphi = W,$$ and unit coordinate vectors $(\vec{r}), (\vec{\theta}), (\vec{\varphi})$, which define vector directions of corresponding equation lines, are in brackets after equations.

Birth Of Elementary Particles From Physical Vacuum

Let's consider a spiral wave of photon (8)

$$w(\xi,t)\vec{m} = w_0 \, e^{i(\omega_* t - k_* \xi)}\vec{m}, \quad \omega_* = k_* c, \quad \xi = (\vec{n}\cdot\vec{r}),$$

propagating with the velocity c in physical vacuum in the direction of a vector $\vec{n}$ and having a wavelength $\lambda = 2\pi / k_*$ and radius of the outer spiral $r_0 = c / \omega_* = 1 / k_*$. Colliding with an obstacle (a field of an atomic nucleus or other photon), the wave is compressed in the direction of the vector $\vec{n}$ and bifurcated into a solution of the system of equations (14), in which the linear speed of rotation of the wave by the angle φ is equal to $W = (c / r_0)r\sin\theta$ (the direction of the axis z in (14) coincides with the direction of the vector $\vec{n}$.) Such a solution of the system (14), describing the compressed or curled photon, as well as all other solutions, describing various elementary particles, we shall search among the solutions with zero coordinate of velocity vector by the angle θ So, we shall put in (14) Ω=0 and result in equation system of elementary particles:

$$\frac{\partial \rho}{\partial t} + \frac{1}{r^2}\frac{\partial(r^2\rho V)}{\partial r} + \frac{1}{r\sin\theta}\frac{\partial(\rho W)}{\partial \varphi} = 0,$$

$$\frac{\partial(\rho V)}{\partial t} + V\frac{\partial(\rho V)}{\partial r} + \frac{W}{r\sin\theta}\frac{\partial(\rho V)}{\partial \varphi} = 0, \; (\vec{r})$$

$$\frac{\partial(\rho W)}{\partial t} + V\frac{\partial(\rho W)}{\partial r} + \frac{W}{r\sin\theta}\frac{\partial(\rho W)}{\partial \varphi} = 0, \; (\vec{\varphi})$$

(15)

The solution for the curled photon we shall find from the system (15), putting in it $W = (c / r_0)r\sin\theta$, $V = 0$.

Then we shall obtain $\rho = \rho_0(1 + q_0(r)\exp(i(\omega_* t - \varphi)))$. That is, at curling the photon is transformed into a longitudinal wave of small compression - stretching of the density of physical vacuum, propagating on parallels inside a sphere of radius r_0 with constant angular velocity $\omega = \omega_* = c / r_0$. Curled photon has no mass and charge, so it can hypothetically apply for the role of neutrino though this hypothesis requires additional check and experimental confirmation. Let's show now that equation system (15) has solutions, which possess all known properties of elementary particles when $r \le r_0$ is small enough. These solutions will be sought as waves propagating with constant angular velocity by the angle Φ under the influence of small-amplitude oscillations of physical vacuum density

$$W = \frac{c}{r_0}r\sin\theta, \quad \rho(r,\varphi,t) = \rho_0 + q(r,\varphi,t)$$

and small-amplitude oscillations of function

$V(r,\varphi,t) \neq 0$ when $r \le r_0$ when $r \le r_0$ is small enough. That is every elementary particle is some bifurcation from curled photon. Under such problem formulation, each elementary particle is a sphere of radius r_0, inside of which waves, created by small-amplitude oscillations of physical vacuum density, propagating along to any parallel (circle with radius $r\sin\theta, r \le r_0$) with constant angular velocity (frequency) c/r_0 , making full roundabout way by angle $0 \le \varphi \le 2\pi$ over equal time $T = 2\pi r \sin\theta / W = 2\pi r_0 / c$. In addition, linear velocity of these waves increases linearly with the radius, reaching its maximum value (velocity of light c) on sphere's equator when $r = r_0,\ \sin\theta = 1$ (Fig.2) (Fig.2).

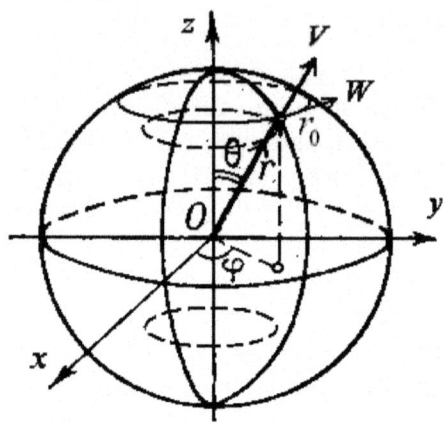

Figure 2: Scheme of any elementary particle.

Substitution of assumed form of solution of (16) into equation system (15), with a drop of second infinitesimal order terms and multiplications of small terms, will result in the following system of equations

$$\frac{\partial q}{\partial t} + \rho_0 (\frac{\partial V}{\partial r} + \frac{2V}{r}) + \frac{c}{r_0}\frac{\partial q}{\partial \varphi} = 0,$$

$$\frac{\partial V}{\partial t} + \frac{c}{r_0}\frac{\partial V}{\partial \varphi} = 0,\ (\vec{r})$$

$$\frac{\partial q}{\partial t} + \rho_0 \frac{V}{r} + \frac{c}{r_0}\frac{\partial q}{\partial \varphi} = 0,\ (\vec{\varphi}) \qquad (17)$$

It is necessary to notice that at such approximation nonlinear term of second infinitesimal order $V \partial(\rho V) / \partial r\ \vec{r}$ has been entirely neglected. The role of this term becomes significant only with relatively large r ® ¥ and, probably, with relatively small 0 r ® . As it will be shown below, this term exactly generate gravitational field of a particle with relatively large r . It is rather probable, that the same term describes nuclear interactions at 0. r ® It's not difficult to get the solutions of equation system (17) in the following form

$$V(r,\varphi,t) \approx \frac{V_0}{r}e^{i(\omega t - kr_0\varphi)}, \quad \omega = ck, \quad \rho(r,\varphi,t) \approx \rho_0(1 - \frac{V_0 r_0}{cr^2}\varphi e^{i(\omega t - kr_0\varphi)}).$$

(18)

However, not every solution in form (16), (18) is an elementary particle. Such solution has to possess properties of charge conservation and universality, as well as quantum properties of mass, momentum and energy. Moreover, over the time of full roundabout way of the wave along the sphere equator, electric field intensity must conserve its sign. Such classical and quantum mechanical terms as electric and magnetic field of elementary particle, its charge, mass, energy, momentum, spin also need correct definitions through the characteristics of physical vacuum.

First, let's give the definition of electric field and electric charge of elementary particle similarly to the case of plane electromagnetic waves propagation, examined above.

Definition. Electric field intensity distribution $\vec{E}$ and charge density distribution ρ_{ch} of elementary particle will be defined as:

$$\vec{E} = E\vec{r} = \frac{W}{r\sin\theta}\frac{\partial(\rho V)}{\partial\varphi}\vec{r}; \quad \rho_{ch} = \frac{1}{4\pi}div(V\frac{\partial(\rho W)}{\partial r}\vec{\varphi}).$$

It follows from (16) and (18) that inside a particle at $r \le r_0$

$$\vec{E} = E\vec{r} = \frac{c}{r_0}\frac{\partial(\rho V)}{\partial\varphi}\vec{r} \approx -\frac{ikr_0 c\rho_0 V_0}{r_0 r}e^{i(\omega t - kr_0\varphi)}\vec{r}; \quad \rho_{ch} \approx -\frac{ikc\rho_0 V_0}{4\pi r^2}e^{i(\omega t - kr_0\varphi)}$$

Let's determine an instant value of the charge q_{ch} of elementary particle. Let $\omega t = 2\pi l + kr_0\varphi_*$, where $0 \le kr_0\varphi_* < 2\pi$. Integrating the density distribution of charge over sphere's volume with radius r_0 we shall obtain

$$q_{ch} = -\int_0^\pi \int_{\varphi_*}^{\varphi_* + 2\pi} \int_0^{r_0} \frac{ikc\rho_0 V_0}{4\pi r^2}e^{i(kr\varphi_* - kr_0\varphi)}r^2\sin\theta\, dr d\varphi d\theta =$$

$$= \frac{c\rho_0 V_0}{2\pi}(e^{-i2\pi kr_0} - 1) = \begin{cases} 0, & kr_0 = n/2, \ n = 2m, \ m = 0,1,... \\ -\frac{c\rho_0 V_0}{\pi}, & kr_0 = n/2, \ n = 2m+1. \end{cases}$$

What follows from formula (21) is that solution (16), (18) of the equation system (15) can be interpreted as an elementary particle only in such case, when wave number kr_0 is an integer or a half-integer value. For integer value of kr_0 the charge is zero, for any half-integer value of kr_0 charge equals common by modulus universal value $q = c\rho_0 V_0/\pi$.

Integrating the density distribution of charge over sphere's volume with radius r_0 for $\varphi_* - 2\pi \leq \varphi \leq \varphi_*$ we shall obtain positive value of particle charge q. Thus there are actually two particles bifurcating from curled photon (particle and antiparticle), which have the same frequencies $\omega = n\omega_*/2$ and charges, which modules are equal to q , but have opposite signs. In that case the wavelengths of created periodic solutions by the angle φ are less than 2π in half-integer value of times. That is time of the wave's full roundabout way by angle $0 \leq \varphi \leq 2\pi$ along any parallel of the sphere with radius r_0 equals integer number $2kr_0$ of half-periods $T_p = \pi / \omega = \pi / kc$ of physical vacuum density and electric field intensity oscillations, which conserves its sign on the last uneven half-period, being equal to the charge's sign. It's important to point out that electric field of elementary particle directed along radius is created by particle's electric charge, but at the same time the charge is divergence of a completely different inner field of the particle, which is represented by second term in the third equation of equation system (15)and directed by the angle φ. Also notice that electric field intensity distribution of elementary particle inside the particle (that is within the sphere of radius r_0) defined by the third term in the second equation of equation system (15), decreases as $1/r$, so it removes the problem of infinite energy and mass of elementary particles.

Other Basic Properties Of Elementary Particles

Let's now determine other properties of an elementary particle: internal energy ε, mass m, momentum p and spin σ. Expressions of Planck constant h , as well as fine structure constant, which can be rightfully called the most mysterious constant of microcosm physics, will also be derived. First, let's determine internal energy formula with a use of expression of work A, executed by field forces of the particle

$$\frac{dA}{dt} = \int_B \Lambda \vec{F} \cdot \vec{W} dB.$$

$$(22)$$

Here B is the volume of elementary particle sphere of radius r_0, $\vec{F}$ is the field, which influences on charges distributed inside a sphere with distribution density Λ and has a nonzero projection on velocity vector $\vec{W}$, that is on direction of vector $\vec{\varphi}$. This field can not be electric field, which is directed along radius $\vec{r}$. This field can only be the summary field directed by angle $\vec{\varphi}$ from the third equation of system(15)

$$\vec{F} = V \frac{\partial(\rho W)}{\partial r} \vec{\varphi} + \frac{W}{r\sin\theta} \frac{\partial(\rho W)}{\partial \varphi} \vec{\varphi} \approx -ik \frac{c\rho_0 V_0 \sin\theta}{r} \varphi e^{i(\omega t - kr_0 \varphi)} \vec{\varphi},$$

and it has to execute the work over not only electric charge with distribution density ρ_{ch}, but also over all other charges determined by divergence of this field. After determination of full charge distribution density

$$\Lambda = div\vec{F} = -\frac{ikc\rho_0 V_0}{r^2} \frac{\partial}{\partial \varphi}(\varphi e^{i(\omega t - kr_0 \varphi)})$$

let's insert it as well as derived expression of internal field $\vec{F}$ into the formula (22) to get the following expression

$$\frac{dA}{dt} = -\int_B \frac{k^2 c^3 \rho_0^2 V_0^2 \sin^2\theta}{2r_0 r^2} \frac{\partial(\varphi e^{i(\omega t - kr_0 \varphi)})^2}{\partial \varphi} dB =$$

$$= -e^{2i\omega t} \int_0^\pi \int_0^{2\pi} \int_0^{r_0} \frac{k^2 c^3 \rho_0^2 V_0^2 \sin^2\theta}{2r_0 r^2} \frac{\partial(\varphi^2 e^{-2ikr_0 \varphi})}{\partial \varphi} r^2 \sin\theta \, dr d\varphi d\theta.$$

Integrating the last equation and taking into account that $\omega = kc$ one can obtain finally

$$A = ie^{2i\omega t} \frac{4kc^2 \rho_0^2 V_0^2 \pi^2}{3}; \quad \varepsilon = |A| = \frac{4\pi^2}{3} kc^2 \rho_0^2 V_0^2.$$

Now, to derive the well-known main formulas and correlations of quantum mechanics, it's suffice to denote the mass of elementary particle and Planck constant as

$$m = \frac{4\pi^2}{3} k\rho_0^2 V_0^2 = \frac{4\pi^2}{3} \omega \rho_0^2 V_0^2 / c; \quad \hbar = \frac{4\pi^2}{3} c\rho_0^2 V_0^2.$$

From this it follows immediately:

- Einstein's formula for internal energy of a particle and formulas of impulse and energy for de Broglie's waves
- formula for spin of a particle
- fine structure constant formula

These formulas, derived exclusively by the methods of classical mechanics, are completely identical to the well-known expressions of quantum mechanics as well as clearly reflect the physical essence of charge, mass, energy and spin of elementary particles, allowing to understand the nature of quantum processes in microcosm. It can be seen that the internal energy of the particle is indeed proportional to the square of velocity of light, and proportionality coefficient (mass of the particle) linearly grows with the increase of wave number k, as well as frequency ω of the parental photon. The Plank constant is indeed a constant value depending only on characteristics of physical vacuum and not on the type of the elementary particle. The spin of the particle indeed has a value of either integer or half-integer number of $\hbar$, which allows to separate all elementary particles in two general categories: bosons and fermions. Still, the most surprising and encouraging fact is the almost precise match of the fine structure constant α with its experimental value of 1/137.

Note also that the simplest particles with the spin of ½ when n = 1 are double period cycles in relation to the initial cycle defined by the motion of curled photon. That brings another proof of the theory introduced in this research – the interpretation of the Pauli principle, the corollary fact of which is that electron returns to the initial state only after the turn of 720, not 360 degrees. According to R. P. Feynman (Feynman & Weinberg, 1987), particle with topology of Moebius band meets the Pauli principle. But in the Feigenbaum-Sharkovskii-Magnitskii universal theory of dynamical chaos (FSM theory) (Magnitskii, 2008a, 2008b, 2010, 2010b, 2011b; Magnitskii & Sidorov, 2006; Evstigneev & Magnitskii, 2010), results of which valid for every nonlinear differential equation system of macrocosm, the solution's difficulty increase starts from double period bifurcation of the original singular cycle. Interesting enough, the newborn cycle of doubled period belongs to the Moebius band around the original cycle! In another words, according to the FSM theory electron and proton are initial and simplest double period bifurcations from the infinite bifurcation cascade. Therefore, FSM theory works not only in macrocosm, but also in microcosm, and elementary particles defined by formulas(16), (18), are not a full infinite set of all elementary particles, which can be born as a result of bifurcations in nonlinear equation system (15). Furthermore, more complex nonperiodic solutions of systems (14) and (15) can be foreseen, which are singular attractors in terms of FSM theory. Thus, any attempts of an experimental detection of the simplest (most elementary), as well as the most complex of elementary particles are essentially futile.

Some Main Classical Equations And Laws

Another proof of validity of the theory presented in this paper is the possibility of a rigorous mathematical conclusion from its unique postulate on existence of physical vacuum of some important phenomenological equations and laws of the modern physics which are widely used by classical electrodynamics and quantum mechanics and not contradicting to common sense interpretation of variables included in them. We consider here the Coulomb's law and Schrodinger's and Dirac's equations.

Coulomb's Law

We assume that outside of a particle of radius r_0 change of density of physical vacuum practically does not occur. Then, neglecting the third equation of (17), we shall obtain, that at $r > r_0$

$$V(r,\varphi,t) \approx \frac{V_0 r_0}{r^2} e^{i(\omega t - k r_0 \varphi)}, \ \omega = ck, \ r > r_0.$$

The vector of electric field intensity distribution of a particle will become

$$\vec{E} = E\vec{r} \approx -\frac{i k r_0 c \rho_0 V_0}{r^2} e^{i(\omega t - k r_0 \varphi)} \vec{r}. \tag{23}$$

Then a vector of electric field intensity of an elementary particle $\vec{E}(r)$ we shall find, averaging instant value of a vector of intensity distribution by the angle φ. Let $\omega t = 2\pi l + k r_0 \varphi_*$, where $0 \le k r_0 \varphi_* < 2\pi$. Then for the particles having a negative charge $-q$, we shall obtain

$$\vec{E}_- = -\frac{1}{2\pi} \int_{\varphi_*}^{\varphi_* + 2\pi} \frac{i k \, r_0 c \rho_0 V_0}{r^2} e^{i(k r \varphi_* - k r_0 \varphi)} \vec{r} d\varphi = -\frac{c \rho_0 V_0}{\pi r^2} \vec{r} = -\frac{q}{r^2} \vec{r}.$$

For the particles having a positive charge $+q$, averaging of instant value of a vector of electric field intensity distribution by the angle φ in the interval $\varphi_* - 2\pi \le \varphi \le \varphi_*$ will give $\vec{E}_+ = (q/r^2)\vec{r}$. Obtained expressions coincide with expressions for intensity of an electric field of a charge in the Coulomb's law, and for a particle having a negative charge, the vector of electric field intensity is directed on radius to the center of a particle, and for a particle having a positive charge, the vector of electric field intensity of a particle is directed on radius from its center.

Schrodinger's Equation

Let's show, that for a free particle of mass m the solution of the Schrodinger's equation

$$i\hbar \frac{\partial \psi}{\partial t} = -\frac{\hbar^2}{2m} \Delta \psi \tag{24}$$

is a scalar function $E^*(r,\varphi,t)$, which is a complex conjugate function to an electric field intensity distribution function of an elementary particle from expression (20). As

$$\frac{\partial E^*}{\partial t} = -i\omega E^*, \ \frac{\partial^2 E^*}{\partial \varphi^2} = -k^2 r_0^2 E^*, \ \frac{\partial E^*}{\partial t} = \frac{i\omega}{k^2 r_0^2}\frac{\partial^2 E^*}{\partial \varphi^2} = \frac{i\omega}{k^2 r_0^2} r^2 \sin^2\theta \, \Delta E^*,$$

then averaging the right part of last expression by the angle θ, we shall obtain in a neighborhood of a sphere of an elementary particle of radius r_0

$$\frac{\partial E^*}{\partial t} \approx (\frac{1}{\pi}\int_0^\pi \frac{i\omega}{k^2}\sin^2\theta \, d\theta)\Delta E^* = i\frac{c^2}{2\omega}\Delta E^*.$$

Multiplying the last expression on iħ we shall obtain

$$i\hbar\frac{\partial E^*}{\partial t} = -\frac{c^2\hbar}{2\omega}\Delta E^* = -\frac{\varepsilon\hbar}{2\omega m}\Delta E^* = -\frac{\omega\hbar^2}{2\omega m}\Delta E^* = -\frac{\hbar^2}{2m}\Delta E^*,$$

that coincides with the equation (24). Thus, it becomes clear a physical sense of ψ - function in the Schrodinger's equation for a free particle - it is the electric field intensity distribution of an elementary particle near the surface of its sphere.

Dirac's Equation

It was already shown in (Magnitskii, 2010a, 2011a) that electric field intensity and charge of elementary particle defined above agree with electromagnetic form of Dirac's equation for electron in bispinor form. Here we shall consider this question in more detail. Dirac's equation in bispinor form has a kind

$$i\hbar\frac{\partial \psi}{\partial t}(\vec{r},t) = (c\sum_{j=1}^{3}\alpha_j p_j + \alpha_0 m_e c^2)\psi(\vec{r},t),$$

(25)

that is a consequence of operator equation

$$\hat{\varepsilon}^2\psi = c^2\vec{p}^2\psi + m_e^2 c^4\psi, \quad \hat{\varepsilon} = i\hbar\frac{\partial}{\partial t}, \quad \vec{p} = -i\hbar\vec{\nabla},$$

where m_e is mass of electron or other fermion, $\hat{\varepsilon}$ and $\vec{p}$ are operators of energy and momentum and α_j – Dirac's matrixes. In the theory of electrodynamics of curvilinear waves (EDCW) of A.Kyriakos (Kyriakos, 2006) the electromagnetic form of Dirac's equation is deduced. It is shown, that if the electromagnetic wave of a photon is propagating in a direction z, then at its hypothetical curling and a birth from it a pair of elementary particles the 4-vector (E_x, E_y, H_x, H_y) of electromagnetic wave of each of particles satisfies the Dirac's equations in bispinor form. So, to show, that the vector function of electric field intensity distribution of an elementary particle in a vicinity of its equator satisfies the equations (25) and (26) we should write down system of the equations of elementary particles (15) in cylindrical system of coordinates which axis z coincides with the axis of rotation of an elementary particle:

$$\frac{\partial \rho}{\partial t} + \frac{1}{r}\frac{\partial (r\rho V)}{\partial r} + \frac{1}{r}\frac{\partial (\rho W)}{\partial \varphi} = 0,$$

$$\frac{\partial (\rho V)}{\partial t} + V\frac{\partial (\rho V)}{\partial r} + \frac{W}{r}\frac{\partial (\rho V)}{\partial \varphi} = 0, \ (\bar{r})$$

$$\frac{\partial (\rho W)}{\partial t} + V\frac{\partial (\rho W)}{\partial r} + \frac{W}{r}\frac{\partial (\rho W)}{\partial \varphi} = 0, \ (\bar{\varphi})$$

(27)

Solution of the system (27), consistent with a solution of the system (19) in the vicinity of the equatorial areas of the elementary particle, has the following kind:

$$W = (c / r_0)r, \ V(r,\varphi,t) \approx V_0 e^{i(\omega t - kr_0\varphi)}, \ \omega = ck, \ \vec{E} = \frac{c}{r r_0}\frac{\partial (\rho V)}{\partial \varphi}\vec{r} = E\vec{r}.$$

Then, as it is easy to verify by the direct substitution, the vector $\vec{E}$ is an approximate solution of the second order equation

$$\frac{\partial^2 \vec{E}}{\partial t^2} - c^2 \nabla^2 \vec{E} + \omega_p^2 \vec{E} = 0.$$

in the vicinity of $r \approx r_0$, where ∇, is Laplace operator in cylindrical coordinate system and the frequency $\omega_p = c / r_0$ is an angular velocity, which can be interpreted as an oscillation frequency of the curled photon electromagnetic wave with a wavelength $\lambda = 2\pi r_0$. Multiplying the obtained equation by $(i\hbar)^2$ and using the relation $\hbar\omega = mc^2$ we obtain for vector $\vec{E}$ an equation

$$\hat{\varepsilon}^2 \vec{E} = c^2 \vec{p}^2 \vec{E} + m_p^2 c^4 \vec{E}.$$

(28)

Equation (28) differs from the equation (26) those, that in it instead of the electron mass m_e there is the mass of the curled photon $m_p = 2m_e$ until the moment of its division into two particles: an electron and a positron. Hence, the vector of electric field intensity distribution of each separate elementary particle after their division is the solution of equations (25) and (26) written in cylindrical system of coordinates.

Therefore, the true physical meaning of wave function ψ from Dirac equation for electron in bispinor form (25) becomes clear – it's a 4-vector (E_x, E_y, H_x, H_y) of particle's electromagnetic wave, but in such elementary particles model, as opposed to the case of plain electromagnetic waves propagation, magnetic field intensity vector is a virtual one, since it is directed on an axis z, while velocity vector component V_z equals to zero. Therefore, there is no real magnetic field of an elementary particle in a considered model.

Electron, Positron, Proton, Antiproton, Neutron And Atom of Hydrogen

It's obvious, that more complex, multi-curled elementary particles correspond to high-frequency perturbation waves with bigger mass and energy. So, it's natural to imply that the simplest half-curled particles with the spin of ½ when n = 1 are pairs "electron-positron" and "proton-antiproton". Both pairs of particles have the same mechanism of a birth. The difference is in the values of frequencies of parental photons and, accordingly, in radiuses of their curling r_0 and in masses of the born particles. Experimental data testify that the mass of proton is in three orders greater than the mass of electron. Consequently, the wave frequency of proton is in three orders greater than the wave frequency of electron and, that is important, the radius of proton is in three orders smaller than the radius of electron. That is, the electron is not a small particle that rotates around the nucleus of an atom, and it is a huge ball which size is comparable to the size of the crystal lattice of substance. This implies that the current in the conductors can not be a movement of free electrons.

It is obvious that the charges of proton and electron should have different signs. Thus, their combinations can form atoms of substance only in the case when the electric field intensity of a particle of smaller radius (proton) is directed to its center, and, accordingly, the electric field intensity of a particle of the greater radius (electron) is directed from its center. That is, proton should have a negative charge in the sense of expression (20), and electron should have a positive charge. Then for instant density of physical vacuum of proton ρ_p inside a sphere with radius of its curling r_p we shall obtain the expression

$$\operatorname{Re}\rho_p = \operatorname{Re}\frac{1}{2\pi}\int_{\varphi_*}^{\varphi_*+2\pi}\rho_0(1-\frac{r_p c V_0}{cr^2}\varphi e^{i(kr\varphi_*-kr_0\varphi)})d\varphi = \rho_0(1+\frac{4V_0 r_p}{\pi c r^2}) > \rho_0.$$

Similar expression we shall receive for instant density of physical vacuum of electron ρ_e inside a sphere with radius of its curling r_e:

$$\operatorname{Re}\rho_e = \operatorname{Re}\frac{1}{2\pi}\int_{\varphi_*-2\pi}^{\varphi_*}\rho_0(1-\frac{r_e c V_0}{cr^2}\varphi e^{i(kr\varphi_*-kr_0\varphi)})d\varphi = \rho_0(1-\frac{4V_0 r_e}{\pi c r^2}) < \rho_0.$$

Consequently, proton is compressed, and electron is rarefied areas of physical vacuum with respect to its stationary density ρ_0.

Elementary antiparticles positron and antiproton are, obviously, in pairs to electron and proton, and have charges of opposite signs, that is their waves are formed by additional half-periods of the waves of double period with respect to the waves of the original photons.

Consider now the possibility of the formation from a pair of proton-electron of the simplest electrically neutral structures, such as neutron and atom of hydrogen. Since the electron has a much larger radius than the radius of a proton, then in the most part of elements of physical vacuum laying inside of the electron, the electric field of the electron directed from its center, less than an electric field of the proton directed to its center. Therefore, an electron having got in area of its capture by an electric field of a proton, should move in its direction until some stable structure in the form of a sphere with a radius of an electron, in which center there is a nucleus as a sphere with a radius of a proton is formed. The electric field intensity outside of an external sphere is equal to zero, as at $r > r_e$

$$\vec{E} = E\vec{r} = \vec{E}_e + \vec{E}_p = \frac{q}{r^2}\vec{r} - \frac{q}{r^2}\vec{r} = 0, \ r > r_e.$$

We can assume that the simplest atom of hydrogen, as well as arbitrary neutron are arranged in this manner. The neutron can differ from the atom of hydrogen in radius and, accordingly, in frequencies of oscillations of waves of its electron and proton. In Fig. 3 a diagram of a hydrogen atom and also a picture of a real hydrogen atom made in Japan (Podrobnosti, 04.11.2010) are presented.

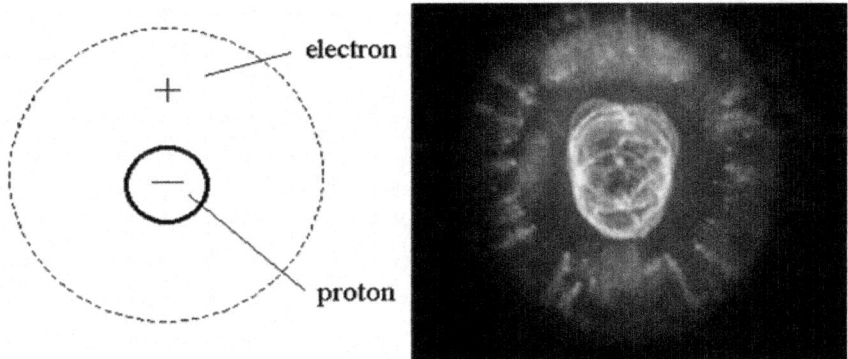

Figure 3: The scheme (at the left) and the photo of a real hydrogen atom.

In this model, the impossibility of formation of atoms of antimatter can be easily explained by the fact that the electric field of the antiproton, which has much smaller radius than the positron, is directed from its center, which prevents the formation of stable structures of antimatter.

GRAVITATION AND GRAVITATIONAL WAVES

Let's demonstrate that the creation of any elementary particle is accompanied by appearance of the gravitation, notably the pressure force in physical vacuum, generated by small periodic perturbations of its density, which in its own turn generate gravitational wave, propagating to the center of newborn particle. It's natural to propose that gravitation works over any distance from the particle, and that when the distance is large, perturbations of physical vacuum density created by the newborn particle depend only on distance r and are independent of angles θ and φ. Based on such assumption, let's seek solutions of system (15) when

r is large in the following form:

$$\rho = \rho_0 + q(r,t), \ V = V(r,t), \ W = 0.$$

Equation system (15) will take a form

$$\frac{\partial \rho}{\partial t} + \frac{1}{r^2}\frac{\partial(r^2\rho V)}{\partial r} = 0, \ \frac{\partial(\rho V)}{\partial t} + V\frac{\partial(\rho V)}{\partial r} = 0, \tag{29}$$

meaning that of all four fields in the initial system (15) only gravitational field $G = V\partial(\rho V) / \partial r$ will remain significant when r is large enough. Furthermore, gravitational field differs from three other previously examined fields since it's severely nonlinear. It can't be linearized basing on the form of velocity W component in analogue with electric and two internal fields of the particle. When r is small and, consequentially, V is small as well, gravitational field can be neglected during the formulation of elementary particles theory. On the contrary, when r is relatively large, all other fields with the exception of gravitational are can be neglected, and that agrees with experimental data. But when 0 r ® V again starts to grow and so we can propose that gravitational term describes also nuclear interactions. again starts to grow and so we can propose that gravitational term describes also nuclear interactions.

Let's seek the solution of equation system (29) in the form of 2 V c/r, that is in the form of a gravitational (radial) wave, which propagates to the center of elementary particle r= 0) with velocity dependant on radius. With the use of function V in equation system (29) and in case of r ® ¥ next expression for small oscillation of physical vacuum density will be derived

$$\rho(r,t) \approx \rho_0(1 + q(r,t)) = \rho_0(1 + q_0 e^{i(\omega t + kr^3/3)}), \ \omega + kc = 0.$$

In this case the pressure force of gravitational wave (gravitational field intensity) expresses as

$$G = V \frac{\partial(\rho V)}{\partial r} \approx \frac{c^2}{r^4} \frac{\partial q}{\partial r} = \frac{i q_0 k c^2}{r^2} e^{i(\omega t + kr^3/3)},$$

and it agrees with the law of universal gravitation. However, the physical essence of gravitation comes in somewhat different light than before. The bodies do not attract each other – each material body creates its own gravitational wave, which propagates from infinity to its center of mass and puts an external pressure on other body with the force, proportional to the mass of the body and inversely proportional to the square of distance between the bodies.

Let's note another significant difference between gravitational and electromagnetic waves. Electromagnetic wave moving with constant velocity has a wavelength, thus, resulting in the existence of electromagnetic wave quant or photon. Gravitational wave moves with velocity dependant on radius, thus, there can be no gravitational wave quant. Traditional parallel between the gravitational wave and its hypothetical carrier, graviton, is apparently the main obstacle for the real discovery of gravitational waves in nature.

CONCLUSION

The theoretical research carried out in the work and its results allow to draw several fundamental conclusions and statements which looks more than plausibly:

- all fields and material objects in the Universe are various perturbations of physical vacuum, microcosm and macrocosm are organized by the same laws – laws of classical mechanics, described by nonlinear differential equation systems in tree-dimensional plane Euclidean space and bifurcations in such systems;

- electromagnetic fields can exist without mass and gravitation, and electromagnetic waves can propagate in any direction with constant velocity (velocity of light) and arbitrary oscillation frequency, which is defined by oscillation frequency of physical vacuum without changes of its density;

- there exist equations, more common than Maxwell equations, deduced from the physical vacuum equations and invariant concerning Galileo transformations, many experimentally established laws of classical and quantum mechanics can be successfully deduced from the physical vacuum equations;

- existence of gravitation, mass and charge inseparably linked with the creation of elementary particles in form of curls of a single gravielectromagnetic field, the attracting force is actually a pressure force in physical vacuum created by gravitational wave, which propagates to the center of the particle with variable velocity and has no wave length;

REFERENCES

1.   N. Evstigneev, N. Magnitskii, 2010 On possible scenarios of the transition to turbulence in Rayleigh-Benard convection. Doklady Mathematics, 82 1 659 662 .

2.   R. Feyman, S. Weinberg, 1987 Elementary Particles and the Laws of Physic. Dirac Memorial lectures. Cambridge University Press, Cambridge, 138

3.   A. Kyriakos, 2006 Theory of the nonlinear quantized electromagnetic waves, adequate of standard model theory. Sp.B. BODlib, 208

4.   N. Magnitskii, S. Sidorov, 2006 New Methods for Chaotic Dynamics. World Scientific, Singapore, 360

5.   N. Magnitskii, 2008 Universal theory of dynamical chaos in nonlinear dissipative systems of differential equations. Commun. Nonlinear Sci. Numer. Simul., 13 416 433 .

6.   N. Magnitskii, 2008 New approach to analysis of Hamiltonian and conservative systems. Differential Equations, 44 12 1682 1690 .

7.   N. Magnitskii, 2010 Mathematical theory of physical vacuum. New Inflow, Moscow, 24

8.   N. Magnitskii, 2010 On topological structure of singular attractors. Differential Equations, 46 11 1552 1560 .

9.   N. Magnitskii, 2011 Mathematical theory of physical vacuum. Commun. Nonlinear Sci. Numer. Simul. 16 2438 2444 .

10.   N. Magnitskii, 2011 Theory of dynamical chaos. URSS, Moscow, 320 (in Russian)

11.   N. Tesla, 2003 Lectures, patents, articles. Tesla Print, Moscow, 198

12.   Podrobnosti 2010 Available from http://podrobnosti.ua/technologies/2010/11/ 04/ 728672.html

Chapter 4

A NEW APPROACH TO CLASSICAL STATISTICAL MECHANICS

Nijalingappa Umakantha

Department of Physics, Karnatak University, Dharwad, India

ABSTRACT

A new approach to classical statistical mechanics is presented; this is based on a new method of specifying the possible "states" of the systems of a statistical assembly and on the relative frequency interpretation of probability. This approach is free from the concept of ensemble, the ergodic hypothesis, and the assumption of equal a priori probabilities.

INTRODUCTION

The object of classical statistical mechanics is to explain the (statistical) properties of an assembly of a large number of identical particles in terms of the (deterministic) laws of classical mechanics. Such a theory has been developed by Gibbs in the first decade of the last century [1]. The Gibbs approach, presented by Tolman [2], has been accepted as the conventional approach to classical statistical mechanics. The theory presented by Tolman and the subsequent authors [3,4] is based on: 1) the concept of "state" of a many-particle system as defined by classical mechanics in the sense that the state of a N-particle system at any instant of time can be specified by a point in 6N dimensional phase space and 2) the notion of probability prevalent at the beginning of the last century according to which probability refers to a manyparticle system "chosen at random from the ensemble" of many-particle systems.

In this paper we present a new approach to classical statistical mechanics based on the significant progress made during and after the third decade of the last century in the theory of probability (a branch of pure mathematics) and the methods of statistical analysis (a branch of applied mathematics) [5,6]. This new approach is based on: 1) a new method of specifying the possible "states" of an assembly of particles which (method) is consistent with the requirements of statistical analysis, and 2) on the relative frequency interpretation of probability. The present approach is an independent approach and should be viewed as such. For the sake of clarity, the distinctive features of the two approaches are also discussed in the text.

PRELIMINARY CONSIDERATIONS

For the sake of clarity we may just mention the main features of statistical analysis and the relation between statistical analysis and probability theory. A large number of physical entities (such as adult men in a population) are said to form a collective, or a population, or a statistical assembly [5], if each entity (or member) of the assembly exists in one of the many (at least two) possible states S_n's (such as the state of parenthood of having n number of children) and the states of the members collected in any systematic manner lead to a random sequence of these possible states, in which each entry belongs to one member. If N_n is the number of times the state S_n occurs in the random sequence having a total number of N_0 entries, then N_n/N_0 is the relative frequency of the state S_n. As there can be many different systematic ways of collecting the data, there would be many random sequences of the same states (relevant to the given assembly). In all these random sequences the relative frequencies of the states would have approximately (in the statistical sense) the same set of values; such sequences are said to be statistically equivalent. One important feature of all statistical properties is that they are independent of such details as: 1) which particular member of the assembly is in which particular state, 2) the regions of space within which the individual members exist in the assembly (such as the places of residence of men), and 3) the total number of members of the assembly. Evidently, the relative frequencies of the possible states in a random sequence possess these properties. In the theory of probability, statistical properties are dealt with in a more abstract and general manner by associating probabilities with the possible states per se; these probabilities are treated as unspecified constants with the understanding that, with reference to any random sequence of states (relevant to a statistical assembly), the numerical values of the relative frequencies of the states in the sequence are approximately equal to the numerical values of the corresponding probabilities. It is extremely important to appreciate the point that (the relative frequencies as well as) the probabilities are associated with the possible states and not with the members of the assembly. With this background we consider classical statistical mechanics. A physical entity having finite non-zero mass bound to a time-independent potential is said to form a conservative system; to the extent the internal structure and the external dimensions of the entity do not play any role in the phenomenon under consideration, we can regard the entity as a particle, a point-mass. We can attribute many physical properties to such a system; the sum-total of all the physical quantities which can be attributed to the system at any instant of time is said to specify fully the "instantaneous state" of the system at that instant. As time passes, the system evolves through a continuous sequence of successive instantaneous states as governed by the laws of classical mechanics; such a sequence may be called a dynamical state of the system. Each dynamical state is determined by one solution of Newton's second law of motion which (solution) is specified by two constants, r_0 and p_0, the position and the momentum of the particle at some "initial" instant of time t_0. A dynamical state may be denoted as $S(r_0, p_0, t_0; r, t)$ which gives the position r of the particle as a function of time t ; once r is given as a function of time, all the physical quantities attributable to the system at any instant t, as well as their variations with t, can be derived mathematically from the function. In the case of a conservative system in a dynamical state, the energy E of the system remains constant so long as the system is in that dynamical state. There can be infinite number of such possible dynamical states for the given system, though a system exists, over a duration of time, in only one possible dynamical state. In some special cases, the system (such as a simple or conical pendulum, or a particle in a closed Kepler orbit) may go repeatedly through the same sequence of successive instantaneous states; such a dynamical state may be called a cyclic dynamical state.

Our object of study in classical statistical mechanics is a statistical assembly consisting of a large number of identical independent conservative systems which obey the laws of classical mechanics. Such systems may be: free particles (helium atoms), rigid rotators (diatomic molecules), harmonic oscillators (atoms in a crystal lattice), etc. In all these cases, first we have to identify the possible states of the systems relevant to the statistical properties of the given assembly and then use the laws of probability to determine the probabilities associated with these states. As our object is to present the new approach, we consider only an assembly of free particles.

AN ASSEMBLY OF FREE PARTICLES IN STATISTICAL EQUILIBRIUM

For the sake of definiteness let us consider an assembly of a large number N_0 of identical particles confined (by what is normally referred to as the walls of a container) within a fixed volume of field-free space of macroscopic dimensions. Evidently, over a duration of time each particle would be in a dynamical state characterized by a pair of constants such as r_0, p_0, with the understanding that all such pairs of constants specifying all the possible dynamical states (of all the particles) correspond to the same initial instant of time t_0. Though we have referred to them as particles, they are indeed physical entities having non-zero spatial extension and hence they would collide with one another exchanging energy and momentum. As a result, a particle would travel along a segment of a straight line belonging to a particular dynamical state, collide with another particle, exchange energy and momentum, make a transition to another particular dynamical state, travel along a segment of a straight line belonging to the new dynamical state to collide again, and so on; every particle would go through such a process incessantly. It is envisaged that as a result of such repeated transitions (from one dynamical state to another) made by all the particles over a sufficiently long initial duration of time, a "state of dynamic equilibrium" would be reached in the sense that the fractions of the total number of particles of the assembly in different possible dynamical states would remain almost constant over subsequent durations of time. Such an assembly is said to be in statistical equilibrium. Our interest is only in the equilibrium state and not in how equilibrium is reached from an initial non-equilibrium state [7].

As mentioned at the outset, first we have to specify the possible states of the particles in a manner that is consistent with the laws of classical mechanics as well as with the methods of statistical analysis. According to classical mechanics, the dynamical state of a (free) particle between two successive collisions is specified by the momentum of the particle and by the coordinates of the particle at the two points of collision. According to the methods of statistical analysis the positions of the particles within the volume of space of the assembly have no relevance to the statistical properties of the assembly. This means that, so far as the statistical properties of the assembly are concerned, we need specify each possible state by momentum only. We refer to them as momentum states. Evidently, the possible values of momentum p have a continuous range (both in magnitude and direction).

We associate probability P(p) dp with the states corresponding to momentum lying between p and p + dp; here dp is an element of constant magnitude dp_x dp_y dp_z. Evidently, P(p) corresponds to unit interval of p values. Our object now is to find this probability distribution using the laws of probability and the methods of statistical analysis. We treat (to begin with) P(p) as an unspecified function of p and then derive a general expression for it by making use of the properties of the equilibrium state of the assembly and the results of probability theory.

If we consider at any one instant of time t_1, the states of different particles (which exist at different points within the volume of the assembly) one after the other in any systematic manner, we get a random sequence of the possible momentum states p's. This random sequence has N_0 number of elements corresponding to N_0 number of particles in the assembly. If the states corresponding to momentum lying between p and p + dp occur $N_1(p)$ number of times in the sequence, the relative frequency $w_1(p) = N_1(p)/N_0$ of these states in the sequence would be approximately (in the statistical sense) equal to the probability P(p) dp. If we consider the states at another instant of time t_2 (after an interval of time long enough for particles to undergo transitions) we get another random sequence of the same states in which also the relative frequency $w_2(p)$ is approximately equal to the probability P(p) dp. Such random sequences are statistically equivalent. Thus because of the dynamic nature of equilibrium, we get a large number of statistically equivalent random sequences of the same states p's, the random sequences being relevant to the "state" of the assembly at the instants of time t_1, t_2, $\cdots$.

Let us consider the random sequence at the instant of time t_1. Now we define (what is referred to in statistical analysis [5,6] as) a binomial random variable R(p) which assumes the value 1 if a state in the random sequence belongs to momentum between p and p + dp and assumes the value 0 if it does not. This leads to a (derived) random sequence of 1's and 0's. Evidently, $\Sigma R(p)$ is the total number $N_1(p)$ of the particles having momentum between p and p + dp at the instant t_1. Corresponding to the random sequence at t_2, we get another value $N_2(p)$. Thus corresponding to random sequences at different instants of time t_1, t_2, $\cdots$, we get different values of N(p). Since a well defined (time-independent) probability P(p) dp is associated with the states between p and p + dp, these values of N(p) would have a well defined distribution characterized by the probability P(p) dp. According to the central limit theorem of the probability theory [5,6], the probability P{N(p)} that the quantity SR(p) has (at the conceptual instant of time) the value N(p) is given by

$$P\{N(p)\} = \frac{\exp\left[-\{N(p) - \bar{N}(p)\}^2 / 2N_0\sigma^2\right]}{\sqrt{2\pi}\sqrt{N_0}\sigma} \qquad (1)$$

where $\bar{N}(p)$ is exactly equal to P(p) dp N_0 and $\sigma^2(p) = P(p)$ dp $\{1 - P(p)$ dp$\}$. Here $\bar{N}(p)$ is the most probable value and, the distribution being symmetric, $\bar{N}(p)$ is also the mean value of N(p) taken over this distribution (which, in effect, is over a duration of time); σ is the standard deviation. This shows that because of the dynamic nature of the equilibrium distribution, the number N(p) of particles having momentum between p and p + dp fluctuates leading to a time-independent Gaussian distribution of values with the mean value $\bar{N}(p)$

and the standard deviation σ. Evidently, this is true of each possible value of p.

It is interesting to consider the state of a particular single particle of the assembly at different instants of time. Because of collisions, the particle would be in different momentum states at different (well separated) instants of time, leading to a random sequence of momentum states; this sequence would have the same number of elements as the number of instants of time at which the state of the particle is considered. Since the probability distribution P(p) dp is associated with the possible states per se (and not with the particles) this random sequence would be statistically equivalent to the random sequence of states of the particles of the assembly at one instant of time. So, the relative frequency distribution w(p) in any long segment of this sequence also would be approximately equal to the probability distribution P(p) dp. We may regard this sequence as being made up of a large number of successive segments each segment having n_0 number of elements. With reference to the first segment, Σ R(p) is the total number $n(p)_1$ of instants of time (out of n_0 number of instants) at which this particle has momentum between p and p + dp; with reference to the second segment we get another number $n(p)_2$; and so on. All these numbers are approximately equal to P(p) dp n_0. These numbers lead to a Gaussian distribution similar to (1). Again, the mean number of instants of time $\bar{v}$ (p) (out of a total number of instants of time n_0) at which the particular particle has momentum between p and p + dp is exactly equal to P(p) dp n_0. This is true of every possible momentum p of this particular particle under consideration. All this is true also of every other particle in the assembly. Significance of this result is that the mean value of a physical quantity taken over the states of a large number of particles of the assembly at any one instant of time is approximately equal to that taken over the states of any one particular particle at equally large number of instants of time.

A particle makes a transition to another momentum state, p' say, as a result of its collision with another particle and this process is independent of the momentum states of all the other particles in the assembly. So the fluctuations in the number N(p') of particles having momentum between p› and p› + dp would be independent of the fluctuations in the numbers relevant to all other momentum values (except for the weak condition that the sum of these numbers should remain constant at N_0). So, the probability P{N(p')}, given by (1), that N(p') number of particles have momentum between p' and p' + dp is independent of the probability P{N(p'')} that N(p'') number of particles have momentum between p'' and p'' + dp for any two distinct values of momentum p' and p''. The probability P{N(p)}for one value of p being independent of that for another value, the probability $P(N_1, N_2, \cdots, N_n, \cdots)$ that N_1 number of particles have momentum between p_1 and p_1 + dp, and N_2 between p_2 and p_2 + dp, $\cdots$, N_n between p_n and p_n + dp, $\cdots$, is given by

$$P(N_1, N_2, \cdots N_n, \cdots) = P(N_1)P(N_2)\cdots P(N_n)\cdots$$
$$= \Pi P(N_n) \qquad (2)$$

This is a multi-dimensional Gaussian distribution and its maximum (which corresponds to each N_n being equal to $\bar{N}_n$, the most probable value) may be specified by the condition

$$\delta P(N_1, N_2, \cdots N_n, \cdots)$$
$$= \left[\{\delta P(N_1)\} P(N_2) \cdots P(N_n) \cdots \right]$$
$$+ \left[P(N_1) \{\delta P(N_2)\} \cdots P(N_n) \cdots \right]$$
$$+ \left[P(N_1) P(N_2) \cdots \{\delta P(N_n)\} \cdots \right]$$
$$= 0. \qquad (3)$$

Here each term has not only the condition for the probability relevant to one value of momentum p being maximum, but also the probabilities relevant to all the other values as well. Thus the condition (3) is not consistent with the fact that the probabilities $P(N_n)$'s are all independent of one another. So we consider, instead, the condition $\delta \ln P(N_1, N_2, \cdots N_n, \cdots) = 0$. We have

$$\delta \ln P(N_1, N_2, \cdots N_n, \cdots)$$
$$= \delta \ln P(N_1) + \delta \ln P(N_2) + \cdots + \delta \ln P(N_n) + \cdots = 0, \qquad (4)$$

where each term refers to one value of momentum only, consistent with $P(N_n)$'s being independent of one another. Thus though both the conditions (3) and (4) look mathematically equivalent, only the condition (4) is physically appropriate. Importance of this result cannot be overemphasized.

All that has been said so far is mere explication, in terms of the laws of probability, of what we should mean by (dynamic) statistical equilibrium, once we assume that time-independent probabilities associated with the possible "states" of the particles exist; in fact, no law physics is involved. A little reflection would show that the above reasoning is so general that it is applicable not only to momentum p but also to energy E. The properties of a statistical assembly are better understood in terms of energy (rather than momentum) of the particles. So we develop the theory by treating energy as the independent variable.

If we associate probability $P(E_n)$ dE with the states corresponding to the energy lying between E_n and $E_n + dE$, then the probability $P(n_1, n_2, \cdots, n_n, \cdots)$ that n_1 number of particles have energy between E_1 and $E_1 + dE$, n_2 between E_2 and $E_2 + dE$, $\cdots$, n_n between E_n and $E_n + dE$, $\cdots$, is given by

$$P(n_1, n_2, \cdots n_n, \cdots) = P(n_1)P(n_2)\cdots P(n_n)\cdots$$
$$= \Pi P(n_n), \qquad (5)$$

which is a multi-dimensional Gaussian distribution and its maximum corresponds to each n_n being equal to $\bar{n}_n$, the most probable value (which is also the mean value). Again, the appropriate condition for this probability being maximum is specified by

$$\delta \ln P(n_1, n_2, \cdots n_n, \cdots)$$
$$= \delta \ln P(n_1) + \delta \ln P(n_2) + \cdots + \delta \ln P(n_n) + \cdots = 0. \tag{6}$$

When we consider the random sequences of states of the systems at a large number of instants of time, the random sequence corresponding to each instant would be characterized by one set of n_n values. Since $\bar{n}_n$'s are the most probable numbers, the number of random sequences having the set of $\bar{n}_n$ values should be larger than the number of those having any other set of possible n_n values. Now we estimate the number of such sequences (having the set of $\bar{n}_n$ values). Let $S_1, S_2, \cdots, S_{N_0}$ be the sequence of particles of the assembly in some order. Each distribution of N_0 number of energy states in this sequence of N_0 number of particles leads to one sequence of energy states of the particles; there are $N_0!$ number of such sequences. But all of them are not distinct because the same states occur many times. For instance, in any sequence the energy states between E_1 and $E_1 + dE$ would occur $\bar{n}_1$ number of times and permutation of these states among themselves does not lead to a new sequence of states; this is so with respect to each of the other states as well. So the total number of distinct sequences of states with $\bar{n}_1$ number of particles in the energy states between E_1 and $E_1 + dE$, $\bar{n}_2$ between E_2 and $E_2 + dE$, $\cdots$, $\bar{n}_n$ between E_n and $E_n + dE$, $\cdots$ is given by

$$\mathcal{N}_m(\bar{n}_1, \bar{n}_2, \cdots \bar{n}_n, \cdots) = N_0! / (\bar{n}_1! \bar{n}_2! \cdots \bar{n}_n! \cdots). \tag{7}$$

Since $\bar{n}_n$'s are the most probable numbers of particles in the relevant states, the number $\mathcal{N}_m$ is larger than the number N corresponding to any other set of n_n values; the subscript m denotes this. So this number $\mathcal{N}_m$ should be proportional to $P_m(\bar{n}_1, \bar{n}_2, \cdots \bar{n}_n, \cdots)$ which is the most probable value of the probability (5). Since it is appropriate to express the equilibrium condition in terms of $\ln P_m$ (rather than P_m), let us consider $\ln \mathcal{N}_m$. We have from (7)

$$\ln \mathcal{N}_m = \ln N_0! - \sum \ln \bar{n}_n!. \tag{8}$$

Using the Sterling approximation for the factorial of a large number n, given by $\ln n! = n \ln n - n$, (8) may be put as

$$\ln \mathcal{N}_m = N_0 \ln N_0 - \sum \bar{n}_n \ln \bar{n}_n. \tag{9}$$

Thus we see that each of the following three conditions characterize the (same) equilibrium state of the statistical assembly. When each n_n assumes its most probable value $\overline{n}_n$: 1) the probability $P(n_1, n_2, \cdots, n_n, \cdots)$ has its maximum value $P(\overline{n}_1, \overline{n}_2, \cdots \overline{n}_n, \cdots)$, 2) the number $\mathscr{N}$ has its largest value $\mathscr{N}_m$, and 3) the quantity $\Sigma n_n \ln n_n$ has its minimum value $\Sigma \overline{n}_n \ln \overline{n}_n$.

In statistical physics we are interested in properties which can be attributed to an assembly over a time scale which is large compared to the time scale relevant to transitions in the "states" of the constituent systems. Such properties depend on (the physical properties relevant to) the possible states of the systems and on the probabilities associated with these states. As mentioned before the states are determined (for the systems of the given assembly) by the laws of physics, and the probabilities (or equivalently the most probable numbers of systems in different states) are to be determined by using the laws of probability (consistent, of course, with the basic properties of the assembly under consideration).

We recognize that the two basic properties of the assembly of particles under consideration are that the total number N_0 and the total energy E_0 of the particles remain constant at all instants of time. Since the most probable numbers $\overline{n}_n$'s are also one set of possible numbers n_n's,

$$\Sigma \overline{n}_n = N_0 \tag{10a}$$

and

$$\Sigma \overline{n}_n E_n = E_0 \tag{10b}$$

Thus $\overline{n}_n$'s should satisfy the three conditions given in (9) and (10). Now $\mathscr{N}_m$ being the largest number, and N_0 and E_0 being constants, we have

$$\delta \ln \mathscr{N}_m = -\Sigma (\ln \overline{n}_n + 1) \delta \overline{n}_n = 0 \tag{11}$$

$$\delta \Sigma \overline{n}_n = \Sigma \delta \overline{n}_n = 0 \text{ and}$$
$$\delta \Sigma \overline{n}_n E_n = \Sigma E_n \delta \overline{n}_n = 0, \tag{12}$$

where $\delta \overline{n}_n$'s are small deviations from the equilibrium values $\overline{n}_n$'s. These three conditions should be satisfied simultaneously. Using Lagrange's method of undetermined multipliers, we may express them as a single equation. We have

$$\Sigma (\ln \overline{n}_n + \alpha + \beta E_n) \delta \overline{n}_n = 0 \tag{13}$$

where a and b are constants. Since (13) is to be satisfied for any arbitrary set of $\delta \overline{n}_n$'s, we should have

$$\ln \overline{n}_n + \alpha + \beta E_n = 0 \tag{14}$$

which means that

$$\bar{n}_n = 1/\left[\exp\left(\alpha + \beta E_n\right)\right] \tag{15}$$

Here $\bar{n}_n$ is the most probable, as well as the mean, number of particles in the energy range between E_n and E_n + dE. Using the well-known arguments we may identify the constant b as $1/kT$, where k is the Boltzmann constant and T the thermodynamic temperature. By virtue of (10a), a can be expressed in terms of other quantities as $Z = N_0 \exp a = \Sigma \exp (-E_n/kT)$; Z is called the partition function. Thus we arrive at the Maxwell-Boltzmann distribution law for energy.

THE CONVENTIONAL APPROACH AND THE NEW APPROACH

The development of statistical physics has been reviewed by Lebowitz [8]. Here we compare the distinctive features of the conventional approach (CA) and the new approach (NA). In CA [2] the object of our study is N_0 number of identical particles confined within a macroscopic volume of space; we refer to it as a many-particle system. The "state" of the many-particle system at any instant of time is specified by the position and momentum of the particles at that instant and changes continuously as a function of time as governed by (deterministic) laws of classical mechanics; the many-particle system per se is not regarded as being in statistical equilibrium. Because of the difficulties in determining the exact state of this many-particle system (due to "our incomplete knowledge" of the system), statistical approach is adopted. This conventional approach is based on three basic assumptions. 1) First we select (rather mentally) a large number of identical many-particle systems (which have "the same structure as the one of actual interest" and "are selected in such a manner as to agree with our partial knowledge as to the precise state of the actual system of interest", and) which exist in all the different accessible states; these are said to form a "representative ensemble" of many-particle systems. The concept of ensemble is the most distinctive feature of CA [9]. 2) Next, the time-averaged values of physical quantities relevant to the actual many-particle system of our interest are assumed to be the same as the respective values averaged, at one instant of time, over the states of the many-particle systems of the ensemble. This is known as the ergodic hypothesis. 3) In taking the average over the accessible states of the many-particle systems of the ensemble, the same "weight" is given to all the states; this is the assumption of equal a priori probabilities. All the physical reasoning and mathematical derivations refer to the representative ensemble of many-particle systems. In this approach probability refers to a (many-particle) system selected at random from among an ensemble of (many-particle) systems; this is consistent with the notion of probability prevalent at the beginning of the last century.

In the new approach also our object of study is N_0 number of identical particles confined within a macroscopic volume of space; we refer to it as an assembly of many particles. The main features of the present approach are: 1) We do accept (following classical mechanics) that the "state" of an assembly of particles is specified by the position and momentum of the particles, but in recognition of the statistical properties being independent of where the particles exist within the assembly, we specify the "state" by momentum of the particles only. 2) At the outset we recognize that collisions induce repeated transitions in the dynamical states of the particles and the assembly per se is identified as being in statistical equilibrium. All the physical reasoning and mathematical derivations refer to this assembly (only). 3) Though collision between two particles is strictly governed by deterministic laws of classical mechanics, when we consider the momentum states of the particles of the assembly (at any instant of time) in any systematic spatial order, they lead to a random sequence of these possible states. This justifies introduction of the concepts of randomness and probability into the theory (within the conceptual framework of classical determinism). 4) Probability of a state is identified as the relative frequency of the state in a random sequence of possible states. 5) The mean number $\bar{n}_n$ of particles in the assembly having energy between E_n and $E_n + dE$, is shown to be the same as the most probable number of particles having energy in that range. The values of physical quantities relevant to a statistical assembly in equilibrium depend on the (time-averaged time-independent) mean numbers of particles in the various energy states, whereas the condition for statistical equilibrium is specified in terms of the most probable numbers of particles in these energy states. If the theory is to be regarded as being logically consistent, these two sets of numbers should be shown to be equal. In the present approach this has been achieved by using the central limit theorem of the probability theory. 6) The mean value of a physical quantity taken over the states of the particles in the assembly at any one instant of time, is shown to be approximately the same as the mean value taken over the different states of any single particle of the assembly at equally large number of instants of time. 7) The reason for maximizing $\ln \mathcal{N}_m(\bar{n}_1, \bar{n}_2, \cdots \bar{n}_n, \cdots)$ is justified, and not just accepted as a matter of convenience. In fact, only as a result of maximizing $\ln \mathcal{N}_m$ (instead of maximizing $\mathcal{N}_m$) do we get the exponential function in (15). A little reflection would show that maximizing $\ln \mathcal{N}_m$ as a matter of convenience is tantamount to assuming what is to be derived. And 8) it is shown that the minimum possible value of the quantity $\Sigma n_n \ln n_n$ given by $\Sigma \bar{n}_n \ln \bar{n}_n$ (also) specifies the equilibrium condition of the assembly. This is "Boltzmann's famous H-theorem" which, according to Tolman, "may be regarded as among the greatest achievements of physical science".

CONCLUDING REMARKS

In conclusion we may state that by making use of the modern theory of probability and statistics, it is possible to develop a new approach which is free from the concept of ensemble, the ergodic hypothesis, and the assumption of equal a priori probabilities (of the conventional approach). However, this is only an alternative approach which leads to the same final results as the well established conventional approach.

REFERENCES

1.   J. W. Gibbs, "Elementary Principles of Statistical Mechanics," Dever, New York, 1960 (Reprint of 1902 Edition).

2.   R. C. Tolman, "The Principles of Statistical Mechanics," Oxford University Press, Oxford, 1938.

3.   D. Ter Haar, "Foundations of Statistical Mechanics," Reviews of Modern Physics, Vol. 27, No. 3, 1955, pp. 289- 338. doi:10.1103/RevModPhys.27.289

4.   D. Chandler, "Introduction to Modern Statistical Mechanics," Oxford University Press, Oxford, 1987.

5.   R. von Mises, "Mathematical Theory of Probability and Statistics," Academic Press, Waltham, 1964.

6.   M. R. Spigel, "Theory and Problems of Probability and Statistics," McGraw-Hill, New York, 1992.

7.   R. Frigg, "Typicality and the Approach to Equilibrium in Boltzmann Statistical Mechanics," Philosophy of Science, Vol. 76, No. 5, December 2009, pp. 997-1008.

8.   J. L. Lebowitz, "Statistical Mechanics," In: H. Stroke, Ed., The First Hundred Years, The Physical Review, AJP Publication, New York, 1995, pp. 465-471.

9.   H. O. Georgii, "The Equivalence of Ensembles for Classical Systems of Particles," Journal of Statistical Physics, Vol. 80, No. 5-6, 1995, pp. 1341-1378. doi:10.1007/BF02179874

Chapter 5

A FATIGUE ANALYSIS OF A HYDRAULIC FRANCIS TURBINE RUNNER

Miriam Flores[1], Gustavo Urquiza[1], José María Rodríguez[2]

[1]Centro de Investigación en Ingeniería y Ciencias Aplicadas, CIICAp, Universidad Autónoma del Estado de Morelos, Cuernavaca, México

[2]Centro Nacional de Investigación y Desarrollo Tecnológico, Cuernavaca, México

ABSTRACT

In this work, the estimation of crack initiation life of a hydraulic Francis turbine runner is presented. The life prediction is based on the local strain approach to predict the initiation life. First, the analysis is carried out in air and in water condition and the runner's natural frequencies were calculated using the finite element (FE) method. The analysis in air is compared with experimental analysis in order to have a representative model of real runner and subsequently the numerical analysis was perform in water. In the case of the runner immersed in water, the added mass effect due to the fluid structure interaction (FSI) is considered. Second, the static and dynamic stresses were calculated according to life estimation. For the calculation of static stresses, the pressure distribution of water and the centrifugal forces were applied to the runner. The dynamic stresses were estimated for interactions between the guide vane and the runner. Lastly, the estimation of the crack initiation life of the runner was obtained.

INTRODUCTION

The tendency of higher power concentration in hydraulic turbines, bring as a consequence an increase in both the load and hydraulic forces in the machine. These conditions produce major stresses in runners and possible vibration problems that could cause fatigue and fracture the blades. The fracture begins in small cracks, brought about by critical operation conditions of the machine over long periods, until failure.

In the most of hydroelectric power plants around the World, turbines have operated for decades and in many of them, the current operating conditions are different from the original design specified, these operations cause vibrations and some have presented cracks in the runners produced by fatigue [1,2].

The fatigue cracks normally are present in regions that have a metallurgical or structural discontinuity and were subjected to higher stresses [3]. The concept of local strains and stresses are the most promising approach to predict the crack initiation growth in a structure subjected to fatigue loads [4]. These concepts are used in this work, and the following procedure is adopted for the estimation of crack initiation growth life of a Hydraulic Francis Turbine Runner that is installed in a Mexican hydroelectric power plant.

MODAL ANALYSIS

In the actual operating conditions, Francis runners are surrounded by water. For the prediction of the dynamical characteristics of the structure at these conditions, it must be taken into account the effect of the fluid that surrounds it. The system had to be treated as a problem of fluid-structure interactions, where the equation of dynamic structure has to be coupled with the fluids equations. It is well known that the equation of the dynamic structure could be formulated as follows:

$$[M_S]\{\ddot{u}\}+[C_S]\{\dot{u}\}+[K_S]\{u\}=\{F_S\} \tag{1}$$

where $[M_s]$ is the structural mass matrix; $[C_s]$, the structural damping matrix; $[K_s]$, the structural stiffness matrix; $\{F_s\}$, the applied load vector and $\{u\}$, the nodal displacement vector. In the case of the coupled model of the structure-water, the behavior of the water pressure could be described with the acoustic wave equation, known as the Helmholtz equation.

$$\nabla^2 P = \frac{1}{c^2}\frac{\partial^2 P}{\partial t^2} \tag{2}$$

where P is the fluid's pressure, c the sound velocity in the fluid's media, t is the time and $\tilde{N}^2$ the Laplace operator. Equation (2) comes from the Navier-Stokes movement equation and the continuity equation considering the following assumptions [5]:

Ÿ The fluid is a compressible fluid (the density change because of the pressure variations)

Ÿ The fluid has no viscosity.

Ÿ There are no flow on the fluid The density and pressure are uniform in the fluid.

In the interface between the solid runner and water, the relation between the normal pressure gradient of the fluid and the normal acceleration of the structure gives the equation [6]:

$$\{n\} \cdot \{\nabla P\} = -\rho_0 \{n\} \cdot \frac{\partial^2 U}{\partial t^2}$$ (3)

where U is the displacement vector of the structure's interface, and r_0 the density of the fluid. Considering the pressure of the fluid that acts in the interface, (1) for the structural dynamics can be described by the form:

$$[M_S]\{\ddot{u}\} + [C_S]\{\dot{u}\} + [K_S]\{u\} = \{F_S\} + \{F_{fS}\}$$ (4)

where $\{F_{fS}\}$ is the load vector because of the fluid's pressure acting in the interface. The finite element discretized equations for the fluid-structure interaction problem were described as:

$$\begin{pmatrix} [M_S] & [0] \\ [M_{fS}] & [M_f] \end{pmatrix} \begin{Bmatrix} \{\ddot{u}\} \\ \{\ddot{p}\} \end{Bmatrix} + \begin{pmatrix} [C_S] & [0] \\ [0] & [C_f] \end{pmatrix} \begin{Bmatrix} \{\dot{u}\} \\ \{\dot{p}\} \end{Bmatrix}$$
$$+ \begin{pmatrix} [K_S] & [K_{fS}] \\ [0] & [K_f] \end{pmatrix} \begin{Bmatrix} \{u\} \\ \{p\} \end{Bmatrix} = \begin{Bmatrix} \{F_S\} \\ \{0\} \end{Bmatrix}$$ (5)

where $[M_{fS}]$ is the mass equivalent matrix in the interface and $[K_{fS}]$ is the stiffness equivalent matrix in the interface. The solution of the finite element modal analysis from the runner-water coupled model gives as a result the natural frequencies and the modal shapes of the structure.

Cyclic Strain in Fatigue

The range of total deformation (De) is the addition of the elastic strain (e_e) and the plastic strain (e_p):

$$\Delta \varepsilon = \Delta \varepsilon_e + \Delta \varepsilon_p$$ (6)

For a stable hysteresis curve, it is suggested [7] that it can be described by a cycle of deformation being the sum of the elastically and plastically ranges, so:

$$\frac{1}{2}\Delta \varepsilon = \frac{1}{2E}\Delta \sigma + \left(\frac{1}{2K'}\Delta \sigma\right)^{1/n'}$$ (7)

where: E is the elastic module, Ds is the real range of stress, K' is the cyclic strength coefficient, and n' is

Neuber's Rule

The Neuber's rule expresses the relation between the nominal stress range DS, and the true stress; and the nominal deformation in the elastic region in the vicinity of the defect in the specimen [8]. This is the nominal deformation in the elastic region in the vicinity of the defect in a specimen. That is:

$$\frac{\left(K_f \Delta S\right)^2}{E} = \Delta\varepsilon\Delta\sigma \qquad (8)$$

Equation (7) is used in the equation of the Neuber's rule (8), obtaining (first approximation):

Fatigue Life

The stress life (S-N) data can be plotted linearly on a log-log scale. The total strain amplitude is the sum of elastic strain amplitude and plastic strain amplitude. The stress life for the elastic part of the strain amplitude is determined by:

$$\frac{1}{2}\Delta\varepsilon_e = \frac{1}{E}\sigma_f'\left(2N_i\right)^b \qquad (10)$$

where s'$_f$ is the fatigue strength coefficient. The plastic strain life in the log-log plot is

$$\frac{1}{2}\Delta\varepsilon_p = \varepsilon_f'\left(2N_i\right)^c \qquad (11)$$

where e'$_f$ is the fatigue ductility coefficient, and c is the fatigue ductility exponent. The total strain amplitude is the strain life equation (include the effect of mean stress s$_m$), as follows:

CASE STUDY

The case study presented was performed for a Francis turbine runner of 38.5 MW with an operation velocity of 180 rpm and it consists of 13 blades. The runner's metallic material is 13.4CrNi stainless steel with elastic module E = 206 GPa, yield strength S$_{ys}$ = 590 MPa, Poisson ratio m = 0.288 and density r = 7700 kg/m³. The adopted procedure for the runner analysis is as follows:

1) The model of the runner was constructed as a FE model to perform the modal and the static stress analysis. The simulations were performed using the commercial software ANSYS.

2) The modal analysis was realized for air and water. The numerical analysis in air is compared with the vibration experimental results obtained for air. The runner's analysis in water considers the interaction of the structure and the fluid to obtain the natural frequencies. Also, the

relation of the frequencies reduction was obtained due to the water that surrounds the runner.

3) The static stress of the runner was calculated taken into account the loads in the operational conditions caused by the centrifugal forces and the fluids static pressure. For the calculation of the dynamical stress, an excitation force was considered for blades passing the guide vanes. The numbers of the guide vanes in this case is 24.

4) With the nominal stresses (from the stress analysis), the estimation of the cracking initiation growth on the runner was calculated.

FE Model

Based on the characteristic of cyclic symmetry for the structure, it was used a runner's sector conformed by a blade and an angle of 360/13 degrees of the crown and the band to run the simulation. The model was discretized with 3D solid structure elements with 20 nodes for the blade, and 3D structural solid elements tetrahedral with 10 nodes for crown and band. In the analysis it was established the conditions of the cyclic analysis. The results were expanded to the whole runner. **Figure 1** shows the model of the runner sector, which is formed by 20894 nodes and 10374 elements. **Figure 2** shows the discretized model of the complete runner. The modal characteristics of runner were obtained using the modal analysis for cyclic geometry of ANSYS using the Block-Lanczos method.

Modal Analysis

The theoretical and experimental study of the structures immersed in water, indicates that the natural frequencies are reduced because of the interaction of this fluid with the structure [9-11]. It is important to determine the natural frequencies of the runner in air and establish if there is a reduction of them when the runner is sur- rounded by water.

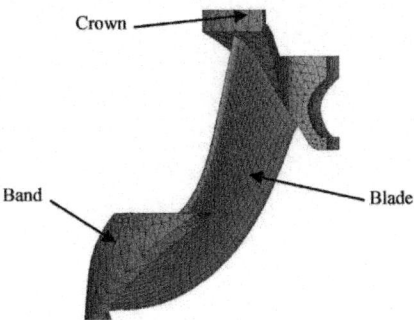

Figure 1: FE model of runner section.

The modal characteristics of a free vibration in air were obtained from a numerical and experimental procedure [12], the results were presented in **Table 1**. The modal shapes of the runner in air were shown in **Figure 3" target="_self"> Figure 3**.

For comparison of the numerical and experimental results in air, the variation (D%) is calculated between them, indicating the concordance between both analyses.

$$\Delta\% = \frac{\left(f_{Sim} - f_{Exp}\right)}{f_{Exp}}(100)$$
(13)

It is observed a variation between ±1.29% and ±3.5%, depending on the frequency, showing a good correlation between the simulation and the experimental results. From this air simulation model, the simulation of the runner in water is performed.

For the runner's simulation in water, the FE model was modified extending the mesh of the structure, the fluid mesh considers that the runner was surrounded by the fluid. Both dominions share the same nodes group in the interface. 3D acoustic fluid elements were used specifying the elements of the fluids-structure interface and 3D infinite acoustic elements for the wall absorption.**Figure 4** showed the mesh model of the complete runner surrounded by water. The water properties under environmental temperature and atmospheric pressure were: density r = 1000 kg/m³ and sound speed in water v = 1483 m/s. The obtained frequencies in this analysis were presented in **Table 2**. The modes of vibration observed

Figure 2: The whole runner model of the Francis turbine.

Table 1. Natural frequencies of the runner in air

Analysis	Frequency, f (Hz)				
	f_1	f_2	f_3	f_4	f_5
Simulation	63.785	76.523	130.42	135.38	148.83
Experimental	66.13	78.12	128.75	131.875	145

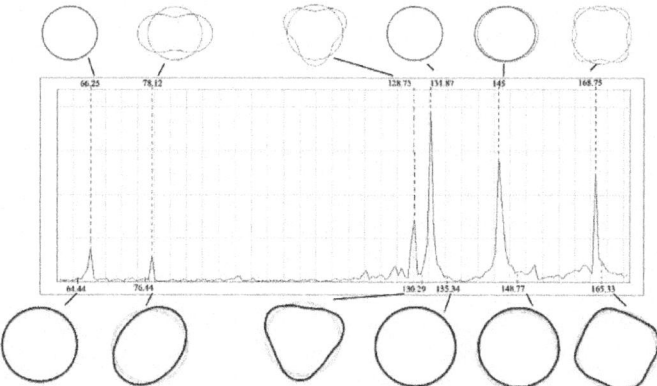

Figure 3: Vibration modes of the runner in air calculated by the experimental analysis (top) and numerical analysis (bottom) of first to sixth natural frequencies.

Table 2: Natural frequencies of the runner in water

Analysis	Frequency, f (Hz)				
	f_1	f_2	f_3	f_4	f_5
Simulation	54.089	61.141	95.076	95.307	100.69

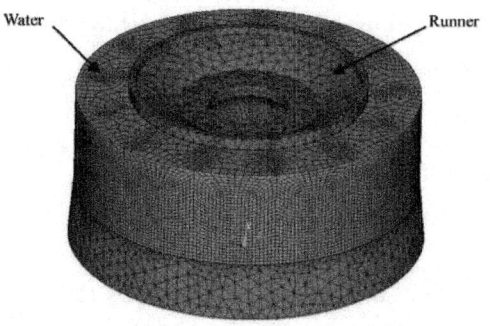

Figure 4: Mesh model of the complete runner surround of water.

were similar to the ones presented before, for the simulation of the runner in air. **Figure 5** shows the natural frequencies of the runner in air and water.

When the obtained results in the runner's simulations in air and submerged in water are compared, it was observed that a decrease in the natural frequencies do exist,

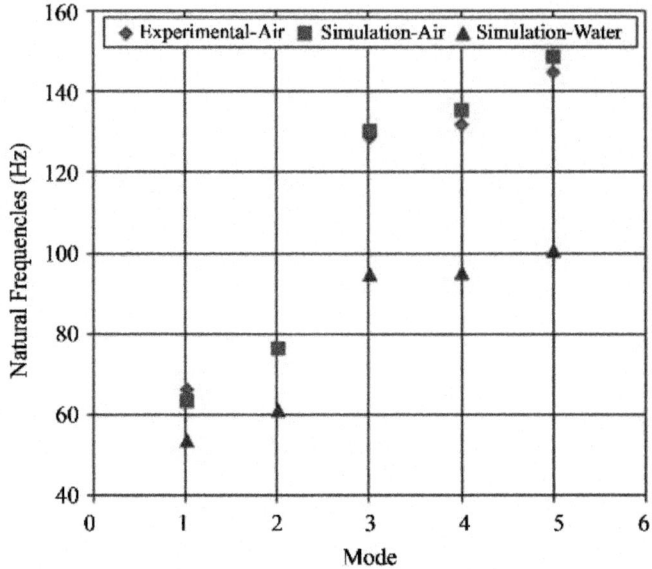

Figure 5. Natural frequencies in air and water.

because of the water that surrounds the structure. The above mentioned is estimated by the ratio of frequencies reduction given by:

$$\delta = \frac{f_{air} - f_{water}}{f_{air}}$$ (14)

where f_{air} and f_{water} are the natural frequencies in air and water respectively. The ratios frequencies reduction d, are shown in **Table 3**.

It was observed that the ratio of frequencies reduction varies from 0.152 to 0.324 depending on the frequencies. These values of d approach to those presented in different works, for example: Tanaka [4] gave an empiric value of 0.2, and Rodríguez [11] and Liang [13] realized a theoretical and experimental investigation modeling the runner in water with simplifications, and presenting for d, values that vary from 0.1 to 0.39.

It was observed that the ratio of frequencies reduction varies from 0.152 to 0.324 depending on the frequencies. These values of d approach to those presented in different works, for example, Tanaka [4] gave an empiric value of

0.2, and Rodríguez [11] and Liang [13] realized a theoretical and experimental investigation modeling the runner in water with simplifications, and presenting for d, values that vary from 0.1 to 0.39.

Stress Analysis

For the static stresses analysis of the runner under operating conditions, it was necessary to include the force of inertia and pressure fluid on the runner. The load in relation with the static pressure of the fluid on the sides of the pressure and suction of the runner's blades were obtained by the Computational Fluids Dynamic (CFD) analysis [12] which uses the finite volume method to solve the Navier-Stokes equations. This load was allocated for the FE model for the stress analysis. The calculation was realized with the Von Mises criteria. The stress distribution was shown in **Figure 6**. The peak stress was found at 56.1 MPa and the average stress was 19.7 MPa. The maximum stress was localized in the blade near the band, close to the runner's axis. The dynamical stresses come from the possible resonance of the harmonic from the guide vanes with the runner's frequency. The calculation of the frequency of the blades passing the guide vanes, f_z, caused by the external force that acts on the blades, having a frequency of:

$$f_z = N_z \times \frac{n}{60} \qquad (15)$$

where N_z is the blades passing the guide vanes number, and n is the operation velocity of the turbine. The corresponding frequencies of the first and second harmonics for the blades passing the guide vanes were 72 and 144 Hz. From these results, it is shown that the second natural frequency of the runner in water was very close to the frequency of the first harmonic for the blades passing the guide vanes, producing a possible resonance effect.

Table 3: Ratio frequencies reduction of the runner

Ratio	Frequency, f (Hz)				
	f_1	f_2	f_3	f_4	f_5
δ	0.152	0.201	0.271	0.296	0.324

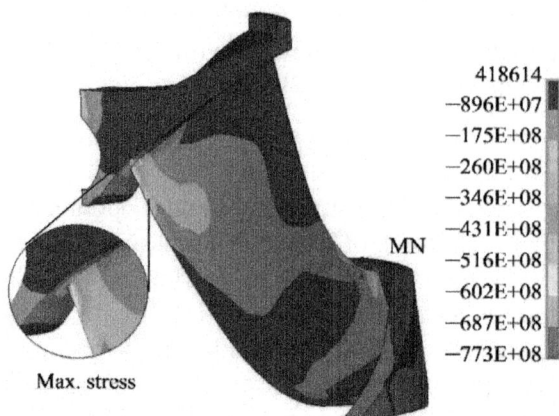

418614
−896E+07
−175E+08
−260E+08
−346E+08
−431E+08
−516E+08
−602E+08
−687E+08
−773E+08

MN

Max. stress

Figure 6: Stress distribution to the centrifuge forces and the static pressure of the fluid.

The force of excitation comes from the first harmonic from the guide vanes and was a maximum at 10% of the stable force, taking the stress of Von Mises in the runner as 19.7 MPa as the stable load. Assuming that the damping ratio was 0.02, the quality factor for the fundamental bending is 25. Because of this, the dynamical stresses were very close to the union zone of the crown and the blade being 49.25MPa.

CRACK INITIATION

For the estimation of the crack initiation growth in the chosen runner's hydraulic turbine, the equations were solved with the values obtained in the previous analysis. The material fatigue properties as: ultimate tensile strength S_u = 735 MPa, cyclic strength coefficient K' = 1730 MPa, cyclic strain hardening exponent n' = 0.14, fatigue strength exponent b = −0.076 and fatigue ductility exponent c = −0.62 were useful for the calculation of crack initiation growth [14,15]. The nominal stress range DS was obtained from the dynamical stresses calculated, with their value as 49.25 MPa, so DS = 98.5 MPa. The real range of stress Ds was determined by Equation (9). The stress concentration factor K equals 5 and was considered that K_f = K, and because of this Ds = 492 MPa. The local strain range from the Neuber's rule (8), was obtained as De = 0.002553.

The crack initiation growth was estimated from the equation for the total amplitude of deformation in the deformation life (12). The mean stress s_m is considered equal to 0. The fatigue strength coefficient $s›_f$, can be approximated equal to the true stress at fracture, $s'_f = s_f$. For steel with Brineel hardness up to 500 it may be approximated as s'_f @ Su + 345 MPa, so $s›_f$ = 1080 MPa.

The fatigue ductility coefficient e'_f, is approximated by the true fracture ductility (13)

$$\varepsilon'_f \cong \varepsilon_f = \ln \frac{1}{1-RA}$$ (16)

where RA is reduction in area. The number of cycles of crack initiation growth is obtained from (12). The life of the blade under resonance conditions is calculated by equation:

$$Ti = \frac{N_i}{Harmonic\ in\ resonance}$$ (17)

In operation conditions, the dynamic stresses induced in the runner are present for possible resonance. In these conditions, the resonance is accumulative and if the machine exceeds the crack initiation life of 23 days, the fracture initiation will occur.

CONCLUSION

A modal analysis was realized and the estimation of crack initiation growth life was calculated, for a Francis turbine runner. A FE model was built for the numerical, modal and static stress analysis. In the modal analysis, the natural frequencies and the modal shapes of the runner in air and surrounded by water, were determined. The simulation results in air as compared with the experimentally obtained, present a maximum variation of ±3.5% and shows a good correlation between them. In the submerged in water runner analysis, it was considered the runner surrounded by this fluid, and also the interface between the fluid and the structure. It was observed a decrease in the natural frequencies of the runner in air and surrounded by water. The modal shapes in both cases were similar. A static analysis was realized in operation loads obtaining the Von Mises stresses. The maximum stress was localized in the blade near to the band, close to the runner axis. The dynamical stresses were calculated from the possible resonance that exists between the second natural frequencies of the runner and the first harmonic of the guide vanes. If the machine operates under these conditions of resonance and dynamical stress, exceeding the time of crack initiation growth of 23 days, the crack initiation growth will occur. The method used in this work for the runner analysis could be used for the dynamical behavior analysis for other turbine runners and crack initiation growth estimation.

ACKNOWLEDGEMENTS

This paper is dedicated to the memory of Dr. Janusz Kubiak Szyszka (Centro de Investigación en Ingeniería y Ciencias Aplicadas, Universidad Autónoma del Estado de Morelos) who left us in April 2009.

REFERENCES

1.   R. Xiao, et al., "Study on Dynamic Analysis of the Francis Turbine Runner," Large Electric Machine and Hydraulic Turbine, Vol. 7, 2001, pp. 41-43.

2.   S. Rao, P. K. Nimbekar, R. Misra and A. K. Singh, "Application of Local Stress-Strain Approach to Predict Fracture Initiation of a Francis Turbine Runner Blade," 7th International Symposium on Transport Phenomena and Dynamics of Rotating Machinery, Hawaii, 22-26 February 1998, pp. 22-26.

3.   S. Rao, "Turbine Blade Life Estimation," Alpha Science International Ltd., Pangbourne, 2000.

4.   H. Tanaka, "Vibration Behavior and Dynamic Stress of Runners of Very High Head Reversible Pump-Turbine," 15th International Association of Hydraulic Engineering & Research, Symposium on Hydraulic Machinery and Systems, Belgrade, 1990, pp. 289-306.

5.   L. E. Kinsler, et al., "Fundamentals of Acoustics," John Wiley and Sons, New York, 1982.

6.   O. C. Zienkiewicz, and R. E. Newton, "Coupled Vibrations of a Structure Submerged in a Compressible Fluid," Symposium on Finite Element Techniques, Stuttgart, 1-15 May 1969, pp. 360-378.

7.   J. F. Martin, T. H. Topper and G. M. Sinclair, "Computer Based Simulation of Cyclic Stress Strain Behavior," T. &A. M. Report No. 326, University of Illinois, Urbana, 1969.

8.   H. Neuber, "Theory of Stress Concentration for ShearStrained Prismatical Bodies with Arbitrary Nonlinear Stress-Strain Law," Journal of Applied Mechanics, Vol. 28, No. 4, 1961, pp. 544-550. doi:10.1115/1.3641780

9.   A. Coutu, H. Aunemo, B. Badding and O. Velagandula, "Dynamic Behavior of High Head Francis Turbine," Hydro 2005, Villach, 17-20 October 2005.

10. C. Monette, A. Coutu and O. Velagandula, "Francis Runner Natural Frequency and Mode Shape Predictions," Waterpower XV, Chattanooga, 23-26 July 2007.

11. C. G. Rodríguez, E. Egusquiza, X. Escaler, M. Farhat, Q. W. Liang and F. Avellan, "Experimental Investigation of Added Mass Effect on a Francis Turbine Runner," Journal of Fluids and Structures, Vol. 22, No. 5, 2006, pp. 699-712.doi:10.1016/j.jfluidstructs.2006.04.001

12. M. Flores, "Fluid-Structure Interaction Study of a Hydraulic Francis Turbine Runner," Ph.D. Dissertation, University Autonomous of Morelos State, Mexico, 2009.

13. Q. W. Liang, C. G. Rodríguez, E. Egusquiza, X. Escaler and F. Avellan, "Modal Response of Hydraulic Turbine Runners," 23th International Association of Hydraulic Engineering & Research, Symposium on Hydraulic Machinery and Systems, Yokohama, October 2006.

14. ASM International, "Mechanical Testing and Evaluation," ASM Handbook, Vol. 8, 2000.

15. D. F. Socie, M. R. Mitchell and E. M. Caulfield, "Fundamentals of Modern Fatigue Analysis," Fracture Control Program Report No. 26, University of Illinois, Urbana, 1977

Chapter 6

DEVELOPMENT OF ANALYTICAL MODEL FOR MODULAR TANK VEHICLE CARRYING LIQUID CARGO

Messaoud Toumi[1], Mohamed Bouazara[1], Marc J. Richard[2]

[1]Department of Applied Sciences, University of Quebec at Chicoutimi, Saguenay, Canada

[2]Department of Mechanical Engineering, Laval University, Quebec, Canada

ABSTRACT

The study of dynamics of tank vehicles carrying liquid fuel cargo is complex. The forces and moments due to liquid sloshing create serious problems related to the instability of tank vehicles. In this paper, a complete analytical model of a modular tank vehicle has been developed. The model included all the vehicle systems and subsystems. Simulation results obtained using this model was compared with those obtained using the popular TruckSim software. The comparison proved the validity of the assumptions used in the analytical model and showed a good correlation under single or double lane change and turning manoeuvers.

INTRODUCTION

In general, numerical models are developed to understand the liquid sloshing phenomenon coupled with tank structure. They are able to determine the coupling behavior, only under specific conditions, such as periodic accelerations. The effects of suspension system, tire and road excitation on a moving vehicle have not been taken into consideration. Regarding the vehicle itself, different simple models for tractors and trailers have been described in literatures to study the dynamic behavior of heavy vehicles during various maneuvers. Ellis [1] developed a simple model for tractor-trailer type bicycle with four degrees of freedom where the load transfer was modeled using an additional degree of freedom (rolling motion). Hyun [2] adopted a model for vehicle with four degrees of freedom for the active control of roll-over of heavy vehicles. While various solid-liquid models have been developed to determine the dynamic behavior of vehicles carrying liquids, few models have been developed to reflect the effects of vehicle systems and subsystems, such as suspension and tire components. The models adopted for the vehicle systems are all based on simplified assumptions.

It is necessary to develop a comprehensive model because a vehicle is composed of various subsystems and the effects of those need to be considered. AutoSim package, one of most popular software for modeling of the behavior of a vehicle, was developed at the University of Michigan [3,4]. Three software applications were created based on the AutoSim package [5]. These software applications are CarSim, TruckSim and BikeSim for cars, heavy vehicles and motorcycles respectively. However, the TruckSim software does not include the effects of motion of a moving load [6-8]. They are easy to use for conventional vehicles only. However, they offer some models for unconventional designs and the models find applications in some specific research projects. Another drawback with these tools is that they work in a closed environment.

VEHICLE KINEMATIC

To develop the model of the vehicle, there are several methods that could be exploited to derive the equations of motion such as Lagrange, Newton and virtual work methods. The popular alternative approach for dynamic modelling of vehicles is to use of simple models having a reasonable excution time. In this study, a new model was developed based on the simplified Ervin model [4]. This model was solved without any mathematical approximation and it took care of the complexity of liquid motion inside the tank. The solutions of the equations were obtained using the mathematical software Maple [9]. The equations were derived based on the principles of Newtonian mechanics and conservation of linear and angular momentums for a solid body.

Coordinate System

The large number of degrees of freedom for translation and rotational motion, required to represent an articulated vehicle, excludes the use of a single coordinate system. In fact, the equations of motion can be written more easily if several coordinate systems are employed. The purpose of this section is to identify the orientation of the various coordinate systems, and specify the variables required to connect the processing unit vectors in the various systems. The inertial coordinate system, the body coordinate system fixed to the sprung mass and the coordinate system fixed to the unsprung mass were used to describe the system. Newton's laws are valid only for a finite acceleration in an inertial coordinate system $[x_n, y_n, z_n]$. The orientations of coordinate axes were expressed in accordance with the Society of Automotive Engineers' standard (SAE), where the positive x axis points anterior, the positive y axis is oriented to the right and the positive z axis points downward. In our model, each sprung mass was represented as a rigid body with six degrees of freedom namely, longitudinal, lateral, vertical, roll, pitch and yaw. For the unsprung mass, there were assigned two degrees of freedom namely, the roll and vertical motions relative to the point of attachment of the sprung mass. The equations were formulated such that there was no limit to the number of sprung and unsprung masses. All the equations were solved, without any mathematical simplification, using the symbolic computational software Maple [9].

Three coordinate systems were used to develop the equations of motion. The first one was attached to the inertial system $[x_n, y_n, z_n]$, the second one was attached to each sprung mass $[x_s, y_s, z_s]$ and the third one was attached to each unsprung mass $[x_u, y_u, z_u]$. Figure 1

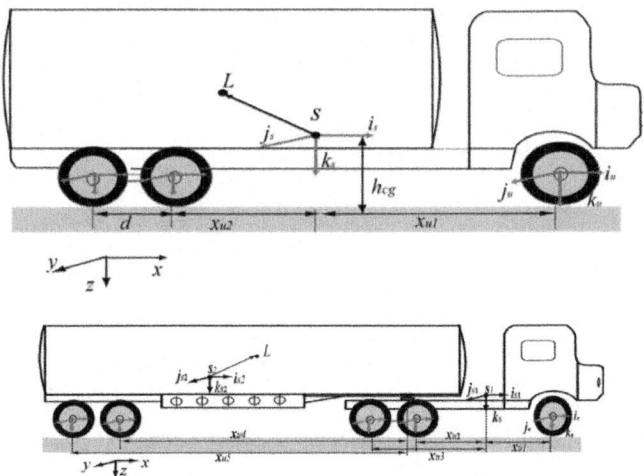

Figure 1: Fixed unit and articulated tank vehicles.

shows the coordinate systems for fixed unit and articulated vehicles.

Coordinate System Fixed to the Sprung Mass

The three rotational motions of the sprung mass were expressed by the three Euler angles: yaw ψ_s (around axis z), pitch θ_s (around axis y) and rolling motion ϕ_s (around axis x) as shown by **Figure 2**.

The transformation matrix between inertial system and the system fixed to the sprung mass was defined separately for the three successive rotations: yaw, pitch and roll.

Yaw $^{\psi_s}$:

$$\left(i_n \quad j_n \quad k_n\right)^{\mathrm{T}} = \lambda_{\psi_s}\left(i_1 \quad j_1 \quad k_1\right)^{\mathrm{T}}$$

$$\lambda_{\psi_s} = \begin{pmatrix} \cos\psi_s & -\sin\psi_s & 0 \\ \sin\psi_s & \cos\psi_s & 0 \\ 0 & 0 & 1 \end{pmatrix}$$

(1)

Pitch $^{\theta_s}$:

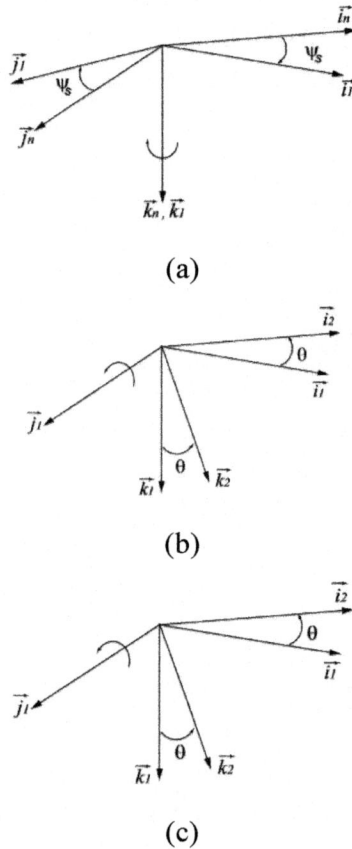

(a)

(b)

(c)

Figure 2: (a) Yaw: $^{w_{sz} = \dot{\psi}_s k_n}$; (b) Pitch: $^{w_\sigma = \dot{\theta}_s j_1}$; (c) Roll: $^{w_{sx} = \dot{\phi}_s i_1}$. Sprung mass orientation defined by euler angles.

$$\lambda_{\theta_s} = \begin{pmatrix} \cos\theta_s & 0 & \sin\theta_s \\ 0 & 1 & 0 \\ -\sin\theta_s & 0 & \cos\theta_s \end{pmatrix} \qquad (2)$$

Roll ϕ_s :

$$\begin{pmatrix} i_2 & j_2 & k_2 \end{pmatrix}^T = \lambda_{\varphi_s} \begin{pmatrix} i_s & j_s & k_s \end{pmatrix}^T$$

$$\lambda_{\varphi_s} = \begin{pmatrix} 1 & 0 & 0 \\ 0 & \cos\varphi_s & -\sin\varphi_s \\ 0 & -\sin\varphi_s & \cos\varphi_s \end{pmatrix} \qquad (3)$$

The transformation matrix, connecting the inertial system and the system fixed to the sprung mass, was obtained by combining the three matrices as follows:

$$\begin{pmatrix} i_n & j_n & k_n \end{pmatrix}^T = R_s'' \begin{pmatrix} i_s & j_s & k_s \end{pmatrix}^T$$

$$R_s'' = \begin{bmatrix} \lambda_{\psi_s} \end{bmatrix}\begin{bmatrix} \lambda_{\theta_s} \end{bmatrix}\begin{bmatrix} \lambda_{\phi_s} \end{bmatrix} \qquad (4)$$

where: (please see Equation (5) below)

$$R_s'' = \begin{bmatrix} C\psi_s C\theta_s & -S\psi_s C\phi_s + C\psi_s S\theta_s S\phi_s & S\psi_s S\phi_s + C\psi_s S\theta_s C\phi_s \\ S\psi_s C\theta_s & C\psi_s C\phi_s + S\psi_s S\theta_s S\phi_s & -C\psi_s S\phi_s + S\psi_s S\theta_s C\phi_s \\ -S\theta_s & C\theta_s S\phi_s & C\theta_s C\phi_s \end{bmatrix} \qquad (5)$$

and

$$\begin{pmatrix} i_s & j_s & k_s \end{pmatrix}^T = \begin{bmatrix} R_s'' \end{bmatrix}^{-1} \begin{pmatrix} i_n & j_n & k_n \end{pmatrix}^T \qquad (6)$$

with index ($C \equiv \cos$, $S \equiv \sin$).

Linear and Angular Velocities of the Sprung Mass

The equations of motion of each sprung mass were developed and written for the system fixed to the sprung mass in terms of linear velocity $[U_s, V_s, W_s]$ and angular velocity $[p_s, q_s, r_s]$ of the center of mass of the sprung mass. In order to calculate the velocity and Euler angles, expressions connecting the linear and angular velocities for both the systems were developed.

$$\begin{pmatrix} \dot{x} & \dot{y} & \dot{z} \end{pmatrix}^T = R_s'' \begin{pmatrix} U_s & V_s & W_s \end{pmatrix}^T \qquad (7)$$

$$p_s i_s + q_s j_s + r_s k_s = \dot{\phi}_s i_s + \dot{\theta}_s j_2 + \dot{\psi}_s k_n \tag{8}$$

Introducing the transformation matrices between the two systems, the relationship between the angular velocities can be calculated by the following equations:

$$\dot{\phi}_s = p_s + \frac{\sin\theta_s \sin\phi_s}{\cos\theta_s} q_s + \frac{\sin\theta_s \cos\phi_s}{\cos\theta_s} r_s$$

$$\dot{\theta}_s = \cos\phi_s q_s - \sin\phi_s r_s \tag{9}$$

$$\dot{\psi}_s = \frac{\sin\phi_s}{\cos\theta_s} q_s + \frac{\cos\phi_s}{\cos\theta_s} r_s$$

Coordinate System Fixed to the Unsprung Mass

As mentioned earlier, two motions were assigned to each unsprung mass, namely, the roll motion and vertical motion relative to the sprung mass. It may be noted that the pitching motion of the unsprung mass, representing the axle of vehicle, is infinitely small and can be neglected [3]. The orientation of the sprung mass relative to the inertial coordinate system was defined by two rotational motions namely, yaw motion ψ_s and roll motion ϕ_u as illustrated in Figure 3.

The transformation matrix between the system fixed to the unsprung mass and the inertial system can be expressed as:

Yaw ψ_s:

$$\begin{pmatrix} i_n & j_n & k_n \end{pmatrix}^{\mathrm{T}} = \lambda_{\psi_s} \begin{pmatrix} i_1 & j_1 & k_1 \end{pmatrix}^{\mathrm{T}}$$

$$\lambda_{\psi_s} = \begin{pmatrix} \cos\psi_s & -\sin\psi_s & 0 \\ \sin\psi_s & \cos\psi_s & 0 \\ 0 & 0 & 1 \end{pmatrix} \tag{10}$$

Roll ϕ_u:

$$\begin{pmatrix} i_u & j_u & k_u \end{pmatrix}^{\mathrm{T}} = \lambda_{\varphi_u} \begin{pmatrix} i_1 & j_1 & k_1 \end{pmatrix}^{\mathrm{T}}$$

$$\lambda_{\varphi_s} = \begin{pmatrix} 1 & 0 & 0 \\ 0 & \cos\varphi_u & -\sin\varphi_u \\ 0 & -\sin\varphi_u & \cos\varphi_u \end{pmatrix} \tag{11}$$

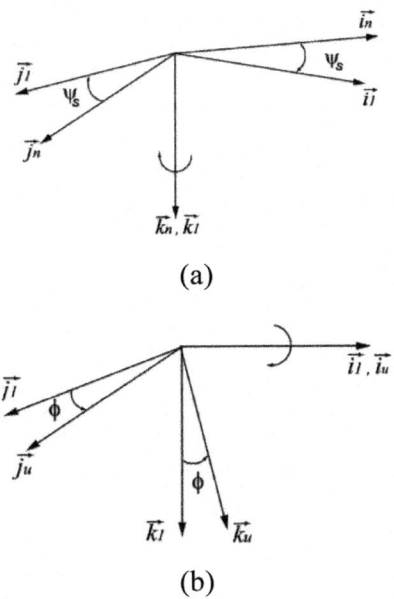

(a)

(b)

Figure 3: (a) Yaw; (b) Roll ϕ_u. Unsprung mass orientation defined by Euler angles.

Therefore, the transformation matrix, that connects the system fixed to the unsprung mass and the inertial system, can be obtained by combining the above two transformations ((10) and (11)).

$$\left(i_u \quad j_u \quad k_u\right)^{\mathrm{T}} = \left[\lambda_{\varphi_u}\right]\left[\lambda_{\psi_s}\right]\left(i_n \quad j_n \quad k_n\right)^{\mathrm{T}} \tag{12}$$

The angular velocity of the unsprung mass can be defined by the following equation:

$$w_u = p_u i_u + r_s k_s \tag{13}$$

By introducing the transformation matrices between the inertial system and the system fixed to the unsprung mass, the angular velocity was expressed in terms of Euler angles as follows:

$$p_u = \dot{\phi}_u \tag{14}$$

On the other hand, the road excitation forces are in contact with the unsprung mass. These forces are transferred to the sprung mass through the suspension system. Therefore, the transformation matrix between the two systems fixed to the sprung and unsprung masses needs to be calculated.

$$\left(i_u \quad j_u \quad k_u\right)^{\mathrm{T}} = \left[\lambda_{\varphi_u}\right]\left[\lambda_{\theta_s}\right]\left[\lambda_{\varphi_s}\right]\left(i_s \quad j_s \quad k_s\right)^{\mathrm{T}} \tag{15}$$

$$R_s^u = \left[\lambda_{\phi_u}\right]\left[\lambda_{\theta_s}\right]\left[\lambda_{\psi_s}\right] \tag{16}$$

Sprung Mass Kinematics

For the derivation of equations of motion of the vehicle it is necessary to calculate the expression for the acceleration of an arbitrary point on the vehicle. Figure 4 shows O_f as the coordinate system fixed to the road (inertial) and O_b as the system of the body coordinate with a

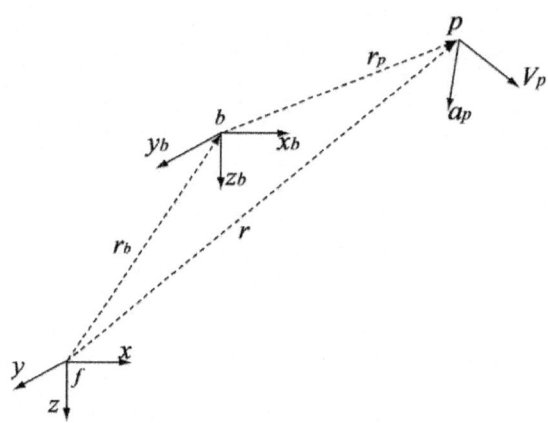

Figure 4: Coordinate systems.

translational velocity v_s and an angular velocity w_s. For a given vector q, the following expression [10] was obtained:

$$\left(\frac{\partial}{\partial t}\right)_f q = \left(\frac{\partial}{\partial t}\right)_b q + w_s \times q \tag{18}$$

The indices f and b indicate that the derivative was calculated with respect to the inertial system and system of the body concerned respectively.

The velocity of point p in the vehicle, relative to inertial system, can be calculated by the following expression:

$$v_p = v_s + \left(\frac{\partial}{\partial t}\right)_f r_p \tag{19}$$

Therefore, substitution of Equation (18) in Equation (19) gives:

$$v_p = v_s + \left(\frac{\partial}{\partial t}\right)_b r_p + w_s \times r_p = v_s + \dot{r}_p + w_s \times r_p \tag{20}$$

The acceleration of the point p can be calculated by differentiating Equation (20) with respect to time:

$$a_p = \left(\frac{\partial}{\partial t}\right)_f v_p$$

$$= \left(\frac{\partial}{\partial t}\right)_b \left(v_s + \dot{r}_p + w_s \times r_p\right) + w_s \times \left(v_s + \dot{r}_p + w_s \times r_p\right)$$

$$= \dot{v}_s + w_s \times v_s + \dot{w}_s \times r_p + w_s \times \left(w_s \times r_p\right) + 2w_s \times \dot{r}_p + \ddot{r}_p \tag{21}$$

Since the center of mass of the sprung mass coincides with the origin of the coordinate system attached to the sprung mass, acceleration of the center of mass of the sprung mass was obtained by replacing $\left(r_p = 0\right)$ in Equation (21):

$a_s = \dot{v}_s + w_s \times v_s$.

$$\begin{pmatrix} a_{sx} \\ a_{sy} \\ a_{sz} \end{pmatrix} = \begin{pmatrix} \dfrac{d}{dt}U + qW - rV \\ \dfrac{d}{dt}V + rU - pW \\ \dfrac{d}{dt}W + pV - qU \end{pmatrix} \tag{22}$$

In this study, it was assumed that the load of the liquid, represented by the center of mass, can move as a material point and can be represented by a remote vector $r_L = [x_L, y_L, z_L]^T$ from the center of mass of the sprung mass with the same angular velocity $[p_s, q_s, r_s]^T$ as that of the sprung mass of the vehicle as shown in Figure 5.

$$R_s^u = \begin{bmatrix} C\theta_s & S\theta_s S\phi_s & S\theta_s C\phi_s \\ -S\theta_s S\phi_u & C\phi_s C\phi_u + S\phi_s S\phi_u C\theta_s & -C\phi_u S\phi_s + S\phi_u C\phi_s C\theta_s \\ C\phi_u S\theta_s & -S\phi_u C\phi_s + C\phi_u S\phi_s C\theta_s & S\phi_u S\phi_s + C\phi_u C\phi_s C\theta_s \end{bmatrix} \tag{17}$$

Hence, the acceleration of the center of mass of the liquid can be obtained by replacing the expression $\left(r_p \equiv r_L\right)$ in Equation (21). Moreover, in this study the interaction between the vehicle and the liquid was modeled as a multi-body system using small time step Δt. As the coordinates of the vector r_L were updated at each time step, the relative velocity and acceleration relative to the coordinate system fixed to the sprung mass were neglected.

$$a_L = \dot{v}_s + w_s \times v_s + \dot{w}_s \times r_L + w_s \times \left(w_s \times r_L\right) \tag{23}$$

$$\begin{Bmatrix} a_{Lx} \\ a_{Ly} \\ a_{Lz} \end{Bmatrix} = \begin{Bmatrix} \dfrac{d}{dt}U + \left(\dfrac{d}{dt}q\right)z_L - \left(\dfrac{d}{dt}r\right)y_L + q\left(W + py_L - qx_L\right) - r\left(V + rx_L - pz_L\right) \\ \dfrac{d}{dt}V + \left(\dfrac{d}{dt}r\right)x_L - \left(\dfrac{d}{dt}p\right)z_L + r\left(U + qz_L - ry_L\right) - p\left(W + py_L - qx_L\right) \\ \dfrac{d}{dt}W + \left(\dfrac{d}{dt}p\right)y_L - \left(\dfrac{d}{dt}q\right)x_L + p\left(V + rx_L - pz_L\right) - q\left(U + qz_L - ry_L\right) \end{Bmatrix}$$

Unsprung Mass Kinematics

The position of the unsprung mass is located in relation to the point where the sprung mass is attached as shown in Figures 5 and 6.

$$\left(r_u\right)_f = r_f + r_r + r_{ru} \tag{24}$$

where:

r_f: represents the position of center of sprung mass from the inertial system.

$r_r = \left(x_r, 0, z_r\right)_s$: represents the position of the roll center relative to the system fixed to the sprung mass. $r_{ru} = \left(0, 0, z_u\right)_u$: represents the position of the roll center relative to the system attached to the unsprung mass. The velocity was calculated by differentiating Equation (24) with respect to time:

$$\begin{aligned}
V_u &= V_s + w_s \times r_r + \dot{r}_r + w_u \times r_{ru} + \dot{r}_{ru} \\
&= \left(V_s + w_s \times r_r\right)_s + \left(w_u \times r_{ru} + \dot{r}_{ru}\right)_u
\end{aligned} \tag{25}$$

The acceleration was calculated by differentiating Equation (25) with respect to time:

$$\begin{aligned}
a_u &= \left(a_s + \dot{w}_s \times r_r + w_s \times \left(w_s \times r_r\right)\right)_s \\
&\quad + \left(\begin{matrix} \dot{w}_u \times r_{ru} + w_u \times \left(w_u \times r_{ru}\right) \\ +2w_u \times \dot{r}_{ru} + \dot{w}_u \times r_{ru} + \ddot{r}_{ru} \end{matrix}\right)_u
\end{aligned} \tag{26}$$

where suffixes (s) and (u) indicate systems fixed to the sprung mass and unsprung mass respectively.

$w_s = \left[p_s, q_s, r_s\right]^T$ is the angular velocity of the sprung mass and $w_u = \left[p_u, 0, r_u\right]^T$ is the angular velocity of the unsprung mass.

As described in Figure 7, suspension forces transmitted to the sprung mass for each axis can be expressed as follows:

$$F_{supi} = \begin{pmatrix} F_{sxi1} + F_{sxi2} \\ F_{syi} \\ F_{szi1} + F_{szi2} \end{pmatrix}_{ui} \tag{27}$$

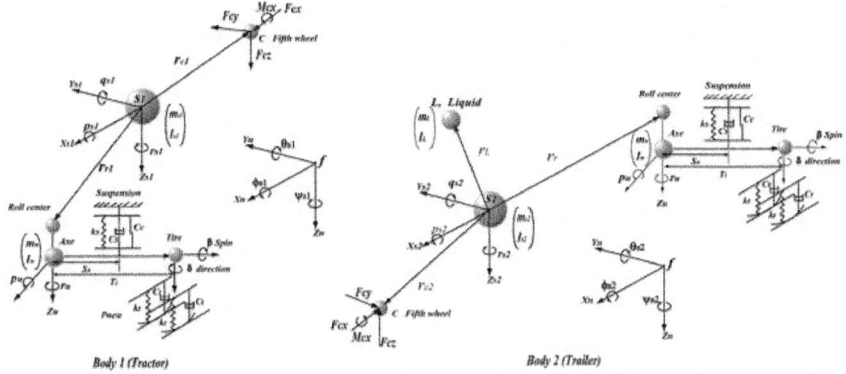

Figure 5: Vehicle mathematical model.

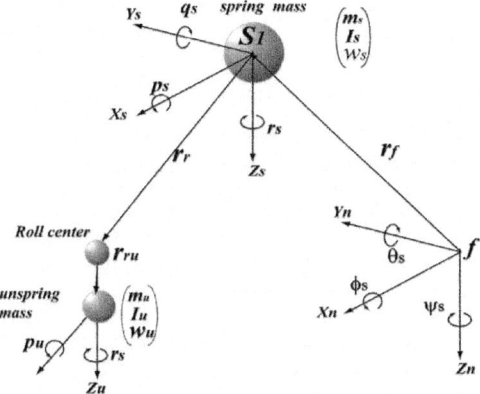

Figure 6: Unsprung mass kinematics.

where:

F_{szi1} and F_{szi2} are the vertical suspension forces at left and right sides respectively.

F_{syi} is the internal lateral force applied to the roll center of each axis. This force is the result of the lateral forces applied to the tires.

F_{sxi1} and F_{sxi2} are the longitudinal suspension forces at the right and left sides respectively.

The suspension forces in the system attached to the sprung mass Si can be defined using the transformation matrix that connects the unsprung mass and sprung mass (Equation (17)).The internal forces can be eliminated according to the dynamic equations of motion for each axis i, as illustrated in Figure 7.

$$F_{supl} = \left(R_{ul}^{si} \begin{pmatrix} F_{sxi1} + F_{xxi2} \\ F_{syi} \\ F_{szi1} + F_{szi2} \end{pmatrix}_{ui} \right)_{si}$$

(28)

$$F_{yui} = -(m_{ui}a_{ui}) \cdot j_{ui} + \left(\sum_{j=1}^{4} F_{wyi} \right) \cos\phi_{ui}$$
$$- \left(\sum_{j=1}^{4} F_{wxi} \right) \sin\phi_{ui} + m_{ui}g \sin\phi_{ui}$$

(29)

$$F_{xsi1} + F_{xsi2} = -(m_{ui}a_{ui}) \cdot i_{ui} + \sum_{j=1}^{4} F_{wxi}$$

(30)

Fifth Wheel Kinematics

The motion of the sprung mass of tractor and trailer are coupled via the fifth wheel joint. Several studies suggested to consider joint connection as a rigid one in the case of translational motion. This allows to consider a joint as a point. With this assumption, the number of degrees of freedom was reduced. Thus we can calculate the expressions for velocity and acceleration of the trailer depending on the velocity and acceleration of the tractor [4]. If the harness is not rigid enough, it can be modeled as an assembly of a spring and a damper in parallel [3]. However, torsional component of the fifth wheel acts in the case of rolling motion. From Figure 8, velocity and acceleration of point C were calculated with respect to the two systems fixed to the sprung masses of tractor and trailer as follows:

$$V_c = V_{s1} + V_{c/s1} = V_{s2} + V_{c/s2}$$

(31)

$$a_c = a_{s1} + a_{c/s1} = a_{s2} + a_{c/s2}$$

with:

$$r_{c/s1} = x_{c1}i_{s1} + z_{c1}k_{s1} \text{ and } r_{c/s2} = x_{c2}i_{s2} + z_{c2}k_{s2}$$

where:

$$V_{c/s1} = w_{s1} \times r_{c/s1}$$

$$V_{c/s2} = w_{s2} \times r_{c/s2}$$

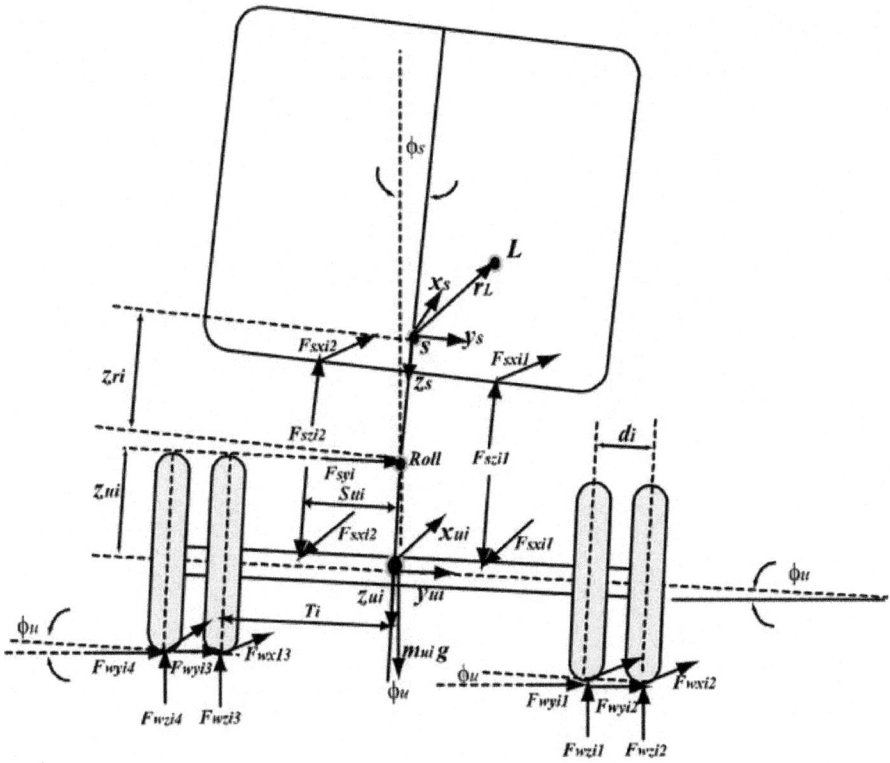

Figure 7: Vehicle model (front view).

and

$$a_{c/s1} = \dot{w}_{s1} \times r_{c/s1} + w_{s1} \times \left(w_{s1} \times r_{c/s1} \right)$$

$$a_{c/s2} = \dot{w}_{s2} \times r_{c/s2} + w_{s2} \times \left(w_{s2} \times r_{c/s2} \right) \tag{32}$$

The following relations can be obtained by introducing the expressions of Equation (32) in Equation (31).

$$V_c = \left(R_{s1}^{s2} \left(\begin{array}{c} U_1 + qz_{c1} \\ V_1 + r_1 x_{c1} - p_1 z_{c1} \\ W_1 - q_1 x_{c1} \end{array} \right)_{s1} \right)_{s2}$$

$$= \left(\begin{array}{c} U_2 + qz_{c2} \\ V_2 + r_2 x_{c2} - p_2 z_{c2} \\ W_2 - q_2 x_{c2} \end{array} \right)_{s2} \tag{33}$$

The transformation matrix R_{s1}^{s2} between the system attached to the sprung mass of trailer $(s2)$ and the system fixed to the sprung mass of tractor $(s1)$ can be calculated through the inertial system as follows:

$$R_{s1}^{s2} = R_n^{s2} \cdot R_{s1}^n.$$

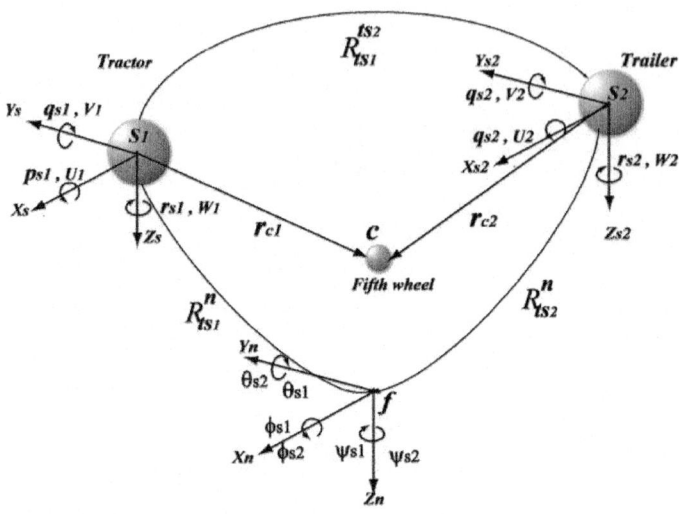

Figure 8: Fifth wheel kinematic.

$$
a_c = \left\{ R_{s1}^{s2} \begin{pmatrix} \frac{d}{dt}U_1 + \left(\frac{d}{dt}q_1\right)z_{c1} + q_1(W_1 - q_1x_{c1}) - r_1(V_1 + r_1x_{c1} - p_1z_{c1}) \\ \frac{d}{dt}V_1 + \left(\frac{d}{dt}r_1\right)x_{c1} - \left(\frac{d}{dt}p_1\right)z_{c1} + r_1(U_1 + q_1z_{c1}) - p_1(W_1 - q_1x_{c1}) \\ \frac{d}{dt}W_1 - \left(\frac{d}{dt}q_1\right)x_{c1} + p_1(V_1 + r_1x_{c1} - p_1z_{c1}) - q_1(U_1 + q_1z_{c1}) \end{pmatrix}_{s1} \right\}_{s2}
$$
$$
= \begin{pmatrix} \frac{d}{dt}U_2 + \left(\frac{d}{dt}q_2\right)z_{c2} + q_2(W_2 - q_2x_{c2}) - r_2(V_2 + r_2x_{c2} - p_2z_{c2}) \\ \frac{d}{dt}V_2 + \left(\frac{d}{dt}r_2\right)x_{c2} - \left(\frac{d}{dt}p_2\right)z_{c2} + r_2(U_2 + q_2z_{c2}) - p_2(W_2 - q_2x_{c2}) \\ \frac{d}{dt}W_2 - \left(\frac{d}{dt}q_2\right)x_{c2} + p_2(V_2 + r_2x_{c2} - p_2z_{c2}) - q_2(U_2 + q_2z_{c2}) \end{pmatrix}_{s2}
$$
$$(34)$$

The simultaneous solution of Equations (33) and (34) gave the final expressions for the velocity and acceleration of the trailer as a function of the velocity and acceleration of the tractor.

The sweep on the roll angle between the tractor and trailer was useful to calculate the constraint of the fifth wheel for the roll motion (roll moment):

$$M_{cs1} = k_{xc}(\phi_{s1} - \phi_{s2})i_{s1}$$

$$M_{cs2} = -R_{s1}^{s2}\left(k_{xc}(\phi_s - \phi_{s2})i_{s1}\right) \qquad (35)$$

VEHICLE KINETICS

This section is devoted to the definition of variables with some algebraic manipulations chosen for the equations of motion. All kinetic parameters were developed for an articulated vehicle. The same settings were applied in the case of a unit vehicle. The free body diagram shown in Figure 7 shows the external and internal forces and moments applied to each subsystem of the vehicle. To obtain the equations of linear and angular motions, it is important to model the rigid body as a set of material points.

Linear Motion

The application of Newton's laws eventually gives the equations of linear motion for the tractor and trailer.

$$\sum F_i = m_i a_i \qquad (36)$$

The equations of translational motion can be obtained by the combination of the Equations (36), (22), (24), (29), and (30). These equations were represented by second order differential equations for sprung mass si:

$$(m_{si} a_{si} + m_{Li} a_{Li}) \cdot i_{si} = \cos\theta_{si} \sum_{j=1}^{k} (F_{sxj1} + F_{sxj2}) - \sin\phi_{uj} \sin\theta_{si} \sum_{j=1}^{k} F_{syj}$$

$$+ \cos\phi_{uj} \sin\theta_{si} \sum_{j=1}^{k} (F_{szj1} + F_{szj2}) - \sin\theta_{si} (m_{si} + m_{Li}) g + \sum (\text{Constraint forces}) \cdot i_{si}$$

(37)

$$(m_{si} a_{si} + m_{Li} a_{Li}) \cdot j_{si} = (\cos\phi_{uj} \cos\phi_{si} + \sin\phi_{uj} \cos\theta_{si} \sin\phi_{si}) \sum_{j=1}^{k} F_{syj} + \sin\phi_{uj} \cos\phi_{si} \sum_{j=1}^{k} (F_{szj1} + F_{szj2})$$

$$- \cos\phi_{uj} \cos\theta_{si} \sin\phi_{si} \sum_{j=1}^{k} (F_{szj1} + F_{szj2}) + \sin\theta_{si} \sin\phi_{si} \sum_{j=1}^{k} (F_{sxj1} + F_{sxj2})$$

$$+ \cos\theta_{si} \sin\phi_{si} (m_{si} + m_{Li}) g + \sum (\text{Constraint forces}) \cdot j_{si} \qquad (38)$$

$$(m_{si} a_{si} + m_{Li} a_{Li}) \cdot k_{si} = (-\cos\phi_{uj} \sin\phi_{si} + \sin\phi_{uj} \cos\theta_{si} \cos\phi_{si}) \sum_{j=1}^{k} F_{syj} - \sin\phi_{uj} \sin\phi_{si} \sum_{j=1}^{k} (F_{szj1} + F_{szj2})$$

$$- \cos\phi_{uj} \cos\theta_{si} \cos\phi_{si} \sum_{j=1}^{k} (F_{szj1} + F_{szj2}) + \sin\theta_{si} \cos\phi_{si} \sum_{j=1}^{k} (F_{sxj1} + F_{sxj2})$$

$$+ \cos\theta_{si} \cos\phi_{si} (m_{si} + m_{Li}) g + \sum (\text{Constraint forces}) \cdot k_{si} \qquad (39)$$

where

$i = 1$: tractor

$i = 2$: trailer.

j: axle number.

k: axle number. $k = 3$ for tractor and $k = 2$ for trailer.

L: liquid.

s_i : sprung mass i .

In this study the constraint forces due to the fifth wheel were eliminated by using the kinematic Equations (33) and (34) developed earlier. It should be noted that all these equations of motion were programmed in the Maple software in a systematic way. Therefore, to obtain the equations of motion in the case of a unit vehicle, only change of the indices $(i=1, k=3)$ was needed.

The equation of vertical motion of the sprung mass for each axis i was given by the folowing expression:

$$
\begin{aligned}
&(m_{ui}a_{ui}) \cdot k_{ui} \\
&= -\left(\sum_{j=1}^{k} F_{wzij}\right)\cos\phi_{ui} - \left(\sum_{j=1}^{k} F_{wyij}\right)\sin\phi_{ui} \\
&\quad + m_{ui}g\cos\phi_{ui} + F_{szi1} + F_{szi2}
\end{aligned}
\tag{40}
$$

where:

i : axle number.

j : number of tires in each axle.

k : $k=2$ fortractor front axle and $k=4$ for the other axles.

Angular Motion

It is important to model the rigid body as a system of material points P with masses m_p to obtain the equation of angular motion. According to Newton's equation, angular momentum relative to the inertial system can be given by the following expression:

$$
M_s = \sum_p M_p = \sum_p r_p \times m_p \ddot{r}_p
\tag{41}
$$

Substituting Equation (21) in (41), the following expression can be obtained:

$$
M_s = \sum_p m_p r_p \times \begin{bmatrix} \dot{v}_s + w_s \times v_s + \dot{w}_s \times r_p \\ + w_s \times (w_s \times r_p) \end{bmatrix}
\tag{42}
$$

$$
\begin{aligned}
M_s &= \sum_p m_p r_p \times [\dot{v}_s + w_s \times v_s] \\
&\quad + \sum_p m_p r_p \times [\dot{w}_s \times r_p] \sum_p m_p r_p \times [w_s \times (w_s \times r_p)]
\end{aligned}
\tag{43}
$$

The first term of Equation (43) can be simplified as:

$$\sum_{p} m_p r_p \times \left[\dot{v}_s + w_s \times v_s \right] = m_s r_s \times \left[\dot{v}_s + w_s \times v_s \right]$$

The second and third terms of Equation (43) can also be simplified [10] as:

$$\sum_{p} m_p r_p \times \left[\dot{w}_s \times r_p \right] = I_s \dot{w}_s$$

$$\sum_{p} m_p r_p \times \left[w_s \times \left(w_s \times r_p \right) \right] = w_s \times I_s w_s$$

Therefore Equation (42) takes the following form:

$$M_s = m_s r_s \times \left[\dot{v}_s + w_s \times v_s \right] + I_s \dot{w}_s + w_s \times I_s w_s \qquad (44)$$

Since the sprung mass center coincides with the origin of the body axis system $(r_s \equiv o)$, the expression of angular motion (44) can be formulated as follows:

$$M_s = I_s \dot{w}_s + w_s \times I_s w_s = \begin{pmatrix} -r_s q_s I_{ys} - \left(\dfrac{d}{dt} r_s \right) I_{xzs} + \left(\dfrac{d}{dt} p_s \right) I_{xs} - p_s q_s I_{xzs} + r_s q_s I_{zs} \\ \left(\dfrac{d}{dt} q_s \right) I_{ys} + \left(p_s \right)^2 I_{xzs} - \left(r_s \right)^2 I_{xzs} - p_s r_s I_{zs} + p_s r_s I_{xs} \\ \left(\dfrac{d}{dt} r_s \right) I_{zs} + p_s q_s I_{ys} + r_s q_s I_{xzs} - p_s q_s I_{xs} - \left(\dfrac{d}{dt} p_s \right) I_{xzs} \end{pmatrix} \qquad (45)$$

The matrix of inertia I_{si} was expressed in the system si as follows:

$$I_{si} = \sum_{s_i} \left(r_p^2 1 - r_p r_p^T \right) m_p = \begin{pmatrix} \sum_{s_i} (y_p^2 + z_p^2) m_p & -\sum_{s_i} (x_p y_p) m_p & -\sum_{s_i} (x_p z_p) m_p \\ -\sum_{s_i} (x_p z_p) m_p & \sum_{s_i} (x_p^2 + z_p^2) m_p & -\sum_{s_i} (y_p z_p) m_p \\ -\sum_{s_i} (z_p x_p) m_p & -\sum_{s_i} (z_p y_p) m_p & \sum_{s_i} (x_p^2 + y_p^2) m_p \end{pmatrix} \qquad (46)$$

Since the tractor body and the trailer body can be modeled as contained bodies, all mathematical expressions can be expressed by integrals $(\int)$ instead of a sums $(\sum)$.

The moments applied to the sprung mass due to the liquid load and the suspension forces expressed in axis system fixed to the sprung mass were calculated as follows:

$$M_{Li} = \begin{pmatrix} x_{Li} \\ y_{Li} \\ z_{Li} \end{pmatrix} \times \left[\begin{pmatrix} m_{Li}(a_{Li}) \cdot i_{si} \\ m_{Li}(a_{Li}) \cdot j_{si} \\ m_{Li}(a_{Li}) \cdot k_{si} \end{pmatrix} + \left[\lambda_{\theta_{si}} \lambda_{\phi_{si}} \right]^{-1} \begin{pmatrix} 0 \\ 0 \\ m_{Li} g \end{pmatrix} \right] \qquad (47)$$

$$M_{susp_i} = \sum_{j=1}^{k}\left[\begin{pmatrix} x_{sj} \\ s_{sj} \\ z_{rj} \end{pmatrix} \times R_{sj}^{si}\begin{pmatrix} F_{sxj1}+F_{sxj2} \\ 0 \\ F_{szj1}+F_{szj2} \end{pmatrix}\right] + \sum_{j=1}^{k}\left[\begin{pmatrix} x_{sj} \\ 0 \\ z_{rj} \end{pmatrix} \times R_{sj}^{si}\begin{pmatrix} 0 \\ F_{syj} \\ 0 \end{pmatrix} + k_{\theta_i}(\phi_{si}-\phi_{sj})\right] \tag{48}$$

Substituting in Equation (45), the terms of the moments due to the liquid charge (47), the moments of the suspension (48) and the moments due to the fifth wheel constraints (35), the final equations of angular motion of the sprung mass (si) can be obtained as follows:

$$(I_{x_{si}}+I_{x_{Li}})\left(\frac{d}{dt}p_s\right) = -r_s q_s\left(I_{z_{si}}+I_{z_{Li}}-I_{y_{si}}-I_{y_{Li}}\right)+\left(p_s q_s+\left(\frac{d}{dt}r_s\right)\right)(I_{xz_{si}}+I_{xz_{Li}})$$
$$+(y_{Li}\cos\theta_{si}\cos\phi_{si}-z_{Li}\cos\theta_{si}\sin\phi_{si})m_{Li}g+y_{Li}(a_{Li})_z-z_{Li}(a_{Li})_y$$
$$+\sum_{j=1}^{k}(k_{\phi_i}(\phi_{si}-\phi_{sj}))+(M_{csi}+M_{supi})\cdot i_{si} \tag{49}$$

$$(I_{y_{si}}+I_{y_{Li}})\left(\frac{d}{dt}q_s\right) = -(p_s^2-r_s^2)(I_{xz_{si}}+I_{xz_{Li}})+(M_{csi}+M_{supi})\cdot j_{si}+p_s r_s\left(I_{z_{si}}+I_{z_{Li}}-I_{x_{si}}-I_{x_{Li}}\right)$$
$$-(z_{Li}\sin\theta_{si}+x_{Li}\cos\theta_{si}\cos\phi_{si})m_{Li}g++z_{Li}(a_{Li})_x-x_{Li}(a_{Li})_z \tag{50}$$

$$\left(I_{z_{si}}+I_{z_{Li}}+\sum_{j=1}^{k}I_{z_{sj}}\right)\left(\frac{d}{dt}r_s\right) = -p_s q_s\left(I_{y_{si}}+I_{y_{Li}}-I_{x_{si}}-I_{x_{Li}}\right)-\left(r_s q_s-\left(\frac{d}{dt}p_s\right)\right)(I_{xz_{si}}+I_{xz_{Li}})$$
$$+(y_{Li}\sin\theta_{si}+x_{Li}\cos\theta_{si}\sin\phi_{si})m_{Li}g+x_{Li}(a_{Li})_y-y_{Li}(a_{Li})_x$$
$$+\sum_{j=1}^{k}M_{Tj}+(M_{csi}+M_{supi})\cdot k_{si} \tag{51}$$

k: axle number k = 3 for tractor and k = 2 for trailer.

The equation for rolling motion of the sprung mass of each axle i can be given by the following expression:

$$I_{xx_{ui}}p_{ui} = (F_{sz_{i1}}-F_{sz_{i2}})s_{ui}-z_{ui}F_{syi}-\sum_{j=1}^{k}(F_{wy_{ij}})(h_{ri}\cos(\phi_{ui})-z_{ui})$$
$$+(F_{wz_{i1}}-F_{sz_{i3}})T_i+(F_{wz_{i2}}-F_{sz_{i4}})(T_i+d_i)+k_{\phi_i}(\phi_{si}-\phi_{ui}) \tag{52}$$

where:

i : axle index.

j : number of tires in each axle.

k : $k=2$ for the front axle of the tractor and $k=4$ for other axles.

Suspension Model

The external forces acting on the vehicle are generated mainly due to the contact forces between wheel and ground. These forces are transmitted to the sprung mass through the suspension system of the vehicle. To simplify the model, the suspension system was represented with a linear spring and a damper assembled in parallel. The vertical force applied on the vehicle through the suspension system was assumed to be equal to the sum of the static equilibrium force and the excitation forces.

$$F_{si} = K_{uj}e_{uj} + C_{uj}\dot{e}_{uj} + F_{static} \qquad (53)$$

where e_{uj} is the suspension deflections and can be calculated based on the geometry of the vehicle.

$$e_{uj} = z_s + \left(-\sin\phi_{uj}\cos\phi_{si} + \cos\phi_{uj}\cos\theta_{si}\sin\phi_{si}\right)s_{uj}$$
$$- \left(\cos\phi_{uj}\sin\theta_{si}\right)x_{uj}$$

$$\dot{e}_{uj} = \frac{d}{dt}e_{uj} \qquad (54)$$

where:

$i = 1$: sprung mass of tractor.

$i = 2$: sprung mass of trailer.

j: axle number ($j = 1, 2, 3$ for tractor and $j = 4, 5$ for trailer).

Tire Model

The tire is an essential element in a vehicle. It represents the contact between wheel and ground. The forces and moments transmitted to the vehicle by the tires due to wheel-ground interaction are complex and nonlinear. These forces and moments depend primarily on normal forces, longitudinal and lateral load transfer, slip rate λ and slip angles α as illustrated in Figure 9. There are several models available for tires. Most studies have used linear model or models based on tables from experimental tests. The forces and moments were characterized according to vehicle velocity, normal force, longitudinal slip ratio and slip angle [4,11-14]. These models usually have a better prediction capability for the traction force of contact. However, their data are specific for the type of tire which reduces their universal use. There are other numerical models which use different analytical approaches [15-17]. The choice of model of

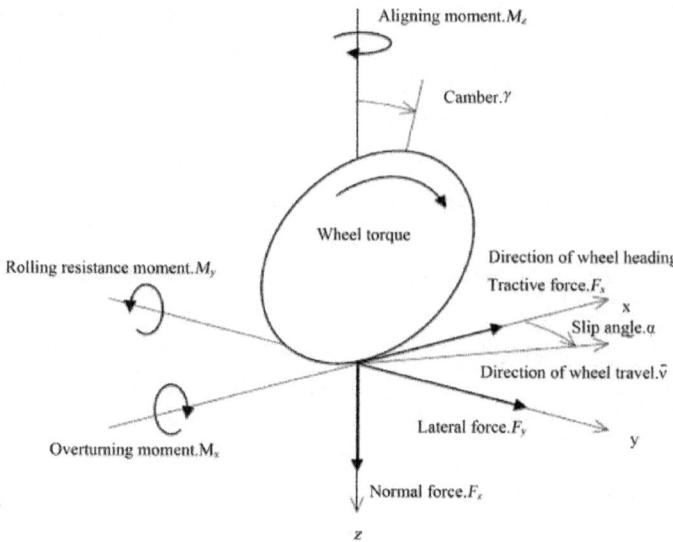

Figure 9: Forces and moments applied on the tire.

the tire affects the calculation of the efforts at the wheelsoil interface. The data from these models are important when one wants to make a dynamic model of a vehicle.

In this study the efforts of the tires were studied with the model called slip circle [11,18]. The model is closely related to the model of friction ellipse shown in Figure 10 [11].

With this model, it is possible to obtain lateral and longitudinal forces in the case of combined motions based only on measured data for separate motions such as, braking/traction alone or direction case, as illustrated in Figure 11.

The calculation was based on the evaluation of friction μ_x and μ_y. The calculation of these coefficients depends on the rate of longitudinal slip λ and slip angle α. The rate of longitudinal slip and slip angle of the tire can be calculated by the formula:

$$\lambda = \frac{r_w \omega_w - U_p}{U_p} \qquad (55)$$

$$\alpha = \tan^{-1} \frac{V_{axe}}{U_p} - \delta$$

$$\lambda = \frac{r_w \omega_w - U_p}{r_w \omega_w} \qquad (56)$$

$$\alpha = \tan^{-1}\frac{V_{axe}}{U_p} - \delta$$

where r_w is the radius of the wheel and ω_w the velocity of rotation of the wheel. V_{axe} is the velocity of lateral translation of the axis and U_p is the longitudinal velocity of the tire as shown inFigure 12.

The expressions can be evaluated from the velocity of center of mass of the vehicle.

$$V_{axej} = \cos\phi_{si}\left(V_{si} + r_{si}x_{uj} - p_{si}z_{rj}\right) - \cos\phi_{uj}p_{uj}H_{uj} \qquad (57)$$

$$U_{pj1} = U_{si} + T_j r_{si}$$

$$U_{pj2} = U_{si} + \left(T_j + d_j/2\right)r_{si} \qquad (58a)$$

$$U_{pj3} = U_{si} - T_j r_{si}$$

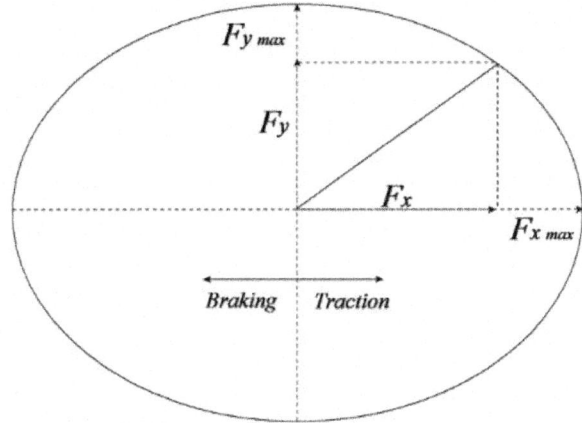

Figure 10: Friction ellipse concept.

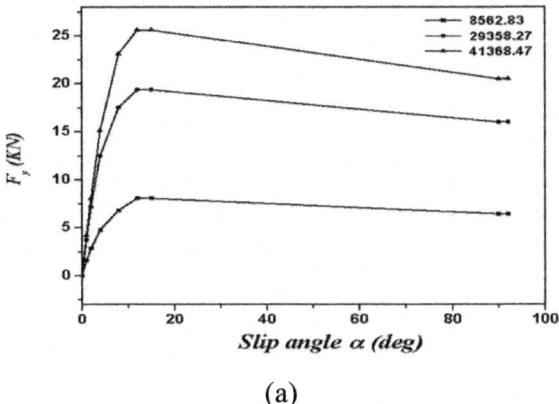

(a)

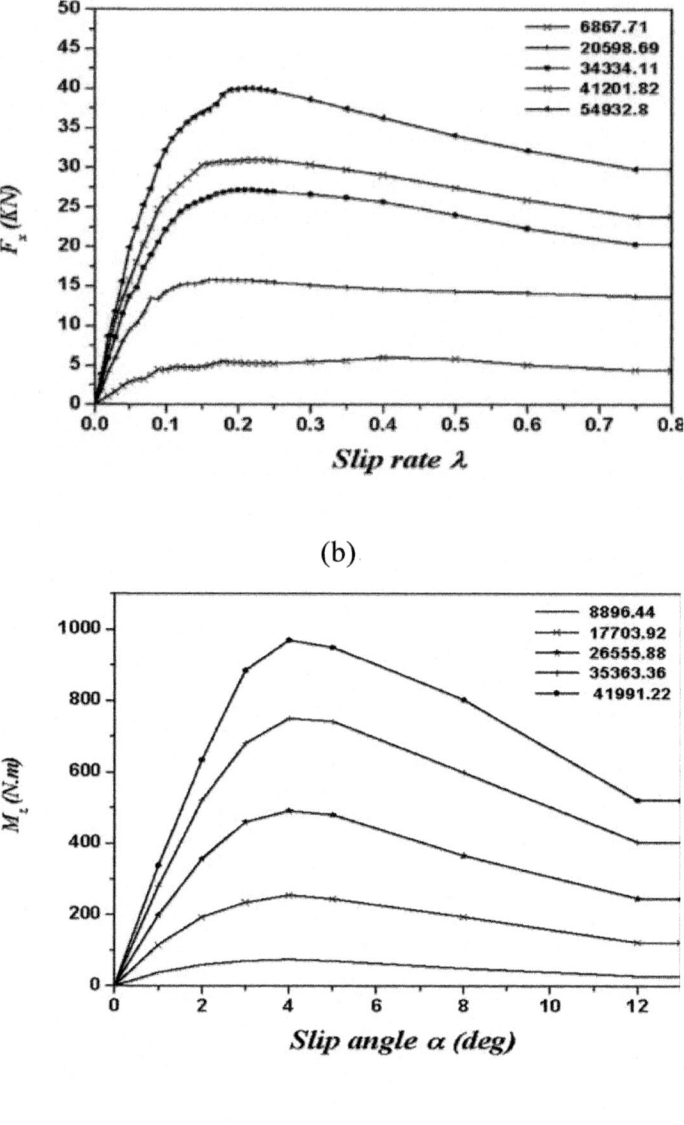

Figure 11: (a) Lateral tire-road contact force; (b) Longitudinal tire-road contact force; (c) Aligning moment generated at tire-road contact. Experimental forces and moments generated at tire-road contact for several vertical load charge [19].

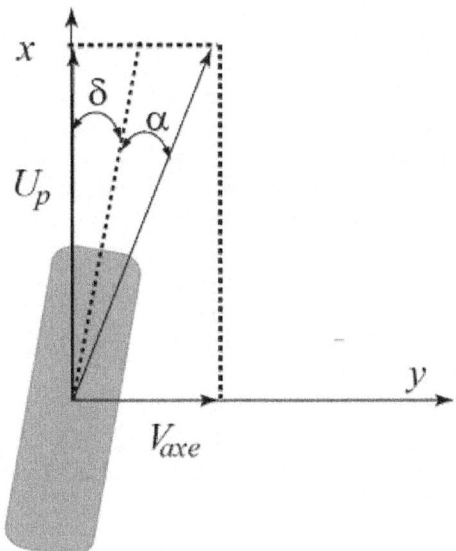

Figure 12. Tire model.

$$U_{pj4} = U_{si} - (T_j + d_j/2)r_{si}$$ (58b)

Tire Vertical Load

In this study, the vertical load of the tire was modeled as a linear spring. Therefore, the vertical force depended on the spring constant.

$$F_{wzij} = Kt_{ij}\Delta_{ij}$$ (59)

The tire deflections were calculated from the geometry of the vehicle as follows:

$$\Delta_{j1} = \Delta_{01} + z_{si} - z_{uj_0} + z_{uj}\left(1 - \cos\phi_{uj}\right)$$
$$- z_{rj} - \sin\theta_{si}x_{uj} + \cos\theta_{si}\cos\phi_{si}z_{rj}$$
$$+ \sin\phi_{uj}T_j + z_{uj}\cos\phi_{uj}$$ (60)

$$\Delta_{j2} = \Delta_{j1} + d_j\cos\phi_{uj}$$

$$\Delta_{j3} = \Delta_{j1} - T_j\cos\phi_{uj}$$ (61)

$$\Delta_{j4} = \Delta_{j1} - \left(T_j + d_j\right)\cos\phi_{uj}$$

To calculate the combined forces, a dimensional vector of slip amplitude γ and direction β was defined [20] as:

$$\gamma = \sqrt{\lambda^2 + (\sin(\alpha))^2}$$

$$\tan(\beta) = \frac{\sin(\alpha)}{\lambda}$$

(62)

The coefficients of friction between the tire and the ground, in the case of the combined forces, took the forms below:

$$\mu(\gamma,\beta) = \mu_x(\gamma)(\cos(\beta))^2 + \mu_y(\gamma))(\sin(\beta))^2$$

$$\mu_x(\gamma,\beta) = \mu(\gamma,\beta)\cos(\beta)$$

(63)

$$\mu_y(\gamma,\beta) = \mu(\gamma,\beta)\sin(\beta)$$

Finally, longitudinal and lateral forces in the case of combined motion were calculated:

$$F_x(\gamma,\beta) = \mu_x(\gamma,\beta)F_z$$

(64)

$$F_y(\gamma,\beta) = \mu_y(\gamma,\beta)F_z$$

Braking Force

The braking force was calculated by taking into account forces and moments developed in the wheel due to rotation (spin) of the wheel as shown in Figure 13. Acceleration of rotation (spin) of the wheels $\ddot{\omega}_{wi}$ were calculated from the rotational motion of the wheels as follows [21]:

$$I_{wi}\ddot{\omega}_{wi} = T_d - (T_{bi} + M_{ri} + F_{xi}r_w)$$

(65)

$$T_d = T_e \eta_{diff} \eta_{trans} \varepsilon$$

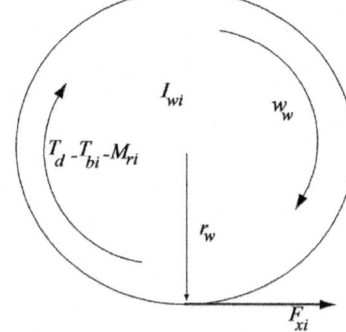

Figure 13: Wheel dynamics.

VEHICLE MODEL VALIDATION

The TruckSim software, developed by the transportation center of the University of Michigan (UMTRI), is specialized in the simulation of heavy vehicles [15,19,22]. The center also developed software applications: CarSim for tourist vehicles and BikeSim for motorcycles. TruckSim, the most popular software in this field, was used to represent and study the dynamics of vehicles in a computer environment. It is possible to analyze a large number of vehicle configurations, since the software has a library of existing models in the transportation industry. However, TruckSim can only add a load that is considered to be fixed on the semi-trailer. This feature does not allow to study the dynamic behavior of liquid sloshing in tanker trucks.

Table 1: Geometric parameters of unit vehicle

(a)

Parameters	Symbols	Values
Spring mass (kg)	m_s	4457 kg
Roll inertia moment (kg·m²)	I_{xx}	2287 kg·m²
Pitch inertia moment (kg·m²)	I_{yy}	35,408 kg·m²
Yaw inertia moment (kg·m²)	I_{zz}	34,823 kg·m²
Height of mass center of gravity of spring (m)	z_{cg}	1.173 m
Distance between the center of mass and the front axis (m)	x_{s1}	1.135 m
Distance between the center of mass and the rear axis 1 (m)	x_{s2}	3.252 m
Distance between the center of mass and the rear axis 2 (m)	x_{s3}	4.522 m

(b)

Parameters	Axle 1	Axle 2	Axle 3
		Unspring mass.	
m_{u} (kg)	527	1007	973
I_{xu} (kg·m²)	612	579	584
I_{zu} (kg·m²)	612	579	584
K_{ϕ} (kg·m²/rad)	432	3389.54	3389.54
$2T_i$ (m)	2.022	2.06	2.06
$2S_i$ (m)	0.828	1.029	1.031
d_i (m)	0	0.31	0.31
H_u (m)	0.533	0.686	0.704

The TruckSim library also has a number of predefined trajectories and maneuvers that can enable researchers to validate the behavior of the vehicle model developed, for difficult maneuvers such as motion in a curve, change of single and double lanes. In the TruckSim environment, the maneuvers are predefined paths, i.e., the excitation is represented by a displacement vector. However, for the model developed in this study, the excitation was defined by the angle of steering or braking torque as an input parameter. From the output vector, the response of the steering angle was recorded. This response was the input parameter for our model that considered the same configuration for a unit or articulated vehicle as defined by the Tables 1 and 2 in the annexure [22]. Two lane change maneuvers (single and double) with a constant speed $v = 70 \, \text{km/h}$ were chosen to compare the two models. Figure 14shows the trajectory and the steer angle during the two maneuvers [22].

Figures 15 and 16 represent the comparison between model simulations for the unit vehicle and the TruckSim software respectively. The vehicle directional responses were evaluated using two difficult motions, such as change of single and double lanes. This comparison was characterized in terms of roll angle, lateral acceleration of the center of mass and trajectory traveled from the center of mass. The simulation showed a good correla

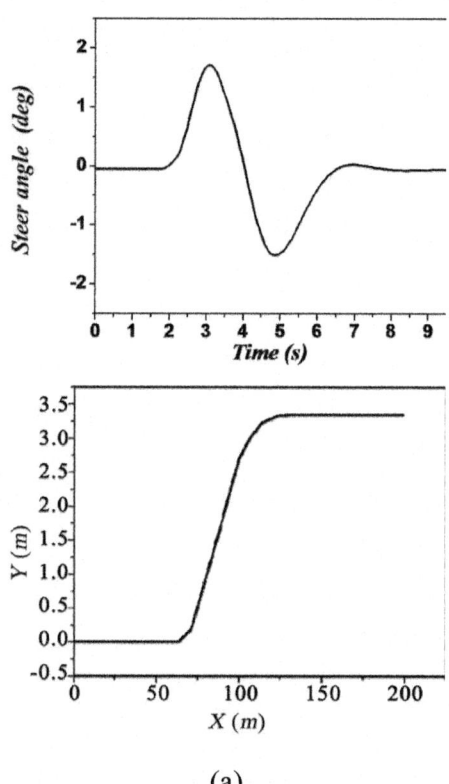

(a)

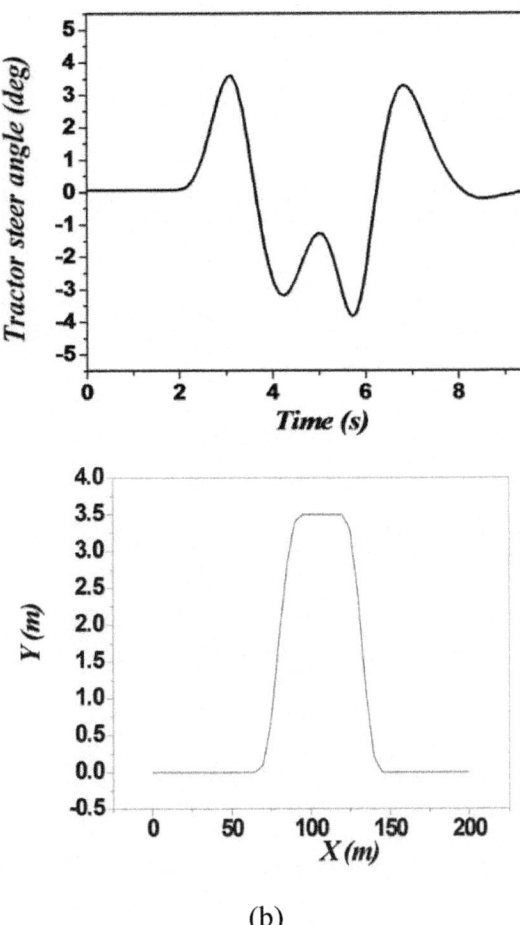

(b)

Figure 14: (a) Simple lane change maneuver and the desired trajectory; (b) Double lane change maneuver and the desired trajectory. Maneuvrer used for the comparaison between the model and TruckSim software. tion between the two models. A small difference was noted for the trajectory of the center of mass.

This difference may be due to the steering angle error of the output vector obtained using TruckSim. In addition, excitement used for the TruckSim software was in closed loop (predefined trajectory).

Table 2: Geometric parameters of articulated vehicle

(a)

Parameters	Tractor $(s1)$	Trailer $(s2)$
Spring mass (kg)	6308	2800
Roll inertia moment (kg·m²)	6879	2400
Pitch inertia moment (kg·m²)	21,711	40,000
Yaw inertia moment (kg·m²)	19,665	40,000
Mixte inertia product Ixz (kg·m²)	130	-
Height of mass (CM) center of gravity of spring (m)	1.02	1.7
Distance between (CM_1) and the fifth wheel (m)	4.601	-
Distance between (CM_2) and the fifth wheel (m)	-	5.5
Distance between (CM_1) and the axle number 1 (m)	1.384	-
Distance between (CM_1) and the axle number 2 (m)	3.242	-
Distance between (CM_1) and the axle number 3 (m)	4.522	-
Distance between (CM_1) and the axle number 4 (m)	-	3.9
Distance between (CM_1) and the axle number 5 (m)	-	5.2

(b)

Parameters	Unsprung mass.				
	Axle 1	Axle 2	Axle 3	Axle 4	Axle 5
m_{sc} (kg)	527	1007	973	735	735
I_{xu} (kg·m²)	612	579	584	586	593
I_{zu} (kg·m²)	612	579	584	586	593
K_s (kg·m²/deg)	1186.3	1581.8	119.8	1468.2	1468.2
$2T_i$ (m)	2.022	2.06	2.06	2.06	2.06
$2S_i$ (m)	0.828	1.029	1.031	1.118	1.118
d_i (m)	0	0.31	0.31	0.31	0.31
H_n (m)	0.553	0.686	0.704	0.717	0.676

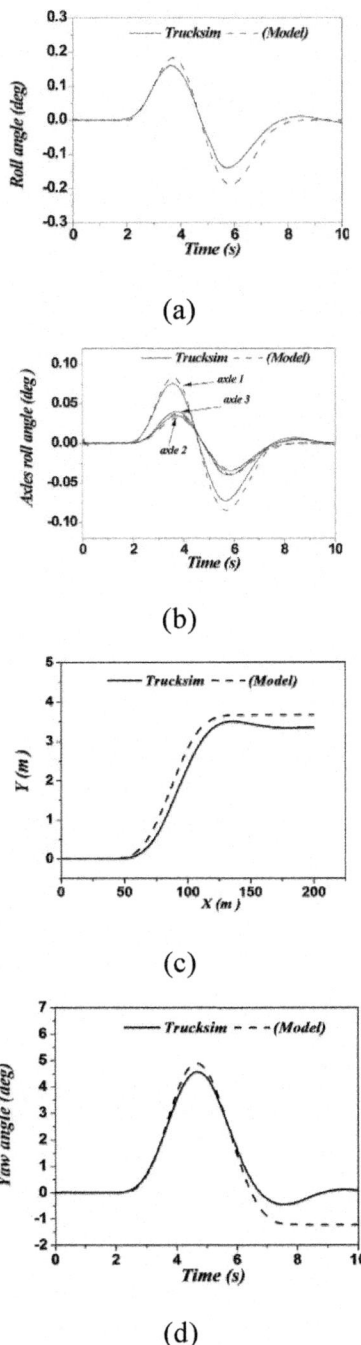

(a)

(b)

(c)

(d)

Figure 15: (a)-(d) Single lane change maneuver for a unit vehicle (Solid: trucksim; Dashed: model).

In case of articulated vehicles, the analysis was more complicated. This difficulty was due to addition of the hinge point where there were additional forces and moments acting between the tractor and trailer. Figures 17 and 18 represent the comparison between the two models for the same difficult issues such as motions during change of single and double lanes. A good overall correlation was observed. However, our model underestimated the response to some extent. This difference in response was observed due to yaw motion. TruckSim software.

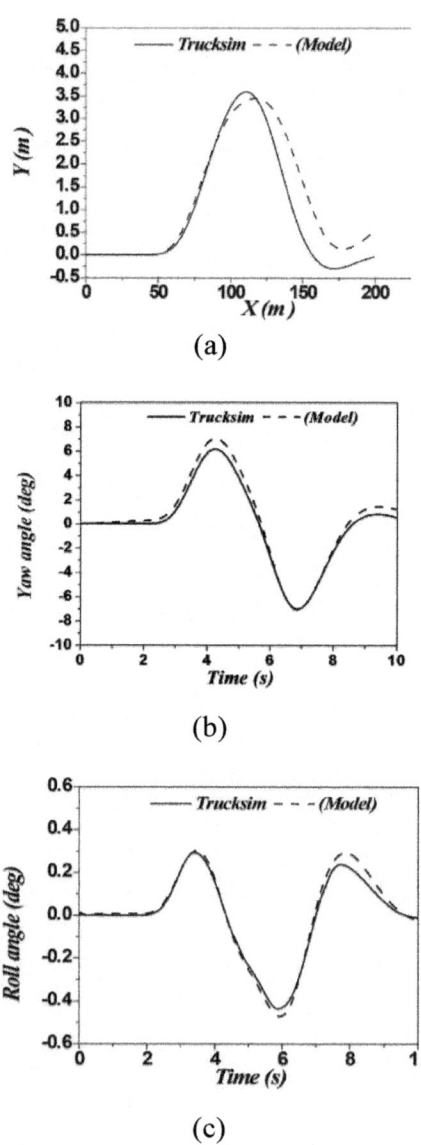

(a)

(b)

(c)

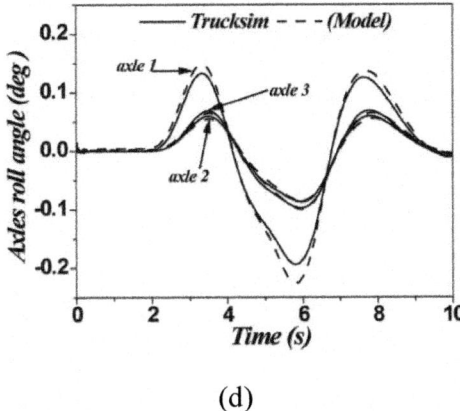

(d)

Figure 16: (a)-(d) Double lane change maneuver for a unit vehicle (Solid: trucksim; Dashed: model).

was able to handle such situation in a better way. This difference can be explained based on the assumption that the fifth wheel was considered as a rigid body for our model. However, in TruckSim it was modeled by a spring-damper combination, which can represent multibody connections between subsystems in a better way. The increase in the response of yaw motion influenced.

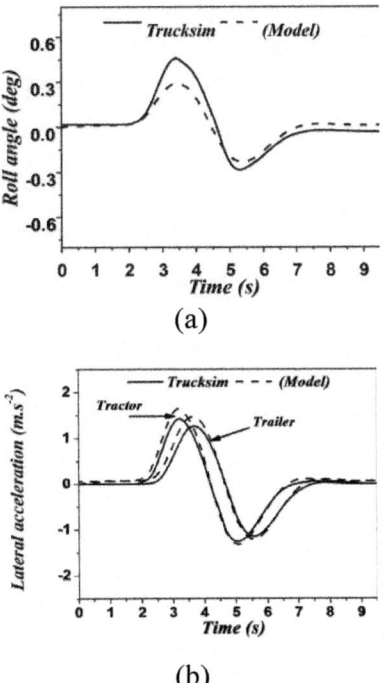

(a)

(b)

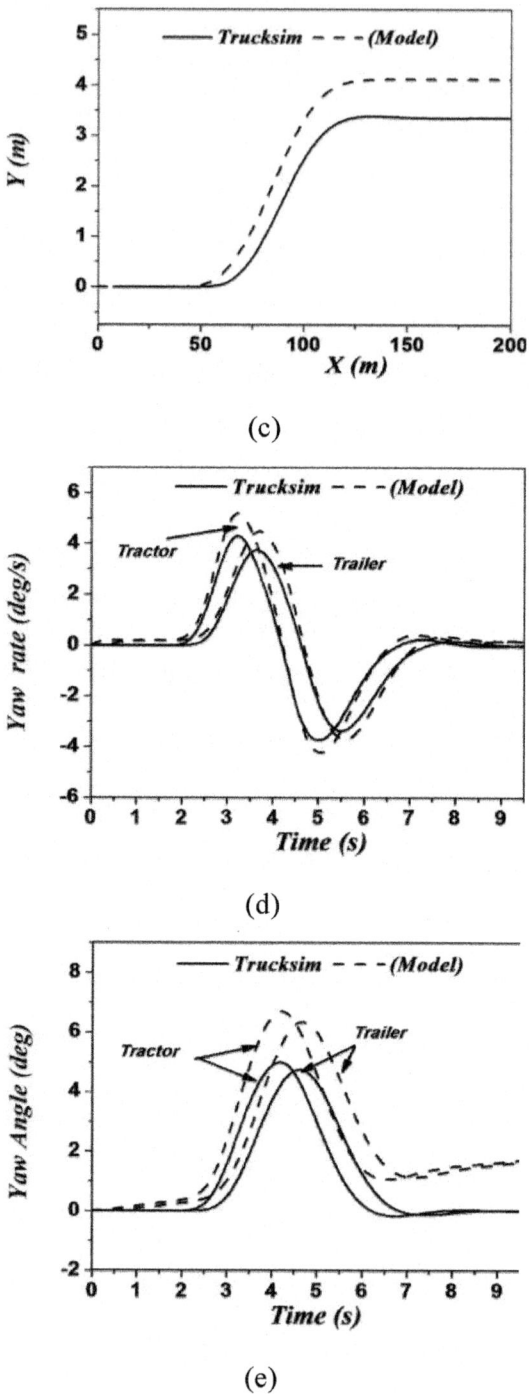

(c)

(d)

(e)

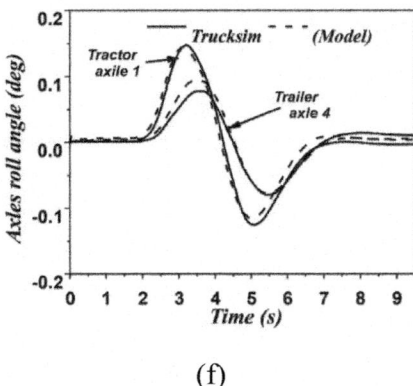

(f)

Figure 17: (a)-(f) Single lane change maneuver for an articulated vehicle (Solid: trucksim; Dashed: model).

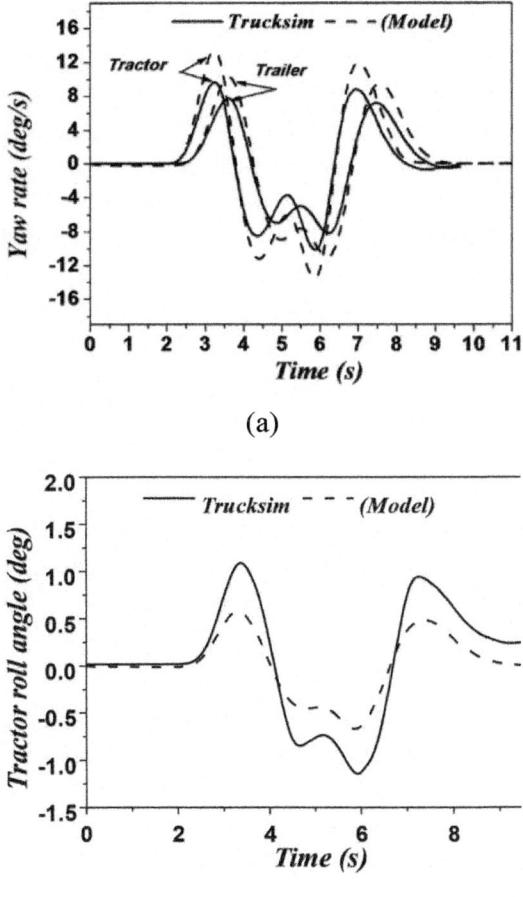

(a)

(b)

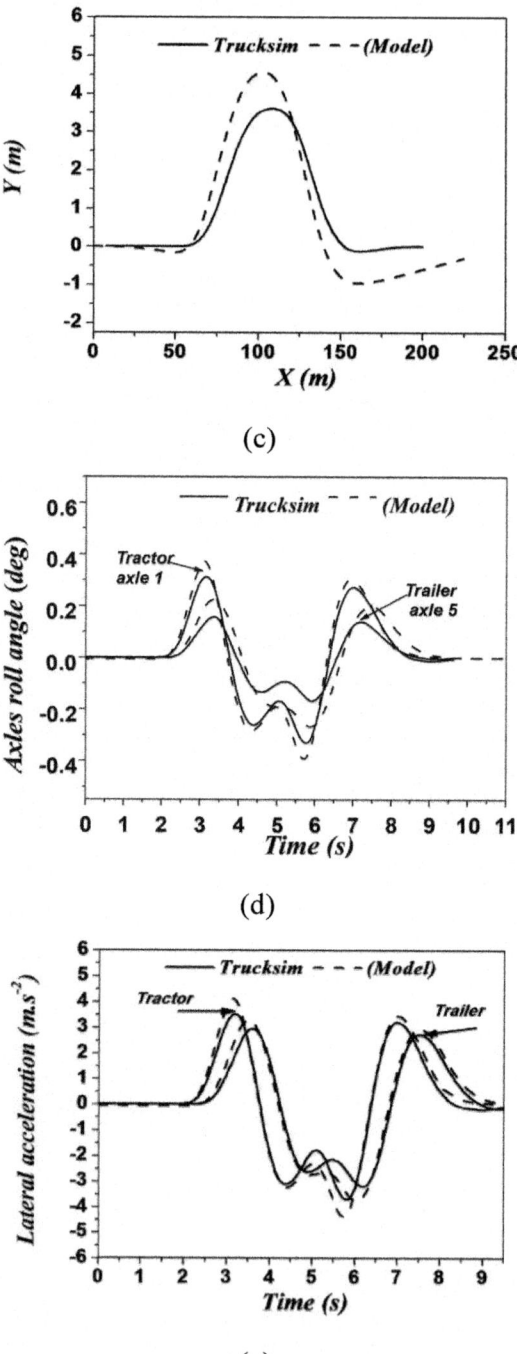

(c)

(d)

(e)

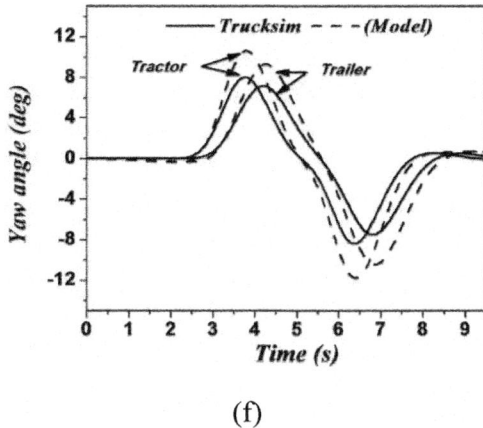

(f)

Figure 18: (a)-(f) Double lane change maneuver for an articulated vehicle (Solid: trucksim; Dashed: model).

the trajectory of the vehicle as illustrated by Figures 17 and 18. This may be due to the error of steering angle recorded from the TruckSim output vector. Still, this difference did not practically affect the very good corre lation between the two models.

CONCLUSIONS

A complete nonlinear three-dimensional vehicle model was developed and validated by Trucksim software. Both unit vehicle and articulated vehicle combination systems were considered in this study. The model gave realistic results in simulation of handling maneuvers near and beyond the adhesion limits.

The loadtransfer for mobile charge due to the liquid was accurately modeled and integrated into the vehicle model as a multibody system. The dynamic responses of tank vehicles were further investigated in view of variations in vehicle maneuvers, fill volume, road condition, and tank configuration [8].

This research, can help better understanding of this kind of complex problem. This will make it possible to answer some queries in the field of safety and the stability of the heavy vehicles, in particular for the tanktruck.

REFERENCES

1. J. Ellis, "Vehicle Handling Dynamics," Mechanical Engineering Publicantions, London, 1994.

2. H. Dongyoon, "Predictive Modeling and Active Control of Rollover in

Heavy Vehicles," Ph.D. Dissertation, Texas University, Austin, 2001.

3.   C. B. Winkler, J. E. Bernard and P. S. Fancher, "A Computer Based Mathematical Method for Predicting the Directional Response of Truck and Tractor-Trailers," Technical Report, Highway Safety Research Institute, University of Michigan, Ann Arbor, 1973.

4.   A. Ann, R. D. Ervin and Y. Guy, "The Influence of Weights and Dimensions on the Stability and Control of Heavy-Duty Trucks in Canada," Volume III Appendices, Final Report, Vehicle Weights and Dimensions Study 2, University of Michigan, Transportation Research Institute, 1986.

5.   M. Sayers, "Symbolic Computer Methods to Automatically Formulate Vehicle Simulations Codes," Ph.D. Dissertation, University of Michigan, Ann Arbor, 1990.

6.   M. Toumi, M. Bouazara and M. J. Richard, "Analytical and Numerical Analysis of the Liquid Longitudinal Sloshing Impact on a Partially Filled Tank-Vehicle with and without Baffles," International Journal of Vehicle System Modeling and Testing, Vol. 3, No. 3, 2009, pp. 229-249. doi:10.1504/IJVSMT.2008.023840

7.   M. Toumi, M. Bouazara and M. J. Richard, "Impact of Liquid Sloshing on the Behavior of Vehicles Carrying Liquid Cargo," European Journal of Mechanical Engineering/Solids, Vol. 28, No. 5, 2009, pp. 1026-1034.

8.   M. Toumi, "Études et Analyse de la Stabilité de Camions Citernes," Ph.D. Dissertation, University of Quebec at Chicoutimi, Canada, 2008, 178 p.

9.   Maple 9. http://www.maplesoft.com/

10.   K. Meriam, "Engineering Mechanics," 5th Edition, Vol. 2, Wiley, Hoboken, 2002.

11.   H. B. Pacejka, "Tire and Vehicle Dynamics," 2nd Edition, 2005.

12.   H. Pacejka and E. Bakker, "The Magic Formula TyreModel," Proceedings of 1st International Colloquium on Tyre Models for Vehicle Dynamics Analysis, 1991, pp 1-18.

13.   J. Y. Wong, "Theory of Ground Vehicles," 3rd Edition, John Wiley and Sons, Hoboken, 2001.

14.   U. Kiencke and L. Nielsen, "Automotive Control System," Springer, Berlin, 2000.

15.   F. Ben Amar, "Modèle de Comportement des Véhicules Tout Terrain pour la Planification Physico-Géométrique des Trajectoires," Ph.D. Dissertation, Université Pierre et Marie Curie, France, 1994.

16.   J. Svendenius and M. Gafvert, "A Semi-Empirical Dynamic Tire Model for Combined-Slip Forces," Vehicle System Dynamics, Vol. 44, No. 2,

2004, pp. 189-208.

17. M. Gafvert, J. Svendenius and J. Andreasson, "Implementation and Application of a Semi-Empirical TireModel in Multi-Body Simulation of Vehicle Handling," Proceedings of the 8th International Symposium on Advanced Vehicle Control, Taipei, 2006.

18. W. Pelz, D. Schuring and M. G. Pottinger, "A Model for Combined Tire Cornering and Braking Forces," SAE Paper, 960180, 1996.

19. M. W. Sayers and S. M. Riley, "Modeling Assumptions for Realistic Multibody Simulations of the Yaw and Roll Behavior of Heavy Trucks," SAE Paper, 960173, 1996.

20. M. Sanfridson, M. Gafvert and V. Claesson, "Truck Model for Yaw Dynamics Control," Technical Report, Lund Institute of Technology, Sweden, 2000.

21. D. Syndey, D. Terry and G. Roberts, "A New Vehicle Simulation Model for Vehicle Design and Safety Research," SAE, Engineering Dynamics Corp., (2001)-01- 0503.

22. Truckism 6. http://www.carsim.com/

Chapter 7

DISCRETE SYMMETRY IN RELATIVISTIC QUANTUM MECHANICS

Guang-jiong Ni[1,2], Suqing Chen[2], Jianjun Xu[2]

[1]Department of Physics, Portland State University, Portland, USA

[2]Department of Physics, Fudan University, Shanghai, China

ABSTRACT

EPR experiment on $K^0 - \bar{K}^0$ system in 1998 [1] strongly hints that one should use operators $\hat{E}_c = -i\hbar \frac{\partial}{\partial t}$ and $\hat{P}_c = i\hbar\nabla$ for the wavefunction (WF) of antiparticle. Further analysis on Klein-Gordon (KG) equation reveals that there is a discrete symmetry hiding in relativistic quantum mechanics (RQM) that $\mathcal{PT} = C$. Here $\mathcal{PT}$ means the (newly defined) combined space-time inversion (with $x \to -x, t \to -t$), while C the transformation of WF ψ between particle and its antiparticle whose definition is just residing in the above symmetry. After combining with FeshbachVillars (FV) dissociation of KG equation $(\psi = \phi + \chi)$ [2], this discrete symmetry can be rigorously reformulated by the invariance of coupling equation of ϕ and χ under either the combined space-time inversion $\mathcal{PT}$ or the mass inversion $(m \to -m)$, which makes the KG equation a self-consistent theory. Dirac equation is also discussed accordingly. Various applications of this discrete symmetry are discussed, including the prediction of antigravity between matter and antimatter as well as the reason why we believe neutrinos are likely the tachyons.

INTRODUCTION

In 1956-1957, the historical discovery of the parity violation [3-6] reveals that both P and C symmetries are violated to maximum in weak interactions. Then in 1964- 1970, both CP and T are experimentally verified to be violated in some cases (though to a tiny degree) [7,8] whereas the product symmetry CPT holds intact to this day [9]. The CPT invariance in quantum field theory (QFT) was first proved by Lüders and Pauli in 1954- 1957 [10-12] via the introduction of the "strong reflection" for proving the CPT theorem. In 1965,

Lee and Wu proposed that the definition of particle $|a\rangle$ versus its antiparticle $|\bar{a}\rangle$ should be [13]

$$|\bar{a}\rangle = \text{CPT}|a\rangle \qquad (1.1)$$

Regrettably, the counterpart of "strong reflection" at the level of RQM went nearly unnoticed in the past decades. In this paper, we are going to study the RQM thoroughly. Not only a discrete symmetry $\mathcal{PT} = \mathcal{C}$ is found in RQM as the counterpart of "strong reflection" in QFT, it is also evolved into the invariance of space-time inversion $(x \to -x, t \to -t)$ or mass inversion $(m \to -m)$, showing that a WF in RQM is always composed of two parts in confrontation inside a particle and then RQM becomes a self-consistent theory. Furthermore, this symmetry can serve as a "theoretical tool" in searching for new applications in today's physics.

The organization of this paper is as follows: In section II, the EPR paradox [14] is discussed together with the $K^0 - \bar{K}^0$ correlation experimental data [1], yielding a strong hint that the energy-momentum operators for antiparticle's WF should be $\hat{E}_c = -i\hbar\frac{\partial}{\partial t}$ and $\hat{p}_c = i\hbar\nabla$ respectively. Section III is focused on a discrete symmetry $\mathcal{PT} = c$, here $\mathcal{PT}$ means the (newly defined) combined space-time inversion (with $\boldsymbol{x \to -x, t \to -t}$), while $\mathcal{C}$ the transformation of WFs between particle and antiparticle, whose definition is just residing in the symmetry. Then after combining with FV dissociation of KG equation [2] in which the WF ψ is composed of two fields: $\psi = \phi + \chi$, the above symmetry can be realized in terms of ϕ and χ rigorously via the invariance of their coupling equation either under the spacetime inversion or a mass inversion $(m \to -m)$ In this way, the probability density is ensured to be positive definite for WFs of either particle or antiparticle. Section IV ascribes various phenomena in the theory of special relativity (SR) to the effects of enhancement of the hidden χ field in a moving particle. In Section V, Dirac equation is discussed accordingly with the importance of helicity being stressed. Section VI contains a brief discussion on the QFT. Sections VII, VIII and IX are devoting to seek for possible applications of the above symmetry in today's physical problems: Why a parity violation phenomenon was overlooked since 1956-1957? Why we believe neutrinos are likely the tachyons? And the prediction of antigravity between matter and antimatter. The last Section X contains a summary. In the Appendix, the Klein paradox is solved for both KG equation and Dirac equation without resorting to the "hole theory".

WHAT THE $K^0 \bar{K}^0$ CORRELATION EXPERIMENTAL DATA ARE TELLING?

To our knowledge, beginning from Bohm and Bell [15,16], physicists gradually turned their research of EPR paradox [14] onto the entangled state composed of electrons, especially photons with spin and achieved fruitful results. However, as pointed out by Guan (1935-2007), EPR's paper [14] is focused on two spinless particles and Guan found that there is a commutation relation hiding in such a system as follows [17]:

Consider two particles in one dimensional space with positions $x_i (i=1,2)$ and momentum operators

$\hat{p}_i = -i\hbar \dfrac{\partial}{\partial x_i}$. Then a commutation relation arises as

$$[x_1 - x_2, \hat{p}_1 + \hat{p}_2] = 0 \tag{2.1}$$

According to QM's principle, there may be a kind of common eigenstate having eigenvalues of these two commutative (i.e., compatible)observables like:

$$p_1 + p_2 = 0, (p_2 = -p_1) \text{ and } (x_1 - x_2) = D \tag{2.2}$$

with D being their distance. The existence of such kind of eigenstate described by Equation (2.2) puzzled Guan, he asked: "How can such kind of quantum state be realized?" A discussion between Guan and one of present authors (Ni) in 1998 led to a paper [18].

Here we are going to discuss further, showing that the correlation experiment on a $K^0 \bar{K}^0$ system (which just realized an entangled state composed of two spinless particles) in 1998 by CPLEAR collaboration [1] actually revealed some important features of QM and then answered the puzzle raised by EPR in a surprising way. First, besides Equation (1), let us consider another three commutation relations simultaneously:

$$\left[t_1 + t_2, \hat{E}_1 - \hat{E}_2 \right] = 0 \tag{2.3}$$

$$[x_1 + x_2, \hat{p}_1 - \hat{p}_2] = 0 \tag{2.4}$$

$$\left[t_1 - t_2, \hat{E}_1 + \hat{E}_2 \right] = 0 \tag{2.5}$$

($E_i = -i\hbar \frac{\partial}{\partial t_i}$ with t_i being the time during which the i-th particle is detected). In accordance with Ref. [1], we also focus on back-to-back events. The evolution of $K^0 \bar{K}^0$'s wavefunction (WF) will be considered in three inertial frames: The center-of-mass system S is at rest in laboratory with its origin x = 0 located at the apparatus' center, where the antiprotons' beam is stopped inside a hydrogen gas target to create $K^0\bar{K}^0$ pairs by $p\bar{p}$ annihilation. The $K^0\bar{K}^0$ pairs are detected by a cylindrical tracking detector located inside a solenoid providing a magnetic field parallel to the antiprotons' beam. For back-to-back events, the space-time coordinates in Equations (1)-(5) refer to particles moving to the right $(x_1 > 0)$ and left $(x_2 < 0)$ respectively. Second, we take an inertial system S' with its origin located at particle 1 (i.e., $x_1' = 0$). S' is moving in a uniform velocity v with respect to s. (For Kaon's momentum of $800\,\text{MeV/c}, \beta = v/c = 0.849$). Another s'' system is chosen with its origin located at particle $2(x_2'' = 0)$. S'' is moving in a velocity $(-v)$ with respect to s. Thus we have Lorentz transformation among the space-time coordinates being

$$\begin{cases} x' = \dfrac{x - vt}{\sqrt{1-\beta^2}}, \\ t' = \dfrac{t - vx/c^2}{\sqrt{1-\beta^2}}, \end{cases} \begin{cases} x'' = \dfrac{x + vt}{\sqrt{1-\beta^2}}, \\ t'' = \dfrac{t + vx/c^2}{\sqrt{1-\beta^2}}, \end{cases}$$

(2.6)

Here t_1' and t_2'' correspond to the proper time t_a and t_b in Ref.[1] respectively. The common time origin $t = t' = t'' = 0$ is adopted.

A $K^0\bar{K}^0$ pair, created in a $J^{PC} = 1^{--}$ antisymmetric state, can be described by a two-body WF depending on time as ([1], see also [19,20])

$$|\Psi(0,0)\rangle^{(\text{antisym})}$$
$$= \frac{1}{\sqrt{2}}\left[|K^0(0)\rangle_a |\bar{K}^0(0)\rangle_b - |\bar{K}^0(0)\rangle_a |K^0(0)\rangle_b\right]$$
$$|\Psi(t_a,t_b)\rangle^{(\text{antisym})}$$
$$= \frac{1}{\sqrt{2}}\left[|K_S(0)\rangle_a |K_L(0)\rangle_b\, e^{-i(\alpha_S t_a + \alpha_L t_b)}\right.$$
$$\left. - |K_L(0)\rangle_a |K_S(0)\rangle_b\, e^{-i(\alpha_L t_a + \alpha_S t_b)}\right]$$

(2.7)

with

$$|K_S\rangle = \frac{1}{\sqrt{2}}\left[|K^0\rangle - |\bar{K}^0\rangle\right], |K_L\rangle = \frac{1}{\sqrt{2}}\left[|K^0\rangle + |\bar{K}^0\rangle\right]$$

(2.8)

where the CP violation has been neglected and

$\alpha_{S,L} = m_{S,L} - i\gamma_{S,L}/2$, $m_{S,L}$ and $\gamma_{S,L}$ being the $K_{S,L}$ masses and decay widths, respectively. From Equation (7), the intensities of events with like-strangeness ($K^0 K^0$ or $\bar{K}^0 \bar{K}^0$) and unlike-strangeness ($K^0 \bar{K}^0$ or $\bar{K}^0 K^0$) can be evaluated as

$$I_{\text{like}}^{(\text{antisy})}\left(t_a,t_b\right)=\frac{1}{8}e^{-2\gamma\tilde{t}}$$

$$\cdot\left\{e^{-\gamma_S|t_a-t_b|}+e^{-\gamma_L|t_a-t_b|}-2e^{-\gamma|t_a-t_b|}\cos\left[\Delta m\left(t_a-t_b\right)\right]\right\} \tag{2.9}$$

$$I_{\text{unlike}}^{(\text{antisy})}\left(t_a,t_b\right)=\frac{1}{8}e^{-2\gamma\tilde{t}}$$

$$\cdot\left\{e^{-\gamma_S|t_a-t_b|}+e^{-\gamma_L|t_a-t_b|}+2e^{-\gamma|t_a-t_b|}\cos\left[\Delta m\left(t_a-t_b\right)\right]\right\} \tag{2.10}$$

where $\Delta m = m_L - m_S, \gamma = (\gamma_S + \gamma_L)/2$ and

$\tilde{t} = t_a\left(\text{for } t_a < t_b\right)$ or $\tilde{t} = t_b\left(\text{for } t_a > t_b\right)$.

Similarly, for $K^0\bar{K}^0$ created in a $J^{PC}=0^{++}$ or 2^{++} symmetric state as:

$$|\Psi(0,0)\rangle^{(sym)}$$
$$=\frac{1}{\sqrt{2}}\left[\left|K^0(0)\right\rangle_a\left|\bar{K}^0(0)\right\rangle_b+\left|\bar{K}^0(0)\right\rangle_a\left|K^0(0)\right\rangle_b\right]$$
$$|\Psi(t_a,t_b)\rangle^{(sym)}$$
$$=\frac{1}{\sqrt{2}}\left[\left|K_L(0)\right\rangle_a\left|K_L(0)\right\rangle_b e^{-i(\alpha_L t_a+\alpha_L t_b)}\right.$$
$$\left.-\left|K_S(0)\right\rangle_a\left|K_S(0)\right\rangle_b e^{-i(\alpha_S t_a+\alpha_S t_b)}\right] \tag{2.11}$$

the predicted intensities read

$$I_{\text{like}}^{(sym)}\left(t_a,t_b\right)$$
$$=\frac{1}{8}\left\{e^{-\gamma_S(t_a+t_b)}+e^{-\gamma_L(t_a+t_b)}-2e^{-\gamma(t_a+t_b)}\cos\left[\Delta m\left(t_a+t_b\right)\right]\right\}$$
$$I_{\text{unlike}}^{(sym)}\left(t_a,t_b\right)$$
$$=\frac{1}{8}\left\{e^{-\gamma_S(t_a+t_b)}+e^{-\gamma_L(t_a+t_b)}+2e^{-\gamma(t_a+t_b)}\cos\left[\Delta m\left(t_a+t_b\right)\right]\right\} \tag{2.12}$$

The experiment [1] reveals that the $K^0\bar{K}^0$ pairs are mainly created in the antisymmetric state shown by Equations (2.9) and (2.10) while the contribution in a symmetric state shown by Equations (2.11) and (2.12) accounts for 7.4%.

What we learn from Ref. [1] in combination with Equations (2.1)-(2.5) are as follows:

(a) Because only back-to-back events are involved in the S system, we denote three commutative operators as: the "distance" operator

$\hat{D}=x_1-x_2=v\left(t_1+t_2\right),\hat{A}=\hat{p}_1+\hat{p}_2$ and $\hat{B}=\hat{E}_1-\hat{E}_2$, Equations (2.1) and (2.3) read

$$\left[\hat{D},\hat{A}\right]=0,\left[\hat{D},\hat{B}\right]=0,\left[\hat{A},\hat{B}\right]=0 \tag{2.13}$$

So they may have a kind of common eigenstate during the measurement composed of $K^0 K^0$ and projected from the symmetric state shown by Equation (11). It is assigned by a continuous eigenvalue $D_j = v(t_1 + t_2)$ (with continuous index j) of operator $\hat{D}$ acting on the WF, $\Psi^{sym}_{K^0 K^0}(x_1, t_1; x_2, t_2)$, as[1]

$$\hat{D}\Psi^{sym}_{K^0 K^0}(x_1, t_1; x_2, t_2)$$
$$= D_j \Psi^{sym}_{K^0 K^0}(x_1, t_1; x_2, t_2)$$
$$= v(t_1 + t_2)\Psi^{sym}_{K^0 K^0}(x_1, t_1; x_2, t_2) \qquad (2.14a)$$

$$\hat{A}\Psi^{sym}_{K^0 K^0}(x_1, t_1; x_2, t_2)$$
$$= A_j^{like}\Psi^{sym}_{K^0 K^0}(x_1, t_1; x_2, t_2)$$
$$= (p_1 + p_2)\Psi^{sym}_{K^0 K^0}(x_1, t_1; x_2, t_2) \qquad (2.15)$$

$$\hat{B}\Psi^{sym}_{K^0 K^0}(x_1, t_1; x_2, t_2)$$
$$= B_j^{like}\Psi^{sym}_{K^0 K^0}(x_1, t_1; x_2, t_2)$$
$$= (E_1 - E_2)\Psi^{sym}_{K^0 K^0}(x_1, t_1; x_2, t_2) \qquad (2.16)$$

where the lowest eigenvalue of $\hat{A}$ is

$A_j^{like} = p_1 + p_2 = 0, (p_2 = -p_1)$, and that of $\hat{B}$ is

$B_j^{like} = E_1 - E_2 = 0, (E_2 = E_1)$ respectively. These eigenstates of like-strangeness events predicted by Equation (11) are really observed in the experiment [1] (these eigenstates of $K^0 K^0$ were overlooked in the Ref. [18]).

(b) The more interesting case occurs for $K^0 \bar{K}^0$ pair created in the antisymmetric state with intensity given by Equation (10) being a function of $(t_a - t_b)$ (not $(t_a + t_b)$ as shown by Equation (12) for symmetric states) which is proportional to $(t_1 - t_2)$ in the S system. In the EPR limit $t_1 = t_2$, $K^0 \bar{K}^0$ events dominate whereas likestrangeness events are strongly suppressed as shown by Equation (9) (see Figure 1 in [1]). So the experimental facts remind us of the possibility that $K^0 \bar{K}^0$ events may be related to common lowest (zero) eigenvalues of some commutative operators (just like what happened in Equations (15) and (16) for operators $\hat{A}$ and $\hat{B}$ (which are applied to symmetric states (due to $\hat{D} = x_1 - x_2 = v(t_1 + t_2)$)) but are not suitable for antisymmetric states), there are another three operators shown by Equations (4) and (5) being: the operator of "flight-path difference" $\hat{F} = x_1 + x_2 = v(t_1 - t_2), \hat{M} = \hat{p}_1 - \hat{p}_2$ and

$\hat{G} = \hat{E}_1 + \hat{E}_2$ with commutation relations as:

$$\left[\hat{F}, \hat{M}\right] = 0, \left[\hat{F}, \hat{G}\right] = 0, \left[\hat{M}, \hat{G}\right] = 0 \tag{2.17}$$

which are just suitable for antisymmetric states. For $K^0 \bar{K}^0$ back-to-back events, assume that one of two particles, say 2, is an antiparticle with its momentum and energy operators being

$$\hat{p}_x^c = i\hbar \frac{\partial}{\partial x}, \hat{E}^c = -i\hbar \frac{\partial}{\partial t} \tag{2.18}$$

(the superscript c means "antiparticle") versus that for particle being

$$\hat{p}_x = -i\hbar \frac{\partial}{\partial x}, \hat{E} = i\hbar \frac{\partial}{\partial t} \tag{2.19}$$

For instance, a freely moving particle's WF reads[2]:

$$\psi(x,t) \sim \exp\left[\frac{i}{\hbar}(px - Et)\right] \tag{2.20}$$

whereas

$$\psi_c(x,t) \sim \exp\left[-\frac{i}{\hbar}(p_c x - E_c t)\right] \tag{2.21}$$

for its antiparticle with $p_c = p$ and $E_c (> 0)$ being momentum and energy of the antiparticle in accordance with Equation (2.18). If using Equations (2.18)-(2.21), we find

$$\hat{F}\Psi_{K^0\bar{K}^0}^{\text{antisym}}(x_1, t_1; x_2, t_2)$$
$$= F_k^{\text{unlike}}\Psi_{K^0\bar{K}^0}^{\text{antisym}}(x_1, t_1; x_2, t_2)$$
$$= v(t_1 - t_2)\Psi_{K^0\bar{K}^0}^{\text{antisym}}(x_1, t_1; x_2, t_2) \tag{2.22}$$

with continuous index k referring to continuous eigenvalues $F_k = v(t_1 - t_2)$. Here, the WF in space-time of this system during measurement reads approximately:

$$\Psi_{K^0\bar{K}^0}^{\text{antisym}}(x_1, t_1; x_2, t_2) \sim e^{i(p_1 x_1 - E_1 t_1)} e^{-i(p_2^c x_2 - E_2^c t_2)} \tag{2.23}$$

with antiparticle 2 moving opposite to particle 1 and $p_2^c = -p_1$.

Now we use $\hat{M}(= \hat{p}_1 - \hat{p}_2) = \hat{p}_1 + \hat{p}_2^c$ on $K^0\bar{K}^0$ system, yielding

$$\hat{M}\Psi^{antisym}_{K^0\bar{K}^0}\left(x_1,t_1;x_2,t_2\right)$$

$$= M^{unlike}_k\Psi^{antisym}_{K^0\bar{K}^0}\left(x_1,t_1;x_2,t_2\right)$$

$$= \left(p_1 + p_2^c\right)\Psi^{antisym}_{K^0\bar{K}^0}\left(x_1,t_1;x_2,t_2\right) \tag{24}$$

Similarly, we have $\hat{G}\left(=\hat{E}_1+\hat{E}_2\right)=\hat{E}_1-\hat{E}_2^c$ and find

$$\hat{G}\Psi^{antisym}_{K^0\bar{K}^0}\left(x_1,t_1;x_2,t_2\right)$$

$$= G^{unlike}_k\Psi^{antisym}_{K^0\bar{K}^0}\left(x_1,t_1;x_2,t_2\right)$$

$$= \left(E_1 - E_2^c\right)\Psi^{antisym}_{K^0\bar{K}^0}\left(x_1,t_1;x_2,t_2\right) \tag{25}$$

Hence we see that once Equations (2.18) and (2.21)

are accepted, the WFs $\Psi^{antisym}_{K^0\bar{K}^0}\left(x_1,t_1;x_2,t_2\right)$ show up in experiments as the only WFs with strongest intensity at the EPR limit $(t_1=t_2)$ corresponding to their three eigenvalues being all zero: $F_k = M^{unlike}_k = G^{unlike}_k = 0$ and they won't change even when accelerator's energies are going up.

If using Equation (2.18), the eigenvalues of $\hat{A}$ and

$\hat{B}$ for the WF $\Psi^{antisym}_{K^0\bar{K}^0}\left(x_1,t_1;x_2,t_2\right)$ are

$A^{unlike}_j = p_1 - p_2^c = 2p_1$ and $B^{unlike}_j = E_1 + E_2^c = 2E_1$ respectively, while that of $\hat{M}$ and $\hat{G}$ for the WF

$\Psi^{antisym}_{K^0K^0}\left(x_1,t_1;x_2,t_2\right)$ are $M^{like}_k = p_1 - p_2 = 2p_1$ and

$G^{like}_k = E_1 + E_2 = 2E_1$, respectively, those eigenvalues are much higher than zero and going up with the accelerator's energy.

Something is very interesting here: If we deny Equation (2.18) but insist on unified operators $\hat{p}$ and $\hat{E}$ for both particle and antiparticle, there would be no difference in eigenvalues between like-strangeness events and unlike-strangeness ones. For example, the M^{unlike}_k and G^{unlike}_k would be $2p_1$ and $2E_1$ too (instead of "0" as in Equations (2.24) and (2.25)). This would mean that three commutative operators $\hat{F},\hat{M}$ and $\hat{G}$ are not enough to distinguish the WF $\Psi^{antisym}_{K^0\bar{K}^0}\left(x_1,t_1;x_2,t_2\right)$ from the WF $\Psi^{antisym}_{K^0K^0}\left(x_1,t_1;x_2,t_2\right)$ even they behave so differently as shown by Equations (9) and (10)), especially at the EPR limit $(t_1=t_2)$.

Equation (2.18) together with the identification of WF

$\Psi^{antisym}_{K^0\bar{K}^0}\left(x_1,t_1;x_2,t_2\right)$ by three zero eigenvalues implies that the difference of a particle from its antiparticle is not something hiding in the "intrinsic space" like opposite charge (for electron and positron) or opposite strangeness (for K^0 and $\bar{K}^0$) but can be displayed in their WFs evolving in space-time at the level of QM.

In summary, instead of one set of WF with its operators (Equations (2.19) and (2.20)), two sets of WFs with operators separately (shown as Equations (2.18)- (2.21)) are strongly supported by the original EPR paradox and its "solution" provided by the $K^0 - \bar{K}^0$ experiment.

To our knowledge, Equation (2.18) can be found at a page note of a paper by Konopinski and Mahmaud in 1953 [21], also appears in Refs. [18,22-28].

HOW TO MAKE KLEIN-GORDON EQUATION A SELF-CONSISTENT THEORY IN RQM? A DISCRETE SYMMETRY

The Negative Energy Solution and the WF of Antiparticle

Let us begin with the energy conservation law for a particle in classical mechanics:

$$E = \frac{1}{2m} p^2 + V(x)$$

(3.1)

Consider the rule promoting observables into operators:

$$E \rightarrow \hat{E} = i\hbar \frac{\partial}{\partial t}, p \rightarrow \hat{p} = -i\hbar \nabla$$

(3.2)

and let Equation (3.1) act on a wavefunction (WF)$^{\psi(x,t)}$, the Schrödinger equation

$$i\hbar \frac{\partial}{\partial t} \psi(x,t) = -\frac{\hbar^2}{2m} \nabla^2 \psi(x,t) + V(x)\psi(x,t)$$

(3.3)

follows immediately. In mid 1920's, considering the kinematical relation for a particle in the theory of special relativity (SR):

$$(E-V)^2 = c^2 p^2 + m^2 c^4$$

(3.4)

and using Equation (3.2) again, the Klein-Gordon (KG) equation was established as:

$$\left(i\hbar \frac{\partial}{\partial t} - V\right)^2 \psi(x,t)$$
$$= -c^2 \hbar^2 \nabla^2 \psi(x,t) + m^2 c^4 \psi(x,t)$$

(3.5)

For a free KG particle, its plane-wave solution reads:

$$\psi(x,t) \sim \exp\left[\frac{i}{\hbar}(p \cdot x - Et)\right]$$

(3.6)

However, two difficulties arose:

(a) The energy E in Equation (6) has two eigenvalues:

$$E = \pm\sqrt{c^2 p^2 + m^2 c^4}$$

(3.7)

In general, $V \neq 0$, the WFs of KG particle's energy eigenstates can always be divided into two parts:

$$\psi \sim \exp\left(-\frac{i}{\hbar}Et\right), E > 0$$

(3.8)

$$\psi \sim \exp\left(-\frac{i}{\hbar}Et\right), E < 0$$

(3.9)

where only the original operators Equation (3.2) are used. But what the "negative energy" means?

(b) The continuity equation is derived from Equation (5) as

$$\frac{\partial \rho}{\partial t} + \nabla \cdot j = 0$$

(3.10)

where

$$\rho = \frac{i\hbar}{2mc^2}\left(\psi^* \frac{\partial}{\partial t}\psi - \psi \frac{\partial}{\partial t}\psi^*\right) - \frac{1}{mc^2}V\psi^*\psi$$

(3.11)

and

$$j = \frac{i\hbar}{2m}\left(\psi\nabla\psi^* - \psi^*\nabla\psi\right)$$

(3.12)

are the "probability density" and "probability current density" respectively. While the latter is the same as that derived from Equation (3.3), Equation (3.11) seems not positive definite and dramatically different from $\rho = \psi^*\psi$ in Equation (3.3). Why?

In hindsight, for a linear equation in RQM, either KG or Dirac equation, the emergence of WFs with both positive and negative energy (E) is inevitable and natural. From mathematical point of view, the set of WFs cannot be complete if without taking the negative energy solutions into account. And physicists believe that these negative-energy solutions might be relevant to antiparticles. However, we physicists admit that both a rest particle's energy $E = mc^2$ and a rest antiparticle's energy $E_c = m_c c^2 = mc^2$ are positive, as verified by numerous experiments like that of pair-creation process $\gamma \rightarrow e^+ + e^-$. The above contradiction constructs socalled "negative-energy paradox" in RQM. For Dirac particle, majority (not all) of physicists accept the "hole theory" to explain the "paradox". But for KG particle, no such kind of "hole theory" can be acceptable. It was this "negative-energy paradox" as well as the four "commutation relations", Equations (2.1)-(2.5), hidden in the two-particle system discussed by EPR [14] gradually prompted us to realize that the root cause of difficulty in RQM lies in an a priori notion—only one kind of WF with one set of operators (like Equation (3.2)) can be acceptable in QM, either for NRQM or RQM.

Once getting rid of the constraint in the above notion and introducing two sets of WFs and operators for particle and antiparticle respectively, we can identify the negative energy solution, Equation (3.9), with the antiparticle's WF directly

$$\psi_c \sim \exp\left(\frac{i}{\hbar}E_c t\right), E_c > 0$$

(3.13)

which implies an antiparticle with positive energy E_c by using Equation (2.18). This claim will be proved rigorously in the next subsection.

One may ask: When you assume the negative energy solution being the WF of antiparticle, how about the difficulty of negative probability density? Below we will see how to solve these two difficulties simultaneously and make KG equation a self-consistent theory at the level of RQM.

The Proof of a Discrete Symmetry $\mathcal{PT} = \mathcal{C}$ for KG Particle

Let us introduce an operator of (newly defined) combined space-time inversion $\mathcal{PT}$ for KG equation. It should change the space-time coordinates as

$$x \to -x, t \to -t$$

(3.14)

then accordingly

$$\hat{p} = -i\hbar\nabla \to \mathcal{PT}\hat{p}(\mathcal{PT})^{-1} = \hat{p}_c = i\hbar\nabla,$$

$$\hat{E} = i\hbar\frac{\partial}{\partial t} \to \mathcal{PT}\hat{E}(\mathcal{PT})^{-1} = \hat{E}_c = -i\hbar\frac{\partial}{\partial t}$$

(3.15)

Because the antiparticle has opposite charge $(-q)$ versus q for particle, so

$$V(x,t) \to \mathcal{PT}V(x,t)(\mathcal{PT})^{-1}$$
$$\equiv V_c(x,t) = -V(x,t)$$

(3.16)

When performing $\mathcal{PT}$ inversion on KG equation, Equation (3.5), from left to right, we meet eventually the WF and define the antiparticle's WF as

$$\mathcal{P}T\psi(x,t) \equiv \mathcal{C}\psi(x,t) = \psi_c(x,t) \tag{3.17}$$

Thus KG particle's equation, Equation (3.5), is transformed into $^{(\hbar=1)}$

$$\left(\hat{E}_c - V_c\right)^2 \psi_c(x,t)$$
$$= -c^2\nabla^2\psi_c(x,t) + m^2\psi_c(x,t) \tag{3.18}$$

or

$$\left(i\frac{\partial}{\partial t} - V\right)^2 \psi_c(x,t)$$
$$= -c^2\nabla^2\psi_c(x,t) + m^2\psi_c(x,t) \tag{3.19}$$

which is formally the same as Equation (3.5) though we should use $\hat{p}_c, \hat{E}_c$ for $\psi_c(x,t)$. Hence the KG equation remains invariant under the $\mathcal{P}T$ operation, Equations (3.14)-(3.17). Notice further that Equation (3.18) is just the "quantized" equation of the kinematical relation for an antiparticle in SR

$$\left(\hat{E}_c - V_c\right)^2 = c^2 p_c^2 + m^2 c^4 \tag{3.20}$$

which is the counterpart of Equation (3.4) for a particle. For example, a KG particle's scattering WF $\psi(x,t;E_1) \sim e^{-iE_1 t} (E_1 > m)$ is attracted by an spherically symmetric potential $V(r) < 0$ and so has a positive phase-shift $\delta_1 > 0$ (in the, say, $S(l=0)$ state). Then physically, its antiparticle's WF

$$\psi_c\left(x,t;E_1^c\right) \sim e^{iE_1^c t}\left(E_1^c = E_1 > m\right)_{\text{is}} \quad \text{repelled} \quad \text{by} \quad \text{the} \quad \text{potential}$$
$V_c(r) = -V(r) > 0$ and has a negative phaseshift $\delta_1^c < 0$.

Note that, however, corresponding to $\psi(x,t;E_1)$, there is another negative energy particle's WF

$$\psi(x,t;-E_1) \sim e^{iE_1 t} \quad \text{satisfying Equation (3.5)}$$

$$\left(i\frac{\partial}{\partial t} - V\right)^2 \psi(x,t;-E_1)$$
$$= \left(-E_1 - V\right)^2 \psi(x,t;-E_1)$$
$$= -c^2\nabla^2\psi(x,t;-E_1) + m^2\psi(x,t;-E_1) \tag{3.21}$$

whose space-time behavior is precisely the same as the antiparticle's WF $\psi_c\left(x,t;E_1^c\right) \sim e^{iE_1^c t}$ with $E_1^c = E_1 > m$

as shown by Equation (3.18) since

$(E_1 + V)^2 = (E_c - V_c)^2$. Thus, for avoiding confusion, we have

$$\mathcal{P}\mathcal{T}\psi(x,t;E_1)$$
$$= \psi(x,t;-E_1) = C\psi(x,t;E_1)$$
$$= \psi_c(x,t;E_1^c) \neq \psi(-x,-t;E_1) \qquad (3.22)$$

and

$$\mathcal{P}\mathcal{T}\psi(x,t)$$
$$= C\psi(x,t) = \psi(-x,-t)$$
$$= \psi_c(x,t)(V=0) \qquad (3.23)$$

achieving the proof of the discrete symmetry $\mathcal{P}\mathcal{T} = C$ for KG particle shown by Equation (3.17). In summary, the "negative-energy paradox" for KG equation is solved in a physical way with following advantages:

a)   By using two sets of WFs and momentum-energy operators for particle and antiparticle respectively, both particle's WF $\psi(x,t)$ and antiparticle's WF $\psi_c(x,t)$ have positive energies $E > 0$ and $E_c > 0$ respectively.

b)   While satisfying the same KG equation with same potential $V(r)$ formally, $\psi(x,t)$ and $\psi_c(x,t)$ are actually subject to opposite "force" for particle and antiparticle respectively.

c)   The space-time behavior of $\psi_c(x,t;E_1^c)$ can be identified with that of a negative energy particle's WF

$$\psi(x,t;-E_1)(E_1 = E_1^c)$$, in a one-to-one correspondence.

Thus from mathematical point of view, all solutions of KG equation form a complete set including both positive and negative energy values of one operator $\hat{E} = i\dfrac{\partial}{\partial t}$ exactly.

By contrast, usually, aiming at finding an anti-particle's WF, one performs the CPT transformation on a particle's WF $\psi(x,t)$, yielding [29-32]

$$\psi(x,t) \rightarrow CPT\psi(x,t) = \psi(-x,-t) \qquad (3.24)$$

whose character can also be summed up as follows:

a')   By using one set of WF and relevant operators for both particle and antiparticle, at the LHS of Equation (3.24), $\psi(x,t)$, and $\psi(-x,-t)$ at RHS must have opposite energies inevitably.

b')   By design in the C transformation, $\psi(x,t)$ and $\psi(-x,-t)$ in Equation (3.24) satisfy different equations with V and $V_c = -V$ respectively. But with opposite energies, they are actually subject to the same (either attractive or repulsive) "force". So one cannot distinguish particle from antiparticle through what their WFs "feel" after the CPT transformation.

c')   From mathematical point of view, we should keep all negative-energy solutions for one equation. However, even facing WFs in doubled numbers, we still don't know how to choose half of them for describing particle and its antiparticle separately in physics.

But we haven't solve the difficulty of negative probability density in KG equation yet, awaiting for another enlightenment which was already there since 1958.

Feshbach and Villars (FV) Dissociation of KG WF $\psi = \phi + \chi$, a Reformulated Symmetry between ϕ and χ under the Space-Time (or Mass) Inversion

In 1958, dividing the WF into $\psi = \phi + \chi$, Feshbach and Villars [2] recast Equation (5) into two coupled Schrödinger-like equations as[3]:

$$\begin{cases} \left(i\hbar \frac{\partial}{\partial t} - V \right)\phi = mc^2\phi - \frac{\hbar^2}{2m}\nabla^2(\phi + \chi) \\ \left(i\hbar \frac{\partial}{\partial t} - V \right)\chi = -mc^2\chi + \frac{\hbar^2}{2m}\nabla^2(\phi + \chi) \end{cases} \qquad (3.25)$$

where

$$\begin{cases} \phi = \frac{1}{2}\left[\left(1 - \frac{1}{mc^2}V \right)\psi + \frac{i\hbar}{mc^2}\dot{\psi} \right] \\ \chi = \frac{1}{2}\left[\left(1 + \frac{1}{mc^2}V \right)\psi - \frac{i\hbar}{mc^2}\dot{\psi} \right] \end{cases} \qquad (3.26)$$

$\left(\dot{\psi} = \frac{\partial \psi}{\partial t} \right)$. Interestingly, the "probability density", Equation (3.11) can be recast into a difference between two positive-definite densities [18,20]:

$$\rho = \phi^*\phi - \chi^*\chi \qquad (3.27)$$

while the probability current density contains interference terms between ϕ and χ:

$$j = \frac{i\hbar}{2m}\left[\left(\phi\nabla\phi^* - \phi^*\nabla\phi \right) + \left(\chi\nabla\chi^* - \chi^*\nabla\chi \right) \right.$$
$$\left. + \left(\phi\nabla\chi^* - \chi^*\nabla\phi \right) + \left(\chi\nabla\phi^* - \phi^*\nabla\chi \right) \right] \qquad (3.28)$$

The expression of $\mathcal{P}$ as shown by Equation (3.27) strongly hints that the $\mathcal{PT} = \mathcal{C}$ symmetry proved in the last subsection may be combined with the FV dissociation of KG equation such that the positive-definite property of $\mathcal{P}$ can be ensured for both particle and antiparticle.

Indeed, after inspecting Equation (3.25) carefully, we do find a hidden symmetry in the sense that it is invariant (in its form) under the following reformulated space-time inversion $(x \to -x, t \to -t)$, i.e., $\mathcal{PT} = \mathcal{C}$ transformation:

$$
\begin{cases}
x \to -x, t \to -t, \\
V(x,t) \to -V(x,t) = V_c(x,t), \\
\psi(x,t) \to PT\psi(x,t) = \psi_c(x,t), \\
\phi(x,t) \to PT\phi(x,t) = \chi_c(x,t), \\
\chi(x,t) \to PT\chi(x,t) = \phi_c(x,t)
\end{cases}
\tag{3.29}
$$

Performing transformation Equation (3.29) on Equation (3.26), we find χ_c satisfying the same equation of χ and ϕ_c satisfying that of ϕ. They read

$$
\begin{cases}
\chi_c = \dfrac{1}{2}\left[\left(1 + \dfrac{1}{mc^2}V\right)\psi_c - \dfrac{i\hbar}{mc^2}\dot\psi_c\right] \\
\phi_c = \dfrac{1}{2}\left[\left(1 - \dfrac{1}{mc^2}V\right)\psi_c + \dfrac{i\hbar}{mc^2}\dot\psi_c\right]
\end{cases}
\tag{3.30}
$$

Remember, for ψ_c, we should use operator Equation (3.15). Accordingly, the probability density for ψ_c is defined as

$$
\begin{aligned}
\rho \to PT\rho &= \rho_c \\
&= \frac{i\hbar}{2mc^2}\left(\psi_c\dot\psi_c^* - \psi_c^*\dot\psi_c\right) + \frac{1}{mc^2}V\psi_c^*\psi_c \\
&= \chi_c^*\chi_c - \phi_c^*\phi_c
\end{aligned}
\tag{3.31}
$$

Similarly, we have $(\nabla\psi \to -\nabla\psi_c)$

$$
j \to PTj = j_c = \frac{i\hbar}{2m}\left(\psi_c^*\nabla\psi_c - \psi_c\nabla\psi_c^*\right)
\tag{3.32}
$$

For simplicity, consider a free KG particle $(V = 0)$ with WF Equation (3.6). Then $|\phi| > |\chi|$

$$
\begin{cases}
\phi = \dfrac{1}{2}\left(1 + \dfrac{E}{mc^2}\right)\psi \\
\chi = \dfrac{1}{2}\left(1 - \dfrac{E}{mc^2}\right)\psi
\end{cases},
\quad
\begin{cases}
\rho = |\phi|^2 - |\chi|^2 > 0 \\
j = \dfrac{1}{m}p|\psi|^2
\end{cases}
\tag{3.33}
$$

But for a free $(V=0)$ KG antiparticle with WF Equation (2.21), it has $|\chi_c| > |\phi_c|$

$$\begin{cases} \phi_c = \dfrac{1}{2}\left(1 - \dfrac{E_c}{mc^2}\right)\psi_c \\ \chi_c = \dfrac{1}{2}\left(1 + \dfrac{E_c}{mc^2}\right)\psi_c \end{cases}, \quad \begin{cases} \rho_c = |\chi_c|^2 - |\phi_c|^2 > 0 \\ j_c = \dfrac{1}{m}\, p_c\,|\psi_c|^2 \end{cases}$$

$$(3.34)$$

Equations (3.33) and (3.34) satisfy all physical conditions we need. If $V \neq 0$, as long as $(E - V) > 0$ for particle or $(E_c - V_c) > 0$ for antiparticle, the situation remains the same. However, once $(E - V) < 0$ or $(E_c - V_c) < 0$, some complications would occur. For further discussion, please see the Appendix.

Therefore, we see that the reformulated space-time inversion, Equation (3.29), reflects the underlying symmetry between a particle's WF ψ and its antiparticle's WF ψ_c. As both E and P in ψ or E_c and P_c in ψ_c are positive definite, all difficulties in KG equation disappear and the latter becomes a self-consistent theory.

Moreover, instead of Equation (3.29), a "mass inversion $(m \to -m)$" can realize the same symmetry, the invariance under a $PT\text{-}c$ transformation, via the following operation on Equation (3.25):

$$\begin{cases} m \to -m_c = -m \\ V(x,t) \to V(x,t) = -V_c(x,t), \\ \psi(x,t) \to \psi_c(x,t), \\ \phi(x,t) \to \chi_c(x,t), \\ \chi(x,t) \to \phi_c(x,t) \end{cases}$$

$$(3.35)$$

Notice that, when $m \to -m$, we have $\hat{p} \to -\hat{p}_c$ and

$\hat{E} \to -\hat{E}_c$, i.e. $-i\hbar\nabla \to -i\hbar\nabla$, $i\hbar\dfrac{\partial}{\partial t} \to i\hbar\dfrac{\partial}{\partial t}$, in contrast to Equation (3.15) [1].[4]

The reason why $V \to -V$ in the space-time inversion Equation (3.29) whereas $V \to V$ in the mass inversion Equation (3.35) can be seen from the classical equation: The Lorentz force F on a particle exerted by an external potential Φ reads: $F = -\nabla V = -\nabla(q\Phi) = ma$. As the acceleration a of particle will change to $-a$ for its antiparticle, there are two alternative explanations: either due to the inversion of charge $q \to -q$ (i.e., $V \to -V$ but keeping m unchanged) or due to the inversion of mass $m \to -m$ (but keeping V unchanged).

REINTERPRETATION OF WF AND THE RELATIVISTIC EFFECTS

The success of FV's dissociation of KG equation should be ascribed to their deep insight that a unified WF ψ is composed of two fields ϕ and χ in confrontation. Note that Equation (3.25) reduces into two equations separately for a static KG particle $(V = 0, \hbar = c = 1)$:

$$\begin{cases} i\dfrac{\partial}{\partial t}\phi = m\phi, \\ i\dfrac{\partial}{\partial t}\chi = -m\chi \end{cases} \tag{4.1}$$

with two separated solutions being:

$$\begin{cases} E = m > 0, \\ \phi \sim e^{-iEt}, \\ \chi = 0 \end{cases}, \begin{cases} E = -m = -E_c < 0, E_c = m > 0 \\ \chi_c \sim e^{iE_c t}, \\ \phi_c = 0 \end{cases} \tag{4.2}$$

Once the particle (antiparticle) is moving with a velocity, $v \neq 0$, ϕ and χ (χ_c and ϕ_c) couple together and the WF $\psi = \phi + \chi$ ($\psi_c = \phi_c + \chi_c$) for a free particle (antiparticle) read (in one-dimensional space)

$$\psi \sim \phi \sim \chi \sim \exp\left[i(px - Et)\right], (|\phi| > |\chi|) \tag{4.3a}$$

$$\psi_c \sim \chi_c \sim \phi_c \sim \exp\left[-i(p_c x - E_c t)\right], (|\chi_c| > |\phi_c|) \tag{4.3b}$$

$(p_c = p > 0, E_c = E > 0)$ respectively. In Equation (4.3a), ϕ dominates $\chi (|\phi| > |\chi|)$. By contrast, in Equation (4.3b) it is χ_c who dominates ϕ_c (The status remains the same for $V \neq 0$ cases as discussed in the last section).

Despite ϕ and ϕ_c (χ and χ_c) having the "intrinsic tendency" to evolve as $\exp\left[i(px - Et)\right]\left(\exp\left[-i(px - Et)\right]\right)$, however, in a WF of particle (antiparticle), $\chi(\phi_c)$ must follow $\phi(\chi_c)$ to evolve like that shown by Equation (4.3a) (Equation (4.3b)), as $|\phi| > |\chi|(|\chi_c| > |\phi_c|)$. So it seems suitable to name ϕ the "hidden particle field" inside a particle while χ the "hidden antiparticle field" (rather than the "negative-energy component") inside the same particle.

Let us try to reinterpret the phenomena displayed in the kinematics of special relativity (SR) via the enhancement of χ field in a particle [22-25]:

(a) Lorentz transformation Consider a particle's WF shown by Equation (4.3a) in an inertial frame S (laboratory). Then take another S' frame resting on the particle, so $p' = 0$ and $E' = E_0 = mc^2$. The WF in S' frame reads:

$$\psi(x',t') \sim \exp\left[\frac{i}{\hbar}(p'x' - E't')\right] = \exp\left[-\frac{i}{\hbar}E_0 t'\right] \qquad (4.4)$$

Here the space-time coordinates (x',t') are introduced and defined in the S' frame via the phase of WF as follows: Based on the assertion that "phase remains invariant under the coordinate transformation" which was named the "law of phase harmony" by de Broglie and was regarded by himself as the fundamental achievement all his life [34], comparing the phase in Equation (4.4) with that in Equation (4.3a) and using

$E = E_0 / \sqrt{1 - v^2/c^2}$, $p = Ev/c^2$, one finds

$$t' = \frac{t - vx/c^2}{\sqrt{1 - v^2/c^2}} \qquad (4.5)$$

Then, all formulas in the Lorentz transformation can be obtained. In some sense, what used here is a particle's wave-packet which serves as a microscopic "ruler", also a "clock" simultaneously.

(b) There is a speed limit c for a massive particle.

For a free KG particle, using Equation (3.33), we may define an "impurity ratio" R for the amplitude of hidden χ field to that of ϕ field and calculate it being

$$R_{free}^{KG} = \frac{|\chi|}{|\phi|} = \left[\frac{1 - \sqrt{1 - (v/c)^2}}{1 + \sqrt{1 - (v/c)^2}}\right] \qquad (4.6)$$

When $v \to 0, |\chi| \to 0$, with the increase of v, $|\chi|/|\phi|$ increases monotonously. The particle becomes more and more "impure" until $|\chi|/|\phi| \to 1$ as a limit of particle being still a particle. As shown by Equation (4.6), the reason why its velocity has a limiting value c (the speed of light) is because ϕ and χ have opposite evolution tendencies in space-time as shown by Equations (4.1)- (4.3) essentially, χ strives to hold ϕ back from going forward until a balance nearly reached when $|\chi| \to |\phi|$ and $v \to c$.

(c) The "length contraction" (FitzGerald-Lorentz contraction) and "time dilation"

As usual, we will show "length contraction" via a wave-packet of KG particle moving at a high-speed (v) but further ascribe it to the enhancement of χ field hidden inside the particle.

First, consider a wave-packet of KG particle at rest [25, 35]

$\psi(x,t)$

$$= \left(4\sigma\pi^3\right)^{-1/4} \int_{-\infty}^{\infty} \exp\left(-\frac{k^2}{2\sigma}\right) \exp\left[i(kx-\omega t)\right] dk \tag{4.7}$$

Assuming $\sqrt{\sigma} \ll \frac{mc}{\hbar}$, we have approximately that

$\psi(x,t)$

$$\cong \frac{(\sigma/\pi)^{1/4}}{(1+i\sigma\hbar t/m)^{1/2}} \exp\left[-\frac{\sigma x^2}{2(1+i\sigma\hbar t/m)} - \frac{imc^2 t}{\hbar}\right] \tag{4.8}$$

If $\sigma\hbar t/m \ll 1$, the diffusion of wave-packet at low speed $(v \ll c)$ can be ignored. Then we perform a "boost transformation"

$$\left(x \to (x-vt)/\sqrt{1-\beta^2}, t \to (t-vx/c^2)/\sqrt{1-\beta^2}, \beta = v/c\right)$$

to push the wave-packet to high velocity$(v \to c)$, yielding

$$\psi_{boost}(x,t) = \left(\frac{\sigma}{\pi}\right)^{1/4} e^{i\alpha\xi}$$

$$\cdot \exp\left(-i\frac{mc^2}{\hbar}\sqrt{1-\beta^2} t\right) \exp\left(-\frac{\xi^2}{2\varpi^2}\right) \tag{4.9}$$

where $\xi = \frac{mc}{\hbar}(x-vt), \alpha = \beta/\sqrt{1-\beta^2}$ and

$$\varpi = \frac{mc\sqrt{1-\beta^2}}{\hbar\sqrt{\sigma}} \propto \sqrt{1-\beta^2} \tag{4.10}$$

Here ϖ is the width of wave-packet measured from its center $\xi=0$. Equations (4.7)-(4.10) show the "length contraction".

Second, we calculate from Equations (4.9) and (3.33)

the values of $|\phi|^2, |\chi|^2$ and the probability density

$\rho = |\phi|^2 - |\chi|^2$ respectively.[5] Their peak values all increase with the increase

of v (boost effect). However, the "intensity" of $|\phi|^2$ or $|\chi|^2$ increases even faster than that of ρ while keeping the constraint $|\phi| > |\chi|$ in the boosting process.

We also calculate the square of "impurity ratio" R for this moving wave-packet:

$$\left[R_{free}^{KG}\right]^2 = \frac{\int_{-\infty}^{\infty} |\chi|^2 \, dx}{\int_{-\infty}^{\infty} |\phi|^2 \, dx} = \left[\frac{1-\sqrt{1-(v/c)^2}}{1+\sqrt{1-(v/c)^2}}\right]^2 \tag{4.11}$$

which is the counterpart of Equation (4.6) for a plane WF of KG particle.

With these calculations, we might intuitively understand the length contraction as an effect of coupling (i.e. entanglement) between ϕ and χ fields due to their opposite evolution tendencies in space as discussed in previous point (b).

Let's turn to the "time dilation" shown by the variation of the mean life

$$\tau = \frac{\tau_0}{\sqrt{1-\beta^2}}$$
(4.12)

of a particle, say, a pion (π^- or π^+) with its velocity v. To understand it, let's return back to Equations (4.1)- (4.3) at $x=0$ and view the WF $\psi(\psi_c)$ on its complex plane with $\mathrm{Re}\psi$ and $\mathrm{Im}\psi$ ($\mathrm{Re}\psi_c$ and $\mathrm{Im}\psi_c$) as abscissa and ordinate. We may see that the time reading of the "inner clock" for a particle (or an antiparticle) is "clockwise" (or "counter clockwise"). Thus with the increase of particle velocity, though the time reading remains clockwise (due to the dominance of ϕ field), it runs slower and slower because of the enhancement of hidden χ field.

(d) WF's group velocity u_g versus phase velocity u_p.

In RQM, a particle's velocity v should be identified with its group velocity u_g. Actually, we have

$$u_g = \frac{d\omega}{dk} = \frac{dE}{dp} = \frac{d}{dp}\sqrt{p^2c^2 + m^2c^4}$$

$$= \frac{pc^2}{E} = v \xrightarrow[E \to \infty]{} c$$
(4.13)

However, the fact that there is an upper bound for particle's velocity doesn't mean that no speed can exceed that of light, c. Indeed, there is another velocity u_p, the phase velocity in the WF

$$u_p = \frac{\omega}{k} = \frac{E}{p}$$
(4.14)

And the relation $E^2 = p^2c^2 + m^2c^4$ implies that

$$u_g u_p = c^2, u_p = \frac{c^2}{u_g} = \frac{c^2}{v}$$
(4.15)

In our opinion, the role of $u_p > c$ here is crucial to maintain the quantum coherence of WF in the space-time globally, we will further discuss this problem elsewhere. In 1923, de Broglie discovered Equation (4.15) in his relativistic theory. However, in the Schrödinger equation of NRQM, the phase velocity remains undefined. See Ref [34].

DIRAC EQUATION AS COUPLED EQUATIONS OF TWO-COMPONENT SPINORS

Let us turn to the Dirac equation describing an electron

$$\left(i\hbar \frac{\partial}{\partial t} - V \right)\psi = H\psi = \left(-i\hbar c\alpha \cdot \nabla + \beta mc^2 \right)\psi \tag{5.1}$$

with α and β being 4×4 matrices, the WF ψ is a four-component spinor

$$\psi = \begin{pmatrix} \phi \\ \chi \end{pmatrix} \tag{5.2}$$

Usually, the two-component spinors ϕ and χ are called "positive" and "negative" energy components. In our point of view, they are the hiding "particle" and "antiparticle" fields in a particle (electron) respectively ([25], see below). Substitution of Equation (2) into Equation (1) leads to

$$\begin{cases} \left(i\hbar \frac{\partial}{\partial t} - V \right)\phi = -i\hbar c\sigma \cdot \nabla\chi + mc^2\phi \\ \left(i\hbar \frac{\partial}{\partial t} - V \right)\chi = -i\hbar c\sigma \cdot \nabla\phi - mc^2\chi \end{cases} \tag{5.3}$$

(σ are Pauli matrices). Equation (3) is invariant under the combined space-time inversion with

$$\begin{cases} x \to -x, t \to -t, \\ \phi(x,t) \to C\phi(x,t) = \chi_c(x,t), \\ \chi(x,t) \to C\chi(x,t) = \phi_c(x,t) \\ V(x,t) \to -V(x,t) = V_c(x,t) \end{cases} \tag{5.4}$$

showing that in its form of two-component spinors, Dirac equation is in conformity with the underlying symmetry Equation (3.29). Note that under the space-time inversion, the σ remain unchanged (However, see Equations (9)- (11) below). Alternatively, Equation (3) also remains invariant under a mass inversion as

$$m \to -m, \phi(x,t) \to \chi_c(x,t), \\ \chi(x,t) \to \phi_c(x,t), V \to V \tag{5.5}$$

In either case of Equation (5.4) or (5.5), we have[6]

$$\psi(x,t) = \begin{pmatrix} \phi(x,t) \\ \chi(x,t) \end{pmatrix} \to \begin{pmatrix} \chi_c(x,t) \\ \phi_c(x,t) \end{pmatrix} = \psi_c'(x,t)$$

(5.6)

For concreteness, we consider a free electron moving along the z axis with momentum $p = p_z > 0$ and having a helicity $h = \sigma \cdot p/|p| = 1$, its WF reads:

$$\psi(z,t) \sim \begin{pmatrix} \phi \\ \chi \end{pmatrix} \sim \begin{pmatrix} 1 \\ 0 \\ \dfrac{p}{E+m} \\ 0 \end{pmatrix} \exp\left[i(pz - Et)\right]$$

(5.7)

with $|\phi| > |\chi|$. Under a space-time inversion

$$(z \to -z, t \to -t, p \to p_c, E \to E_c)$$ or mass inversion

$$(m \to -m, p \to -p_c, E \to -E_c),$$ it is transformed into a WF for positron (moving along z axis)

$$\psi_c'(z,t) \sim \begin{pmatrix} \chi_c \\ \phi_c \end{pmatrix} \sim \begin{pmatrix} 1 \\ 0 \\ \dfrac{p_c}{E_c + m} \\ 0 \end{pmatrix} \exp\left[-i(p_c z - E_c t)\right]$$

(5.8)

with $|\chi_c| > |\phi_c|, (p_c > 0, E_c > 0)$. However, the positron's helicity becomes $h_c = \dfrac{\sigma_c \cdot p_c}{|p_c|} = -1$. This is because the total angular momentum operator for an electron reads

$$\hat{J} = \hat{L} + \frac{\hbar}{2}\sigma$$

(5.9)

Under a space-time inversion, the orbital angular momentum operator is transformed as

$$\hat{L} = r \times \hat{p} = r \times (-i\hbar\nabla)$$
$$\to -r \times (i\hbar\nabla) = -r \times \hat{p}_c = -\hat{L}_c$$

(5.10)

To get $\hat{J} \to -\hat{J}_c$ with $\hat{J}_c = \hat{L}_c + \dfrac{\hbar}{2}\hat{\sigma}_c$, we should have

$$\hat{\sigma}_c = -\hat{\sigma}$$

(5.11)

Hence the values of matrix element for positron's spin operator σ_c is just the negative to that for σ in the same matrix representation.

Notice that Equation (7) describes an electron with positive helicity, i.e., $\Sigma \cdot \hat{p}\psi = p_z\psi = p\psi_7$. Under a space-time inversion, it is transformed into

$$\left(-\Sigma_c\right)\cdot\hat{\boldsymbol{p}}_c\psi'_c = \Sigma_z\left(i\hbar\frac{\partial}{\partial z}\right)\psi'_c = p_c\psi'_c$$ in Equation (8), i.e.,

$\Sigma_c\cdot\hat{\boldsymbol{p}}_c\psi'_c = -p_c\psi'_c$, meaning that Equation (8) describes a positron with negative helicity.

In its form of four-component spinor, Dirac equation, Equation (5.1) with $V=0$, is usually written in a covariant form as (Pauli metric is used:

$$x_4 = ict, \gamma_k = -i\beta\alpha_k, \gamma_4 = \beta, \gamma_5 = \gamma_1\gamma_2\gamma_3\gamma_4 = -\begin{pmatrix} 0 & I \\ I & 0 \end{pmatrix}$$, see Ref. [25]):

$$\left(\gamma_\mu\partial_\mu + m\right)\psi = 0 \tag{5.12}$$

Under a space-time (or mass) inversion, it turns into an equation for antiparticle:

$$\left(-\gamma_\mu\partial_\mu + m\right)\psi'_c = 0 \tag{5.13}$$

with an example of ψ'_c shown in Equation (8). Let us perform a representation transformation:

$$\psi'_c \to \psi_c = \left(-\gamma_5\right)\psi'_c = \begin{pmatrix} \phi_c \\ \chi_c \end{pmatrix} \tag{5.14}$$

and arrive at

$$\left(\gamma_\mu\partial_\mu + m\right)\psi_c = 0 \tag{5.15}$$

due to $\{\gamma_5, \gamma_\mu\}=0$. Since ψ_c and ψ'_c are essentially the same in physics, (this is obviously seen from its resolved form, Equation (5.3)), it is merely a trivial thing to change the position of χ_c in the 4-component spinor (lower in Equation (5.14) and upper in Equation (5.8)).

What important is $|\chi_c|>|\phi_c|$ for characterizing an antiparticle versus $|\phi|>|\chi|$ for a particle. Therefore, if a particle with energy E runs into a potential barrier $V=V_0>E+m$, its kinetic energy

$(T=E-V_0<0)$ becomes negative, and its WF's third component in Equation (5.7) suddenly turns into

$$\frac{p'}{E-V_0+m} = \frac{-p'}{V_0-E-m}, \left(p' = \sqrt{\left(E-V_0\right)^2 - m^2}\right)$$
,

whose absolute magnitude is larger than that of the first component. This means that it is an antiparticle's WF satisfying Equation (5.15) (with $E_c = V_0 - E(> m)$ and $|\chi_c| > |\phi_c|$) and will be crucial for the explanation of Klein paradox in Dirac equation (For detail, please see Appendix). However, we need to discuss the "probability density" ρ and "probability current density" j for a Dirac particle versus ρ_c and j_c for its antiparticle. Different from that in KG equation, now we have

$$\rho = \psi^\dagger \psi = \phi^\dagger \phi + \chi^\dagger \chi$$
$$\rightarrow \rho_c = \psi_c^\dagger \psi_c = \chi_c^\dagger \chi_c + \phi_c^\dagger \phi_c \qquad (5.16)$$

which is positive definite for either particle or antiparticle. On the other hand, we have

$$j = c\psi^\dagger \alpha \psi = c\left(\phi^\dagger \sigma \chi + \chi^\dagger \sigma \phi\right)$$
$$\rightarrow j_c = c\psi_c^\dagger \alpha \psi_c = c\left(\chi_c^\dagger \sigma \phi_c + \phi_c^\dagger \sigma \chi_c\right) \qquad (5.17)$$

(we prefer to keep σ rather than σ_c for antiparticle). For Equations (5.7), (5.8) and (5.14), we find $(c = \hbar = 1)$

$$j_z \sim \frac{2p}{E + m} > 0 \rightarrow j_z^c \sim \frac{2p_c}{E_c + m} > 0 (V = 0) \qquad (5.18)$$

which means that the probability current is always along the momentum's direction for either a particle or antiparticle.

Above discussions at RQM level may be summarized as follows: The first symptom for the appearance of an antiparticle is: If we perform an energy operator

$(E = i\hbar \partial / \partial t)$ on a WF and find a negative energy

$(E < 0)$ or a negative kinetic energy $(E - V < 0)$, we'd better doubt the WF being a description of antiparticle and use the operators for antiparticle, Equation (2.18). Then for further confirmation, two more criterions for ρ and j are needed (see Appendix).

THE STRONG REFLECTION INVARIANCE IN CPT THEOREM AND QFT

In QFT, the starting point is the field operator which is constructed for free complex boson field as [36]

$$\begin{cases} \hat{\psi}(x,t) = \sum_p \dfrac{1}{\sqrt{2V\omega_p}}\left\{\hat{a}_p \exp\left[i(p\cdot x - Et)\right] + \hat{b}_p^\dagger \exp\left[-i(p\cdot x - Et)\right]\right\} \\[2mm] \hat{\psi}^\dagger(x,t) = \sum_p \dfrac{1}{\sqrt{2V\omega_p}}\left\{\hat{a}_p^\dagger \exp\left[-i(p\cdot x - Et)\right] + \hat{b}_p \exp\left[i(p\cdot x - Et)\right]\right\} \end{cases}$$

(6.1)

Similarly, the field operator for free Dirac field reads:

$$\left.\hat{b}_p^{(h)\dagger}v^{(h)}(p)e^{-i(p\cdot x - Et)}\right]\quad \begin{cases}\hat{\psi}(x,t) = \dfrac{1}{\sqrt{V}}\sum_p\sum_{h=\pm1}\sqrt{\dfrac{m}{E}}\left[\hat{a}_p^{(h)}u^{(h)}(p)e^{i(p\cdot x - Et)} + \right.\\[2mm]\left. + \hat{b}_p^{(h)}v^{(h)\dagger}(p)e^{i(p\cdot x - Et)}\right]\quad \hat{\psi}^\dagger(x,t) = \dfrac{1}{\sqrt{V}}\sum_p\sum_{h=\pm1}\sqrt{\dfrac{m}{E}}\left[\hat{a}_p^{(h)\dagger}u^{(h)\dagger}(p)e^{i(p\cdot x - Et)}\right.\end{cases}$$

(6.2)

In Equation (6.1), the annihilation operator $\hat{a}_p$ for particle and the creation operator $\hat{b}_p^\dagger$ for antiparticle in Fock space are introduced. In Equation (6.2), instead of index s $(=\pm1/2$, the spin's projection along the fixed z axis in space), the helicity h is used. See Ref. [37].

Let us return back to the CPT theorem proved by Lüders and Pauli in 1954-1957 [10-12]. The proof of CPT theorem contains a crucial step being the construction of so-called "strong reflection", consisting in a reflection of space and time about some arbitrarily chosen origin, i.e. $r \to -r, t \to -t$.

Pauli proposed and explained the strong reflection in Ref. [12] as follows: When the space-time coordinates change their sign, every particle transforms into its antiparticle simultaneously. The physical sense of the strong reflection is the substitution of every emission (absorption) operator of a particle by the corresponding absorption (emission) operator of its antiparticle. And there is no need to reverse the sign of the electric charge when the sign of space-time coordinates is reversed.

What Pauli claimed, in our understanding, means that under the strong reflection for boson field, one has

$$\begin{cases} x \to -x, t \to -t, \\ \hat{a}_p \leftrightarrows \hat{b}_p^\dagger, \hat{a}_p^\dagger \leftrightarrows \hat{b}_p \end{cases}$$

(6.3)

The mutual transformation, Equation (6.3), in Fock space ensures the field operators, Equation (6.1), invariant under the strong reflection in the sense of (see also [25,26]):

$$\hat{\psi}(x,t) \to \left(\widehat{PT}\right)\hat{\psi}(x,t)\left(\widehat{PT}\right)^{-1}$$
$$= \hat{\psi}(-x,-t) = \hat{\psi}(x,t)$$
$$\hat{\psi}^{\dagger}(x,t) \to \left(\widehat{PT}\right)\hat{\psi}^{\dagger}(x,t)\left(\widehat{PT}\right)^{-1}$$
$$= \hat{\psi}^{\dagger}(-x,-t) = \hat{\psi}^{\dagger}(x,t) \tag{6.4}$$

Here let us introduce the notation $\widehat{PT}$ to represent the strong reflection so that the presentation could be easier and clearer as shown above. Similarly, for Dirac field, under the strong reflection one has

$$\begin{cases} x \to -x, t \to -t, \\ \hat{a}_p^{(h)} \leftrightarrows \hat{b}_p^{(-h)\dagger}, \hat{a}_p^{(h)\dagger} \leftrightarrows \hat{b}_p^{(-h)} \end{cases} \tag{6.5}$$

Here it is important to notice that the helicity, h, will be reversed before and after the strong reflection for a particle and its antiparticle respectively as discussed in Section V. Because Equation (6.2) is written in 4 component spinor covariant form, the invariance of Dirac field operator under the strong reflection should be expressed rigorously as

$$\hat{\psi}(x,t) \to \left(\widehat{PT}\right)\hat{\psi}(x,t)\left(\widehat{PT}\right)^{-1}$$
$$= -\gamma_5\hat{\psi}(-x,-t) = \hat{\psi}(x,t)$$
$$\hat{\psi}^{\dagger}(x,t) \to \left(\widehat{PT}\right)\hat{\psi}^{\dagger}(x,t)\left(\widehat{PT}\right)^{-1}$$
$$= \hat{\psi}^{\dagger}(-x,-t)(-\gamma_5) = \hat{\psi}^{\dagger}(x,t) \tag{6.6}$$

$$\hat{\psi}(-x,-t) = -\gamma_5\hat{\psi}(x,t),$$
$$\hat{\psi}^{\dagger}(-x,-t) = \hat{\psi}^{\dagger}(x,t)(-\gamma_5) \tag{6.7}$$

which are useful in proving the "spin-statistics connection" by strong reflection invariance.

QFT is a successful theory just because it is established on sound basis with the field operator being one of its cornerstones. Historically, through various trials and checks, Equations (6.1)-(6.2) were eventually found (see Section 3.5 of Ref. [36]). Why they are correct and why one would fail otherwise? In our understanding, it is just because they are invariant under the strong reflection as shown by Equations (6.4) and (6.6).

However, as emphasized by Pauli [12] and further stressed by Lüders [11], at least two more rules should be added in doing calculations:

(a)   The order of an operator product in Fock space has to be reversed under the strong reflection, e.g.,

$\left(\widehat{PT}\right)\hat{A}\hat{B}\left(\widehat{PT}\right)^{-1} = \left(\widehat{PT}\right)\hat{B}\left(\widehat{PT}\right)^{-1}\left(\widehat{PT}\right)\hat{A}\left(\widehat{PT}\right)^{-1}$. So is the order of a process occurred in a many-particle system.

(b)   Another rule is: One should always take the normal ordering when dealing with quadratic forms like $\hat{\bar{\psi}}(x)\hat{\psi}(x)$ etc.

Then Pauli and Lüders were able to prove that the Hamiltonian density $\mathcal{H}(x,t)$ for a broad kind of model in relativistic QFT is invariant under an operation of "strong reflection", i.e.,

$$\hat{\mathcal{H}}(x,t) \to \widehat{PT}\hat{\mathcal{H}}(x,t)\left(\widehat{PT}\right)^{-1}$$
$$= \hat{\mathcal{H}}(-x,-t) = \hat{\mathcal{H}}(x,t) \qquad (6.8)$$

The Hamiltonian density is also invariant under a Hermitian conjugation (H.C.) as:

$$\hat{\mathcal{H}}(x,t) \to \hat{\mathcal{H}}^{\dagger}(x,t) = \hat{\mathcal{H}}(x,t) \qquad (6.9)$$

Furthermore, they proved the CPT theorem via the identification of the product of T, C, and P in QFT with the combined operation of the strong reflection and a Hermitian conjugation.

The validity of CPT invariance, i.e. Equations (6.8) and (6.9) has been verified experimentally since the discovery of parity violation ([3-8] etc.) and the establishment (and development) of standard model ([38] etc.) in particle physics till this day. See the excellent book, Ref. [19] and the Review of Particle Physics, Ref. [9].

After restudying the historical contribution of PauliLüders strong reflection invariance, we feel good in understanding that what we claim in RQM (Sections III-V) is essentially the same as or very close to their idea.

In fact, this paper is the direct continuation of our first one in 1974 [22], which was inspired jointly by the discoveries of violations in P, C, CP, T symmetries individually (but CPT invariance holds), also by Lee-Wu's proposal in 1965 that the relationship between a particle $|a\rangle$ and its antiparticle $|\bar{a}\rangle$ should be [13]:

$$|\bar{a}\rangle = CPT|a\rangle \qquad (6.10)$$

and especially by Pauli's invention of the strong reflection in 1955 [12].

Below, we would like to show that WFs for a particle and its antiparticle given in Equations (5.7) and (5.8) are precisely that derived from QFT as expected.

Using Equation (6.2) for Dirac field, we find the WF of an electron being

$$\psi_{e^-}(x,t) = \langle 0|\hat{\psi}(x,t)|e^-, p_1, h_1\rangle$$
$$= \langle 0|\hat{\psi}(x,t)|\hat{a}_{p_1}^{(h_1)\dagger}|0\rangle$$
$$= \frac{1}{\sqrt{V}}\sqrt{\frac{m}{E_1}}u^{(h_1)}(p_1)e^{-i(p_1 \cdot x - E_1 t)}$$

(6.11)

but the hermitian conjugate of a positron's WF is given by

$$\psi_{e^+}^{\dagger}(x,t) = \langle 0|\hat{\psi}^{\dagger}(x,t)|e^+, p_c, h_c\rangle$$
$$= \langle 0|\hat{\psi}^{\dagger}(x,t)|\hat{b}_{p_c}^{(h_c)\dagger}|0\rangle$$
$$= \frac{1}{\sqrt{V}}\sqrt{\frac{m}{E_c}}v^{(h_c)\dagger}(p_c)e^{i(p_c \cdot x - E_c t)}$$

(6.12)

which leads to positron's WF being

$$\psi_{e^+}(x,t) = \frac{1}{\sqrt{V}}\sqrt{\frac{m}{E_c}}v^{(h_c)}(p_c)e^{-i(p_c \cdot x - E_c t)}$$

(6.13)

Similarly, Equations (2.20) and (2.21) can be derived from Equation (6.1) as expected.

AN OVERSIGHT IN QFT (HELICITY STATES OR SPIN STATES?)—WHY A PARITY-VIOLATION PHENOMENON WAS OVERLOOKED SINCE 1956-1957?

Through analysis in RQM till QFT, we stress the necessity of using helicity (h) to describe a fermion or antifermion. Here is an interesting example. Since 2002, Shi and Ni [39-43] predicted a parity-violation phenomenon as follows:

An unstable (decaying) fermion (e.g., neutron or muon) has different mean lifetimes for being right-handed (RH) or left-handed (LH) polarized during its flight with the same speed $v(\beta = v/c)$

$$\tau_R = \frac{\tau}{1-\beta}, \tau_L = \frac{\tau}{1+\beta}$$

(7.1)

where $\tau = \tau_0/\sqrt{1-\beta^2}$, τ_0 the mean lifetime when it is at rest. Similarly, for its antifermion, their lifetimes will be

$$\bar{\tau}_R = \frac{\tau}{1+\beta}, \bar{\tau}_L = \frac{\tau}{1-\beta}$$

(7.2)

Hence, the lifetime asymmetry can be defined as

$$A = \frac{\tau_R - \tau_L}{\tau_R + \tau_L} = \beta \tag{7.3}$$

This is not a small effect. For instance, in Fermilab, physicists consider to build a muon collider [44]. The collision of μ^- and μ^+ beams must happen before the muons decay. It was estimated that if a muon rings along at 1.5 TeV, the time dilation of SR stretches its lifetime to 30 milliseconds—up from 2 microseconds when it's still. That's time enough for 500 circuits in the final ring. However, as discussed in Ref. [43], if the prediction of life asymmetry Equation (7.1) is correct, the lifetime of RH μ^- will be stretched to 146 days while that of LH μ^- only 15 milliseconds. The lifetime asymmetry of μ^+ will be just the opposite as shown by Equation (7.2). Therefore, it seems necessary to take Equations (7.1)- (7.2) into account in the design of a muon collider.

The problem is: How can such a parity-violation phenomenon be overlooked since 1956-1957? One theoretical reason is: in the past, for describing a fermion in flight $(v \neq 0)$, instead of helicity states, the "spin-states" assigned by S (spin's projection along the fixed Z axis in space) were often incorrectly used (see [40-42]). So previous calculations on the lifetime always led to a prediction that $\tau = \tau_0 / \sqrt{1 - \beta^2}$ without parity-violation in contrast to Equations (7.1)-(7.3).[8] The interesting thing is: While Equations (7.1) and (7.2) display the violation of P or C symmetry to its maximum, their "cross-symmetry", $\tau_R = \bar{\tau}_L$ and $\tau_L = \bar{\tau}_R$, reflects the symmetry of $\mathcal{PT} = C$ shown by Equation (6.5) exactly.

DIRAC PARTICLES CONSERVE THE PARITY WHEREAS NEUTRINOS ARE LIKELY THE TACHYONS

Why Dirac Equation Respects the Parity Symmetry?

In the standard representation of Dirac equation for free particle $(\hbar = c = 1)$

$$i \frac{\partial}{\partial t} \psi^{(D)} = -i \alpha \cdot \nabla \psi^{(D)} + \beta m \psi^{(D)} \tag{8.1}$$

Let us choose

$$\alpha = -\begin{pmatrix} 0 & \sigma \\ \sigma & 0 \end{pmatrix}, \beta = \begin{pmatrix} I & 0 \\ 0 & -I \end{pmatrix}, \psi^{(D)} = \begin{pmatrix} \phi^{(D)} \\ \chi^{(D)} \end{pmatrix} \quad \text{, then}$$

$$\begin{cases} i\dfrac{\partial}{\partial t}\phi^{(D)} = i\sigma \cdot \nabla \chi^{(D)} + m\phi^{(D)} \\ i\dfrac{\partial}{\partial t}\chi^{(D)} = i\sigma \cdot \nabla \phi^{(D)} - m\chi^{(D)} \end{cases}$$

(8.2)

As discussed in section V, Equations (8.1) and (8.2) are invariant under the space-time inversion:

$$\begin{cases} x \to -x, t \to -t \\ \phi^{(D)}(x,t) \to \phi^{(D)}(-x,-t) = \chi_c^{(D)}(x,t) \\ \chi^{(D)}(x,t) \to \chi^{(D)}(-x,-t) = \phi_c^{(D)}(x,t) \end{cases}$$

(8.3)

with subscript "c" meaning the antiparticle. After transforming $\psi^{(n)}$ into the "Weyl representation" (chiral representation) as

$$\psi^{(D)} \to \frac{1}{\sqrt{2}}\begin{pmatrix} I & I \\ I & -I \end{pmatrix}\begin{pmatrix} \phi^{(D)} \\ \chi^{(D)} \end{pmatrix} = \begin{pmatrix} \xi^{(D)} \\ \eta^{(D)} \end{pmatrix}$$

(8.4)

we have

$$\begin{cases} i\dfrac{\partial}{\partial t}\xi^{(D)} = i\sigma \cdot \nabla \xi^{(D)} + m\eta^{(D)} \\ i\dfrac{\partial}{\partial t}\eta^{(D)} = -i\sigma \cdot \nabla \eta^{(D)} + m\xi^{(D)} \end{cases}$$

(8.5)

If $m = 0$, Equation (8.5) reduces into two Weyl equations describing two kinds of permanently LH and RH polarized massless fermions respectively. So we may name $\xi^{(D)}$ and $\eta^{(D)}$ (which are usually called as chirality states or chiral fields in 4-component covariant form) as the "hidden LH and RH spinning fields" inside a Dirac particle, which can be either LH or RH polarized (with helicity $h = -1$ or 1) explicitly. See below.

A new symmetry is hidden in Equation (8.5), which remains invariant under the pure space inversion $(x \to -x, t \to t)$ transformation, i.e., the parity operation as

$$\begin{cases} \xi^{(D)}(x,t) \to \xi^{(D)}(-x,t) = \eta^{(D)'}(x,t) \\ \eta^{(D)}(x,t) \to \eta^{(D)}(-x,t) = \xi^{(D)'}(x,t) \end{cases}$$

(8.6)

Here we add " ' " in the superscript of RHS to stress that the WF after the space inversion may be different from that at the LHS (before the space inversion). We knew that the WF in Dirac representation after a space inversion reads

$$\hat{P}\psi^{(D)}(x,t) - \gamma_4 \psi^{(D)}(-x,t)$$

(8.7)

Using Equation (8.6), the RHS of Equation (8.7) turns out to be

$$\frac{1}{\sqrt{2}}\gamma_4\begin{pmatrix}\xi^{(D)}(-x,t)+\eta^{(D)}(-x,t)\\\xi^{(D)}(-x,t)-\eta^{(D)}(-x,t)\end{pmatrix}$$

$$=\frac{1}{\sqrt{2}}\begin{pmatrix}\xi^{(D)'}(-x,t)+\eta^{(D)'}(-x,t)\\\xi^{(D)'}(-x,t)-\eta^{(D)'}(-x,t)\end{pmatrix}=\psi^{(D)'}(x,t)$$

$$(8.8)$$

Hence, we understand the reason why a Dirac particle respects the parity symmetry as shown by Equation (8.7) is because it enjoys the symmetry Equation (8.6) hiding in the 2-component spinor form (in Weyl representation).

For concreteness, let's write down the solution of Equation (8.1)

$$\psi^{(D)}(x,t)=\begin{pmatrix}\phi^{(D)}\\\chi^{(D)}\end{pmatrix}\sim\begin{pmatrix}\phi_0\\\dfrac{-\sigma\cdot p}{E+m}\phi_0\end{pmatrix},$$

$$\left(E=\sqrt{p^2+m^2}>0\right)$$

$$(8.9)$$

Furthermore, we choose a simplest "spin state" with

$$\hat{p}\psi^{(D)}=p_z\psi^{(D)}\text{ and }\hat{\sigma}_z\psi^{(D)}=\psi^{(D)}:$$

$$\psi^{(D)}_{s_z=1/2}(z,t)=\begin{pmatrix}\phi^{(D)}\\\chi^{(D)}\end{pmatrix}\sim\begin{pmatrix}1\\0\\\dfrac{-p_z}{E+m}\\0\end{pmatrix}e^{i(p_z z-Et)}(E>0)$$

$$(8.10)$$

While Equation (8.10) is an eigenfunction of $\hat{\sigma}_z$ with eigenvalue $s_z=1/2$, its helicity h remains unfixed, depending on the value of p_z being positive or negative. Only after $p_z=p>0$ is fixed, can we have a "helicity state" describing a RH particle with $h=1$:

$$\psi^{(D)}_{RH}(z,t)=\begin{pmatrix}\phi^{(D)}\\\chi^{(D)}\end{pmatrix}\sim\begin{pmatrix}1\\0\\\dfrac{-p}{E+m}\\0\end{pmatrix}e^{i(pz-Et)}$$

$$(p>0,E>0)$$

$$(8.11)$$

Looking at Equation (8.11) in the Weyl representation, we see that

$$\xi^{(D)} = \frac{1}{\sqrt{2}}\left(\phi^{(D)} + \chi^{(D)}\right) \sim \frac{1}{\sqrt{2}}\begin{pmatrix} 1 - \dfrac{p}{E+m} \\ 0 \end{pmatrix},$$

$$\eta^{(D)} = \frac{1}{\sqrt{2}}\left(\phi^{(D)} - \chi^{(D)}\right) \sim \frac{1}{\sqrt{2}}\begin{pmatrix} 1 + \dfrac{p}{E+m} \\ 0 \end{pmatrix}$$

(8.12)

$\left|\xi^{(D)}\right| < \left|\eta^{(D)}\right|$. So Equation (8.11) describes a RH particle just because the $\eta^{(D)}$ field dominates the $\xi^{(D)}$ field. Now we perform a space inversion on Equation (8.11), according to the rule Equation (8.7), yielding

$$\hat{P}\psi_{RH}^{(D)}(z,t) \sim \begin{pmatrix} 1 \\ 0 \\ \dfrac{p}{E+m} \\ 0 \end{pmatrix} e^{i(-pz-Et)} = \begin{pmatrix} \phi^{(D)'} \\ \chi^{(D)'} \end{pmatrix} = \psi^{(D)'}(z,t)$$

$$\xi^{(D)'} = \frac{1}{\sqrt{2}}\left(\phi^{(D)'} + \chi^{(D)'}\right) \sim \frac{1}{\sqrt{2}}\begin{pmatrix} 1 + \dfrac{p}{E+m} \\ 0 \end{pmatrix},$$

$$\eta^{(D)'} = \frac{1}{\sqrt{2}}\left(\phi^{(D)'} - \chi^{(D)'}\right) \sim \frac{1}{\sqrt{2}}\begin{pmatrix} 1 - \dfrac{p}{E+m} \\ 0 \end{pmatrix}$$

$$\xi^{(D)'} = \frac{E+p}{m}\eta^{(D)'}, \left|\xi^{(D)'}\right| > \left|\eta^{(D)'}\right|$$

(8.13)

Hence we see that the reason why $\psi^{(D)'}(z,t)$ becomes a LH WF, i.e.,

$$\hat{P}\psi_{RH}^{(D)}(z,t) = \psi_{LH}^{(D)}(z,t)$$

(8.14)

is just because of the dominance of $\xi^{(D)'}$ field over

$\eta^{(D)'}$ field after the P-operation. Before and after the operation, $p \to -p$, the dominant (subordinate) field is transformed into dominant (subordinate) field:

$\eta^{(D)} \to \xi^{(D)'}, \left(\xi^{(D)} \to \eta^{(D)'}\right)$, as shown by Equation (8.6).

In summary, Dirac equation is invariant under a space inversion whereas its concrete solution of WF may be not. The latter may change from that for a RH particle to a LH one or vice versa, but with the same mass m, showing the law of parity conservation exactly.

Tachyon Equation as a Counterpart of the Dirac Equation

Now a question arises: Can we find an equation which violates the symmetry of pure space inversion?

The answer is "yes". Let's introduce a new equation in Weyl representation from Equation (8.5) by erasing the superscript (D), replacing the mass term by $m \rightarrow m_s$ and changing its sign from "+" to "−" in the first equation of Equation (8.5) only [46]

$$\begin{cases} i\dfrac{\partial}{\partial t}\xi = i\sigma \cdot \nabla \xi - m_s\eta \\ i\dfrac{\partial}{\partial t}\eta = -i\sigma \cdot \nabla \eta + m_s\xi \end{cases}$$

$$(8.15)$$

where m_s (real and positive) refers to the mass of a hypothetical particle. We will see immediately that it is a "superluminal particle" or "tachyon".

Indeed, substituting a plane-wave solution

$$\xi \sim \eta \sim \exp\left[i\left(p_z z - Et\right)\right]\begin{pmatrix} 0 \\ 1 \end{pmatrix}$$

$$(8.16)$$

with the particle's helicity $h = -1$ into Equation (8.15), we find that $\left(p_z = p > 0, E > 0\right)$

$$E^2 = p^2 - m_s^2$$

$$(8.17)$$

$$\xi = \frac{1}{m_s}(p+E)\eta, |\xi| > |\eta|$$

$$(8.18)$$

Since $E = \hbar\omega$ and $p = \hbar k$, from Equation (8.17), the dispersion-relation of wave reads

$$\omega^2 = k^2 - m_s^2$$

$$(8.19)$$

As in Section IV, we define the wave's phase velocity u_p as

$$u_p = \frac{\omega}{k}$$

$$(8.20)$$

while its group velocity u_g

$$u_g = \frac{d\omega}{dk} = v$$

$$(8.21)$$

being identical with the particle's velocity v. Equation (8.19) yields a relation between them coinciding with Equation (4.15) exactly:

$$u_p u_R = c^2 \qquad (8.22)$$

However, the relations among E, p and v are dramatically different

$$E = \frac{m_s c^2}{\sqrt{\frac{v^2}{c^2} - 1}}, p = \frac{m_s v}{\sqrt{\frac{v^2}{c^2} - 1}} \qquad (8.23)$$

which dictate $v > c$ such that E, p are real and $E > 0$.

Like Equation (8.4), we define:

$$\phi = \frac{1}{\sqrt{2}}(\xi + \eta), \chi = \frac{1}{\sqrt{2}}(\xi - \eta) \qquad (8.24)$$

and find from Equation (8.15) that (in Dirac representation)

$$i\frac{\partial}{\partial t}\psi = -i\boldsymbol{\alpha}\cdot\nabla\psi + \beta_s m_s \psi \qquad (8.25)$$

$$\begin{cases} i\frac{\partial}{\partial t}\phi = i\boldsymbol{\sigma}\cdot\nabla\chi + m_s\chi \\ i\frac{\partial}{\partial t}\chi = i\boldsymbol{\sigma}\cdot\nabla\phi - m_s\phi \end{cases} \qquad (8.26)$$

$$\left(\psi = \begin{pmatrix} \phi \\ \chi \end{pmatrix}, \beta_s = \begin{pmatrix} 0 & I \\ -I & 0 \end{pmatrix}\right).$$ Despite the difference between Equation (8.26) and Dirac equation, Equation (8.2), both of them respect the combined space-time inversion $(\mathcal{PT})$ symmetry like Equation (8.3)

$$\begin{cases} \phi(x,t) \to \phi(-x,-t) = \chi_c(x,t) \\ \chi(x,t) \to \chi(-x,-t) = \phi_c(x,t) \end{cases} \qquad (8.27)$$

with

$$\begin{cases} i\frac{\partial}{\partial t}\chi_c = i\boldsymbol{\sigma}\cdot\nabla\phi_c - m_s\phi_c \\ i\frac{\partial}{\partial t}\phi_c = i\boldsymbol{\sigma}\cdot\nabla\chi_c + m_s\chi_c \end{cases} \qquad (8.28)$$

Similarly, we define the WF in Weyl representation after $\mathcal{PT}$ inversion as:

$$\begin{cases} \xi(x,t) \to PT\xi(x,t) = \xi(-x,-t) = \eta_c(x,t) \\ \eta(x,t) \to PT\eta(x,t) = \eta(-x,-t) = \xi_c(x,t) \end{cases} \qquad (8.29)$$

Based on Equations (8.27)-(8.29), we find

$$\begin{cases} \eta_c(x,t) = \dfrac{1}{\sqrt{2}}\big[\chi_c(x,t) + \phi_c(x,t)\big] \\ \xi_c(x,t) = \dfrac{1}{\sqrt{2}}\big[\chi_c(x,t) - \phi_c(x,t)\big] \end{cases} \tag{8.30}$$

$$\begin{cases} i\dfrac{\partial}{\partial t}\eta_c = i\boldsymbol{\sigma}\cdot\nabla\eta_c + m_s\xi_c \\ i\dfrac{\partial}{\partial t}\xi_c = -i\boldsymbol{\sigma}\cdot\nabla\xi_c - m_s\eta_c \end{cases} \tag{8.31}$$

which can also be obtained via the $\mathcal{PT}$ operation on Equation (8.15). Equations (8.15) and (8.31) are better to be compared in the following form:

$$\begin{cases} \hat{E}\xi = -\boldsymbol{\sigma}\cdot\hat{p}\xi - m_s\eta \\ \hat{E}\eta = \boldsymbol{\sigma}\cdot\hat{p}\eta + m_s\xi \end{cases} \tag{8.32}$$

$$\begin{cases} \hat{E}_c\eta_c = \boldsymbol{\sigma}_c\cdot\hat{p}_c\eta_c - m_s\xi_c \\ \hat{E}_c\xi_c = -\boldsymbol{\sigma}_c\cdot\hat{p}_c\xi_c + m_s\eta_c \end{cases} \tag{8.33}$$

$\left(\hat{E}_c = -i\dfrac{\partial}{\partial t}, \hat{p}_c = i\nabla, \sigma_c = -\sigma\right)$. Interestingly, Equation

(8.33) can also be reached from Equation (8.32) via a "mass inversion" like that in Sections III and V:

$$\begin{cases} m_s \to -m_s, \\ \hat{E} \to -\hat{E}_c = i\dfrac{\partial}{\partial t}\,(\text{no change in } t) \\ \hat{p} \to -\hat{p}_c = -i\nabla\,(\text{no change in } x) \\ \sigma \to -\sigma_c = \sigma\,(\text{no change in } \sigma) \\ \xi(x,t) \to \eta_c(x,t), \eta(x,t) \to \xi_c(x,t) \end{cases} \tag{8.34}$$

Furthermore, the probability density and probability current density before and after the $\mathcal{PT}$ inversion can be derived as:

$$\begin{aligned} \rho &= \phi^\dagger\chi + \chi^\dagger\phi = \xi^\dagger\xi - \eta^\dagger\eta \xrightarrow{\mathcal{PT}} \\ \rho_c &= \chi_c^\dagger\phi_c + \phi_c^\dagger\chi_c = \eta_c^\dagger\eta_c - \xi_c^\dagger\xi_c \end{aligned} \tag{8.35}$$

and

$$\begin{aligned} j &= -\big(\phi^\dagger\sigma\phi + \chi^\dagger\sigma\chi\big) = -\big(\xi^\dagger\sigma\xi + \eta^\dagger\sigma\eta\big) \xrightarrow{\mathcal{PT}} \\ j_c &= -\big(\chi_c^\dagger\sigma\chi_c + \phi_c^\dagger\sigma\phi_c\big) = -\big(\eta_c^\dagger\sigma\eta_c + \xi_c^\dagger\sigma\xi_c\big) \end{aligned} \tag{8.36}$$

respectively. It is the sharp contrast between Equation (8.35) and Equation (5.16) for Dirac equation (i.e.,

$$\rho^{(D)} = \xi^{(D)\dagger}\xi^{(D)} + \eta^{(D)\dagger}\eta^{(D)}$$), that makes Equation (8.15)

so unique as shown below.

Let us look at the example of WF for tachyon, Equations (8.16)-(8.18), with $E > 0, p_z > 0$ and $h = -1$. It is allowed just because $|\xi| > |\eta|$ and so $\rho > 0$. Second choice of Equation (8.16) with

$$p_z = -p < 0 (p > 0), h = +1 \text{ but}$$

$$\xi = \frac{1}{m_s}(-p+E)\eta, |\xi| < |\eta|$$

(8.37)

should be fobidden due to its $\rho < 0$. Another two possible WFs with $\xi \sim \eta \sim \binom{1}{0}$ have $p_z = p$ and $p_z = -p$ respectively, only the last one with $p_z = -p < 0 (p > 0), h = -1$ is allowed due to its

$$|\xi| > |\eta| \text{ and } \rho > 0.$$

Let us turn to the solution of Equation (8.31) for antitachyon with $E_c > 0$ by just performing $\mathcal{PT}$ operation on Equation (8.16) yielding:

$$\eta_c \sim \xi_c \sim \exp\left[-i\left(p_z^c z - E_c t\right)\right]\binom{0}{1}$$

(8.38)

Now if $p_z^c = p_c \left(= |p_c|\right) > 0$, since

$$\sigma_z^c = -\sigma_z, \sigma_z^c \binom{0}{1} = 1$$, so helicity $h_c = 1$. Substitution of Equation (8.38) into Equation (8.33) yields:

$$\eta_c = \frac{1}{m_s}(p_c + E_c)\xi_c, |\eta_c| > |\xi_c|$$

(8.39)

which is allowed due to $\rho_c > 0$. Second choice of Equation (8.38) with

$$p_z^c = -p_c < 0 (p_c > 0), h = -1 \text{ but}$$

$$\eta_c = \frac{1}{m_s}(-p_c + E_c)\eta, |\eta_c| < |\xi_c|$$

(8.40)

should be forbidden due to its $\rho_c < 0$. In another two possible WFs with $\eta_c \sim \xi_c \sim \binom{1}{0}$, only that with

$p_z^c = -p_c < 0, h_c = +1$ is allowed due to $\rho_c > 0$.

Hence we see that: The tachyon can only exist in a left-handed (LH) polarized state (with helicity $h = -1$) whereas antitachyon only in a right-handed (RH) polarized state (with $h_c = 1$). We tentatively link this strange feature with that found in neutrinos—only ν_L and $\bar{\nu}_R$ exists in nature whereas ν_R and $\bar{\nu}_L$ are strictly forbidden.

Furthermore, at first sight, although Equation (8.15) certainly has no symmetry under the space inversion $(x \to -x, t \to t)$, it seems to enjoy a pure "timeinversion" $(x \to x, t \to -t)$ symmetry like

$$\begin{cases} \xi(x,t) \to \xi(x,-t) = \eta_c'(x,t) \\ \eta(x,t) \to \eta(x,-t) = \xi_c'(x,t) \end{cases}$$
(8.41)

$$\begin{cases} i\dfrac{\partial}{\partial t}\eta_c' = -i\sigma \cdot \nabla \eta_c' + m_s \xi_c' \\ i\dfrac{\partial}{\partial t}\xi_c' = i\sigma \cdot \nabla \xi_c' - m_s \eta_c' \end{cases}$$
(8.42)

We add "$'$" in the superscript of η_c' to stress that $\eta_c'(x,t)$ (being a time reversed WF), though looks like some antitachyon's WF, is obviously different from $\eta_c(x,t)$ gained through the PT inversion, Equation (8.31). Actually, based on Equations (8.29)-(8.31) and (8.41)-(8.42), we have:

$$\begin{cases} \eta_c'(x,t) = \eta_c(-x,t), \xi_c'(x,t) = \xi_c(-x,t) \\ \eta_c'(-x,t) = \eta_c(x,t), \xi_c'(-x,t) = \xi_c(x,t) \end{cases}$$
(8.43)

Interestingly, we cannot find from Equation (8.42) the "physical solution" of $\eta_c'(x,t)$ with $|\eta_c'| > |\xi_c'|$ (so $\rho_c > 0$) and $h_c = 1$ (for $\bar{\nu}_R$) simultaneously. Only $\eta_c'(-x,t)$ makes physical sense, but it is just $\eta_c(x,t)$ like that discussed in Equation (8.39). Notice that the sign change $x \to -x$ in the phase of WF makes a change in the direction of momentum $p_c \to -p_c$. But a WF is always composed of two fields in confrontation, like η_c versus ξ_c here. And the explicit helicity h_c is determined by which one of these two hidden fields being in charge. So the change of $x \to -x$ in these four equalities of Equation (8.43) does reverse the status of η_c versus ξ_c (or η_c' vs ξ_c'), rendering helicity reversed explicitly. The subtlety of tachyon equation, unlike Dirac equation, lies in the fact that only V_L and $\overline{V}_R$ exist whereas V_R and $\overline{V}_L$ are strictly forbidden, i.e., the parity symmetry is violated to maximum. Hence, in strict sense, there is also no physically meaningful WF after the operation of pure "time inversion" on Equation (8.15). We will insist on Equation (8.31) rather than Equation (8.42)—there is only one correct way leading from tachyon to antitachyon via the $\mathcal{PT}$ inversion essentially.

In 2000, Equation (8.25) was first proposed by Tsao Chang and then collaborated with Ni in Ref. [46] (see also [47-52] and the Appendix 9B in Ref. [25]). At first sight, the difference between Equations (8.25) and (8.1) amounts to substituting the mass term βm by $\beta_s m_s$

with $\beta_s = \begin{pmatrix} 0 & I \\ -I & 0 \end{pmatrix}$ being an antihermitian matrix.

Usually, for an equation with nonhermitian Hamiltonian, there is no guarantee for the completeness of its mathematical solutions. In other words, the unitarity of its physical states is at risk. Sometimes, however, a nonhermitian Hamiltonian can be accepted in physics. For example, in the optical model for nuclear physics, an imaginary part of potential, $V = V_0 + iV_1$, is used to describe the absorption of incident particles successfully. The interesting thing for "tachyonic neutrino" is: Solutions of Equation (8.15) for $E > 0$ $(E_c > 0)$ are coinciding with that for $|\xi| > |\eta|$ $(|\eta_c| > |\xi_c|)$ whereas another would-be solutions with $E > 0$ but $|\xi| < |\eta|$ $(E_c > 0$ but $|\eta_c| < |\xi_c|)$ are forbidden, see Equations (8.37) and (8.40). It seems like half of would-be solutions disappear automatically. Equivalently, from physical point of view, only half of states with $\rho > 0$ or $\rho_c > 0$ are allowed in nature whereas another half with $\rho < 0$ or $\rho_c < 0$ are not. Hence one unique feature of "tachyon" equation, like Equation (8.15) or (8.26), lies in its strange realization of unitarity violation that half of would-be states (being tentatively identified with V_R and $\overline{V}_L$) are absolutely forbidden whereas another half (V_L and $\overline{V}_R$) are stabilized. The permanently longitudinal polarization property of neutrino and antineutrino like that analysed above was first predicted by Lee and Yang in 1957 [3-5] and had been verified by GGS experiment in 1958 [53]. Further discussion on this topic is currently in preparation.

ANTIGRAVITY BETWEEN MATTER AND ANTIMATTER

In hindsight, there are two Lorentz invariants in the kinematics of SR:

$$c^2 (t_1 - t_2)^2 - (x_1 - x_2)^2$$
$$= c^2 (t_1' - t_2')^2 - (x_1' - x_2')^2 = \text{const} \tag{9.1}$$

$$E^2 - c^2 p^2 = E'^2 - c^2 p'^2 = m^2 c^4 \tag{9.2}$$

It seems quite clear that Equation (9.1) is invariant under the space-time inversion $(x \to -x, t \to -t)$ and Equation (9.2) remains invariant under the mass inversion $(m \to -m)$ We believe that these two discrete symmetries are deeply rooted at the SR's dynamics via its combination with QM and developing into RQM and QFT—the particle and its antiparticle are treated on equal footing and linked by the symmetry $\mathcal{PT} = \mathcal{C}$ essentially. Hence we can perform a mass inversion on Equation (9.2) in each of two inertial frames with arbitrary relative velocity V in the sense of

$$m \to -m_c = -m, E \to -E_c, p \to -p_c \text{, yielding:}$$

$$E_c^2 - c^2 p_c^2 = E_c'^2 - c^2 p_c'^2 = m_c^2 c^4 = m^2 c^4 \tag{9.3}$$

The invariance of Equation (9.2) under mass inversion as a whole reflects the experimental fact that particle and antiparticle are equally existing in nature even at the level of classical physics.

Example: The motion equation for a charged particle (say, electron with charge $q = -e < 0$) in the external electric and magnetic fields, E and B, is given by the Lorentz formula:

$$ma = q \left(E + \frac{1}{c} v \times B \right) \tag{9.4}$$

Then the operation of either $q \to q_c = -q$ or

$m \to -m_c = -m$ on Equation (9.4) will realize the transformation from particle into its antiparticle (say, positron with charge $q_c = -q = e > 0$) with the acceleration change from $a \to a_c = -a$ as

$$ma_c = -q \left(E + \frac{1}{c} v \times B \right) \tag{9.5}$$

Based on what we learn from RQM (Sections III-V) as well as Equations (9.1)-(9.5), we may conjecture that for a classical theory being capable of treating matter and antimatter on an equal footing, it must be invariant under a mass inversion $m \to -m_c = -m$.

Notice that, however, Equation (9.4) (Equation (9.5)) is only valid for particle (antiparticle) moving at low speed, it must be modified to adapt to high-speed cases through the invariance of continuous Lorentz transformation. So we need "double checks" for testing a classical theory being really "relativistic" or not.

Let us restudy the theory of general relativity (GR). In a $(-,+,+,+)$ metric, the Einstein field equation (EFE) reads (see, e.g. , Refs. [54-56])$(c=1)$,

$$G_{\mu\nu} \equiv R_{\mu\nu} - \frac{1}{2} g_{\mu\nu} R = -8\pi G T_{\mu\nu} \tag{9.6}$$

Of course, Equation (9.6) is covariant with respect to the Lorentz transformation. But could it withstand the test of mass inversion?

On the LHS of Equation (9.6), the Einstein tensor $G_{\mu\nu}$ contains no any mass and no charge as well. But on the RHS, the energy-momentum current density tensor $T_{\mu\nu}$ is proportional to particle's mass m and so changes its sign under an operation of $m \to -m$. Hence as a whole, Equation (9.6) cannot remain invariant under the mass inversion. The reason seems rather clear that antimatter was not taking into account when GR was established in 1915. To modify EFE such that it can preserve the invariance of mass inversion, in 2004, one of us (Ni) proposed to add another term with $T_{\mu\nu}^c$ for antimatter, yielding [27]

$$R_{\mu\nu} - \frac{1}{2} g_{\mu\nu} R = -8\pi G \left(T_{\mu\nu} - T_{\mu\nu}^c \right) \tag{9.7}$$

which remains invariant under a mass inversion since:

$$T_{\mu\nu} \to -T_{\mu\nu}^c, T_{\mu\nu}^c \to -T_{\mu\nu} \, (m \to -m) \tag{9.8}$$

In a weak-field (or the post-Newtonian) approximation, this modified EFE, MEFE, Equation (9.7), will lead to modified Newton gravitational law as

$$F_{\text{grav}}(r) = \mp G \frac{mm'}{r^2} \tag{9.9}$$

where the " $-$ " sign means attractive force between m and m' being both matter or antimatter whereas the "+" sign means repulsive force between m and m' (both positive) if one of them is antimatter.

If we define the "gravitational mass" for matter and antimatter separately

$$m_{grav} = \begin{cases} m > 0, & (\text{matter}) \\ -m_c = -m < 0, & (\text{antimatter}) \end{cases} \qquad (9.10)$$

Then Equation (9.9) can be recast into one equation

$$F_{grav}(r) = -G\frac{m_{grav}m'_{grav}}{r^2} \qquad (9.11)$$

which bears a close resemblance to the Coulomb law in classical electrodynamics (CED)

$$F_{Coul}(r) = \frac{qq'}{r^2} \qquad (9.12)$$

In 1986, within the framework of classical field theory (CFT) plus some assumptions, Jagannathan and Singh derived the potential energy of two static point sources as [57]

$$U(r) = (-1)^{n+1} ee' \times (\text{a positive number}) \times \frac{e^{-\mu r}}{r} \qquad (9.13)$$

where n and μ are spin and mass of the mediating field, e is the "charge" of the source. For CED, $n = 1$ whereas $n = 2$ for gravitational field ($\mu \to 0$ in both cases). So Equation (9.13) is in conformity with Equations (9.11) and (9.12) for the case of "like sources" (with $ee' > 0$) [57], where the case for "unlike sources" $(ee' < 0)$ hadn't been discussed. Here Equation (9.11) has been generalized to the case for "unlike sources", but at a price that the "equivalence principle" in GR ceases to be valid when matter and antimatter coexist as shown by Equation (9.10).

In 2011, the antigravity between matter and antimatter was also claimed by Villata in Ref.[58], where the argument seems different from that explained above. But theory is theory, only fact will have the final say. So we are anxiously waiting for the outcome from the AEGIS experiment [59] (at CERN), which is designed to compare the Earth gravitational acceleration on hydrogen and antihydrogen atoms.

SUMMARY

1) Being the combination of SR and QM, RQM is capable of dealing with particle and antiparticle on an equal footing. As long as we admit that the antiparticle's momentum and energy operators should be $\hat{P}_c = i\hbar\nabla$

and $\hat{E}_c = -i\hbar\dfrac{\partial}{\partial t}$ versus $\hat{p} = -i\hbar\nabla$ and $\hat{E} = i\hbar\dfrac{\partial}{\partial t}$ for particle, it can be proved that the "negative-energy" WF ψ of particle corresponds to a "positive-energy" WF ψ_c of antiparticle precisely.

2) In general, an equation in RQM always has a discrete symmetry $PT - C$ which shows up as a transformation between a particle's WF ψ and its antiparticle's WF ψ_c: $\psi(x,t) \leftrightarrows \psi_c(x,t)$. For a free particle, it simply means $\psi(-x,-t) = \psi_c(x,t)$. This is in conformity with the "strong reflection" in QFT invented by Pauli and Lüders, showing that the intrinsic property of a particle cannot be detached from the space-time.

3) Following Feshbach-Villars' deep insight, we are able to divide each and every WF ψ in RQM into two parts, $\psi = \phi + \chi$. Then the above symmetry is further rigorously expressed by an invariance of motion equation in RQM through the transformations $\phi \leftrightarrows \chi_c$ and $\chi \leftrightarrows \phi_c$ under either the space-time inversion

$(x \to -x, t \to -t)$ or a mass inversion $(m \to -m)$. Since $|\phi| > |\chi|$ in ψ whereas $|\chi_c| > |\phi_c|$ in ψ_c, we may name ϕ as the (dominant) hidden particle field in ψ while χ the (subordinate) hidden antiparticle field in ψ. In this way, both the "probability density" ρ for a particle and ρ_c for an antiparticle can be proved to be positive definite. Now we may say that the RQM is ensured to be self-consistent and can be regarded as a sound basis for QFT.

4) All kinematical effects in SR can be ascribed to the enhancement of the magnitude of χ field in a particle's WF accompanying with the increase of particle's velocity.

5) As proved for Dirac particle with spin, the helicity of a particle is just opposite to that of its antiparticle under a space-time (or mass) inversion. Therefore, the experimental tests for the CPT invariance should include not only the equal mass and lifetime of particle versus antiparticle, but also the following fact: A particle and its antiparticle with opposite helicities must coexist in nature with no exception. A prominent example is the neutrino —A neutrino ν_L (antineutrino $\bar{\nu}_R$) is permanently lefthanded (right-handed) polarized whereas the fact that no ν_R exists in nature must means no $\bar{\nu}_L$ as well (as verified by the GGS experiment [53]). See also Section VII.

6) Based on the invariance of space-time inversion or mass inversion (at the level of RQM) and the latter's generalization to the classical physics,

we tentatively discuss some interesting problems in today's physics, including the prediction of antigravity between matter and antimatter, as well as the reason why we believe neutrinos are likely the tachyons.

ACKNOWLEDGEMENTS

We thank E. Bodegom, T. Chang, Y. X. Chen, T. P. Cheng, X. X. Dai, G. Tananbaum, V. Dvoeglazov, Y. Q. Gu, F. Han, J. Jiao, A. Kellerbauer, T. C. Kerrigan, A. Khalil, R. Konenkamp, D. X. Kong, J. S. Leung, P. T. Leung, Q. G. Lin, S. Y. Lou, D. Lu, Z. Q. Ma, D. Mitchell, E. J. Sanchez, Z. Y. Shen, Z. Q. Shi, P. Smejtek, X. T. Song, R. K. Su, G. Tananbaum Y. S. Wang, Z. M. Xu, X. Xue, J. Yan, F. J. Yang, J. F. Yang, R. H. Yu, Y. D. Zhang and W. M. Zhou for encouragement, collaborations and helpful discussions.

APPENDIX: KLEIN PARADOX FOR KLEIN-GORDON EQUATION AND DIRAC EQUATION

We will discuss the Klein paradox [60] for both KG equation and Dirac equation based on Sections III and V, without resorting to the "hole" theory.

AI: Klein Paradox for KG Equation

Consider that a KG particle moves along z axis in onedimensional space and hits a step potential

$$V(z) = \begin{cases} 0, & z < 0; \\ V_0, & z > 0. \end{cases}$$

(A.1)

Its incident WF with momentum $p(>0)$ and energy $E(>0)$ reads

$$\psi_i = a \exp\left[i\left(pz - Et\right)\right], (z < 0)$$

(A.2)

If $E = \sqrt{p^2 + m^2} < V_0$, we expect that the particle wave will be partly reflected at $z = 0$ with WF ψ_r and another transmitted wave ψ_t emerged at $z > 0$:

$$\psi_r = b \exp\left[i\left(-pz - Et\right)\right], (z < 0)$$

(A.3)

$$\psi_t = b' \exp\left[i\left(p'z - Et\right)\right], (z > 0)$$

(A.4)

with $p'^2 = \left(E - V_0\right)^2 - m^2$. See **Figure 1**(a).

Two continuity conditions for WFs and their space derivatives at the boundary $z = 0$ give two simple equations

$$\begin{cases} a+b=b' \\ (a-b)p=b'p' \end{cases}$$

(A.5)

The Klein paradox happens when $V_0 > E + m$ because the momentum $p' = \pm\sqrt{(V_0-E)^2 - m^2}$ is real again and the reflectivity R of incident wave reads

$$R = \left|\frac{b}{a}\right|^2 = \left|\frac{p-p'}{p+p'}\right|^2, \begin{cases} R<1, \text{if } p'>0 \\ R>1, \text{if } p'<0 \end{cases}$$

(A.6)

(See Ref. [18] or § 9.4 in Ref. [25], where discussions are not complete and need to be complemented and corrected here). Because the kinetic energy E' at $z>0$ is negative: $E' = E - V_0 < 0$, what does it mean? Does the particle still remain as a particle?

As discussed in Section III, for a KG particle (or its antiparticle), two criterions must be held: its probability density ρ (or ρ_c) must be positive and its probability current density j (or j_c) must be in the same direction of its momentum p (or p_c).

See **Figure 1**(b), after making a shift in the energy scale, i.e., basing on the new vacuum at $z>0$ region, we redefine a WF $\tilde{\psi}_t$ (which is actually the WF in the "interaction picture", $\tilde{\psi}_t = \psi_t e^{iV_0 t} (z>0)$)

$$\psi_t \to \tilde{\psi}_t = b' \exp[i(p'z - E't)], (z>0)$$

(A.7)

$(E' = E - V_0 < 0)$. From now on we will replace KG WF $\tilde{\psi}_t$ by $\tilde{\phi}_t$ and $\tilde{\chi}_t$ according to Equation (3.26), if $\tilde{\psi}_t$ still describes a "particle", whose probability density ρ_t should be evaluated by Equation (27) with

$V \to \tilde{V}(z) = 0(z>0)$ yielding:

$$\rho_t = |\tilde{\phi}_t|^2 - |\tilde{\chi}_t|^2 = \frac{E'}{m}|b'|^2 < 0, (z>0)$$

(A.8)

And its probability current density j_t should be given by Equation (3.12), yielding:

$$j_t = \frac{p'}{m}|b'|^2, (z>0)$$

(A.9)

Equation (A.8) is certainly not allowed. So to consider a "particle" with momentum $p'>0$ moving to the right makes no sense. Instead, we should consider $p' < 0$ (which also makes no sense for a particle due to the boundary condition) and regard $\tilde{\psi}_t$ as an antiparticle's WF by rewriting it as:

$$\tilde{\psi}_t = \psi_c = b' \exp\left[-i\left(p_c z - E_c t\right)\right], (z > 0) \tag{A.10}$$

Now using Equation (2.18) we see that Equation (A.10) does describe an antiparticle with momentum

$$p_c = -p' = |p'| = \sqrt{E_c^2 - m^2} > 0 \text{ and energy}$$

$$E_c = |E'| = V_0 - E > 0$$. In the mean time, from the antiparticle's point of view (i.e., with $E_c > m$), the potential becomes $V_c(z) = -\tilde{V}(z)$ (comparing Equation (2.21) with Equation (A.10) as shown by **Figure 1**(c).

It is easy to see from Equations (3.30), (3.31) and (A.10) that

$$\begin{cases} \rho_t^c = |\tilde{\chi}_t^c|^2 - |\tilde{\phi}_t^c|^2 = \dfrac{E_c}{m}|b'|^2 > 0, \\[2mm] j_t^c = \dfrac{p_c}{m}|b'|^2 \end{cases} (z > 0) \tag{A.11}$$

So the reflectivity, Equation (A.6), should be fixed as:

$$R_{KG} = \left|\frac{b}{a}\right|^2 = \left|\frac{p + p_c}{p - p_c}\right|^2 = \left(\frac{1 + \gamma'}{1 - \gamma'}\right)^2,$$

$$\gamma' = \frac{p_c}{p} > 0 \tag{A.12}$$

And the transmission coefficient can also be predicted as:

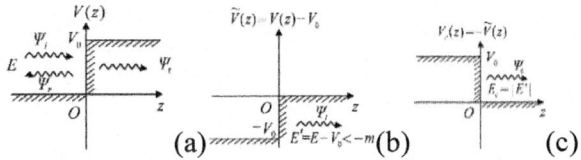

(a)   (b)   (c)

Figure 1. Klein paradox: (a) If $V_0 > E + m$, there will be a wave ψ_t at $z > 0$; (b) Just look at $z > 0$ region, making a shift $V(z) \to \tilde{V}(z) = V(z) - V_0, E \to E' = E - V_0 < -m$; (c) An antiparticle (at $z > 0$) appears with its energy $E_c = |E'| > m$ and the potential is $V_c(z) = -\tilde{V}(z)$.

$$T_{KG} = \frac{j_t^c}{j_i} = \frac{p_c}{p}\left|\frac{b'}{a}\right|^2 = \frac{p_c}{p}\left|1 + \frac{b}{a}\right|^2$$

$$= \frac{4pp_c}{(p - p_c)^2} = \frac{4\gamma'}{(1 - \gamma')^2} \tag{A.13}$$

$$R_{KG} - T_{KG} = 1 \tag{A.14}$$

The variation of T_{KG} seems very interesting:

$$T_{KG} = \begin{cases} 0, \gamma' \to 0 \left(p_c \to 0, E_c \to m \right) \\ \infty, \gamma' \to 1 \left(p_c = p, E_c = E = V_0/2 \right) \\ 0, \gamma' \to \infty \left(p_c \to \infty, E_c = V_0 - E \to \infty \right) \\ 0, \gamma' \to \infty \left(p \to 0, E \to m \right) \end{cases} \tag{A.15}$$

Above equations show us that the incident KG particle triggers a process of "pair creation" occurring at $z = 0$, creating new particles moving to the left side (to join the reflected incident particle) so enhancing the reflectivity $R_{KG} > 1$ and new antiparticles (with equal number of new particles) moving to the right.

To our understanding, this is not a stationary state problem for a single particle, but a nonstationary creation process of many particle-antiparticle system. It is amazing to see the Klein paradox in KG equation being capable of giving some prediction for such kind of process at the level of RQM. Further investigations are needed both theoretically and experimentally.[10]

AII: Klein Paradox for Dirac Equation

Beginning from Klein [60], many authors e.g. Greiner et al. [61,62], have studied this topic. We will join them by using the similar approach like that for KG equation discussed above.

Based on similar picture shown in **Figure 1**, now we have three Dirac WFs under the condition $V_0 > E + m$:

$$\psi_i = a \begin{pmatrix} 1 \\ 0 \\ \dfrac{p}{E+m} \\ 0 \end{pmatrix} e^{i(pz - Et)},$$

$$\psi_r = b \begin{pmatrix} 1 \\ 0 \\ \dfrac{-p}{E+m} \\ 0 \end{pmatrix} e^{i(-pz - Et)} \left(z < 0 \right) \tag{A.16}$$

$$\psi_t = b' \begin{pmatrix} 1 \\ 0 \\ \dfrac{p'}{E - V_0 + m} \\ 0 \end{pmatrix} e^{i(p'z - Et)} = b' \begin{pmatrix} 1 \\ 0 \\ \dfrac{-p'}{V_0 - E - m} \\ 0 \end{pmatrix} e^{i(p'z - Et)}$$

$$= \begin{pmatrix} \phi_t \\ \chi_t \end{pmatrix} (z > 0)$$

(A.17)

where $p' = \pm\sqrt{(V_0 - E)^2 - m^2}$. Unlike Equation (A.8)

for KG equation, the probability density for Dirac WF ψ_t is positive definite (see Equation (5.16))

$$\rho_t = \psi_t^\dagger \psi_t = \phi_t^\dagger \phi_t + \chi_t^\dagger \chi_t$$

(A.18)

Hence we will rely on two criterions: First, the probability current density and momentum must be in the same direction for either a particle or antiparticle.

For ψ_i and ψ_r, their probability current density are $(c = 1)$

$$\begin{cases} j_i = \psi_i^\dagger \alpha_z \psi_i = \phi_i^\dagger \sigma_z \chi_i + \chi_i^\dagger \sigma_z \phi_i = \dfrac{2p}{E+m}|a|^2 > 0 \\ j_r = \psi_r^\dagger \alpha_z \psi_r = \dfrac{-2p}{E+m}|b|^2 < 0 \end{cases} (z < 0)$$

(A.19)

as expected. However, for ψ_t, we meet difficulty similar to that in Equation (A.9)

$$j_t = \psi_t^\dagger \alpha_z \psi_t = \dfrac{-2p'}{V_0 - E - m}|b'|^2 \ (z > 0)$$

(A.20)

the direction of j_t is always opposite to that of p'! The second criterion is: while $|\phi| > |\chi|$ for particle, we must have $|\chi_c| > |\phi_c|$ for antiparticle. Now in ψ_i (or ψ_r), $|\phi_i| > |\chi_i|$ (or $|\phi_r| > |\chi_r|$), but the situation in ψ_t

is dramatically changed, the existence of V_0 renders $|\chi_t| > |\phi_t|$!

The above two criterions, together with the experience in KG equation, prompt us to choose $p' < 0$ and regard ψ_t as an antiparticle's WF. So we rewrite:

$$\psi_t = \psi_t^c e^{-iV_0 t} \quad \text{(A.21a)}$$

$$\psi_i^c = b' \begin{pmatrix} 1 \\ 0 \\ \dfrac{p_c}{E_c - m} \\ 0 \end{pmatrix} e^{-i(p_c z - E_c t)} = \begin{pmatrix} \phi_i^c \\ \chi_i^c \end{pmatrix},$$

$$\tilde{\psi}_i^c = b'_c \begin{pmatrix} 1 \\ 0 \\ \dfrac{p_c}{E_c + m} \\ 0 \end{pmatrix} e^{-i(p_c z - E_c t)} = \begin{pmatrix} \chi_i^c \\ \phi_i^c \end{pmatrix} (z > 0)$$

$$(A.21b)$$

where $\tilde{\psi}_i^c = (-\gamma^5)\psi_i^c$ (with new normalization constant

b'_c replacing b') describes an antiparticle with momentum
$p_c = |p'| = -p' = \sqrt{E_c^2 - m^2} > 0$, energy

$E_c = V_0 - E > 0$ and $|\chi_i^c| > |\phi_i^c|$. Using Equation (5.17) we find

$$j_i^c = \frac{2p_c}{E_c + m}|b'_c|^2 > 0, (z > 0)$$

$$(A.22)$$

as expected. Now it is easy to match Dirac WFs at the boundary $z = 0$, (
$(\psi_i + \psi_r)\big|_{z=0} = \tilde{\psi}_i^c\big|_{z=0}$, yielding[11]

$$\begin{cases} a + b = b'_c \\ \dfrac{(a-b)p}{E+m} = \dfrac{b'_c p_c}{E_c + m} \end{cases} \rightarrow \begin{cases} \dfrac{b}{a} = \dfrac{\xi - \eta}{\xi + \eta} \\ \dfrac{b'_c}{a} = 1 + \dfrac{b}{a} = \dfrac{2\xi}{\xi + \eta} \end{cases}$$

$$(A.23)$$

where $\xi = p(E_c + m) > 0, \eta = p_c(E + m) > 0$. The reflectivity R_D and
transmission coefficient T_D follow from Equations (A.19) and (A.22) as:

$$R_D = \frac{|j_r|}{j_i} = \left|\frac{b}{a}\right|^2 = \left(\frac{1-\gamma}{1+\gamma}\right)^2$$

$$(A.24)$$

$$T_D = \frac{j_i^c}{j_i} = \left|\frac{b'_c}{a}\right|^2 \frac{p_c(E+m)}{p(E_c+m)} = \frac{4\gamma}{(1+\gamma)^2}$$

$$(A.25)$$

$$R_D + T_D = 1$$

$$(A.26)$$

where

$$\gamma = \frac{\eta}{\xi} = \sqrt{\frac{(E_c - m)(E+m)}{(E-m)(E_c+m)}} \geq 0$$

$$(E_c = V_0 - E \geq m)$$

$$(A.27)$$

and

$$T_D = \begin{cases} 0, & \gamma \to 0 (p_c \to 0, E_c \to m) \\ 1, & \gamma = 1 (p_c = p, E_c = E = V_0/2) \\ & \text{(resonant transmission)} \\ \dfrac{2p}{E+p}, \gamma \to \sqrt{\dfrac{E+m}{E-m}} & (E_c = V_0 - E \to \infty) \\ 0, & \gamma \to \infty (p \to 0, E \to m) \end{cases}$$

(A.28)

The variation of T_D bears some resemblance to Equation (A.15) for KG equation but shows striking difference due to sharp contrast between Equations (A.24)- (A.28) and Equations (A.12)-(A.15).

To our understanding, in the above Klein paradox for Dirac equation, there is no "pair creation" process occurring at the boundary $z = 0$. The paradox just amounts to a steady transmission of particle's wave ψ_i into a high potential barrier $V_0 > E+m$ at $z > 0$ region where ψ_t shows up as an antiparticle's WF propagating to the right. In some sense, the existence of a potential barrier V_0 plays a "magic" role of transforming the particle into its antiparticle. Because the probability densities of both particle and antiparticle are positive definite, the total probability can be normalized over the entire space like that for one particle case:

$$\int_{-\infty}^{\infty} \left[\rho(z)\Theta(-z) + \rho_c(z)\Theta(z) \right] dz = 1$$

(A.29)

($\Theta(z)$ is the Heaviside function) and the probability current density remains continuous at the boundary $z = 0$. In other words, the continuity equation holds in the whole space just like what happens in a one-particle stationary state.

It is interesting to compare our result with that in Refs. [61] and [62]. In Ref. [61], Equations (13.24)-(13.28) are essentially the same as ours. But the argument there for choosing $\bar{p} < 0$ in Equation (13.23) is based on the criterion of the group velocity v_{gr} being positive (for the transmitted wave packet moving toward $z = \infty$). And the v_{gr} is stemming from Equation (13.16) which is essentially the probability current density in our Equations (A.21)-(A.22).

However, the author in Ref. [61] also considered the other choice $\bar{p} > 0$ in an example (pp. 265-267 in [61]) based on the hole theory, ending up with the prediction as:

$$R = \left(\frac{1+\gamma}{1-\gamma} \right)^2, T = \frac{4\gamma}{(1-\gamma)^2}, R - T = 1$$

(A.30)

where

$$\gamma = \frac{p_2}{p_1} \frac{E+m}{V_0 - E - m}$$

$$= \sqrt{\frac{(V_0 - E + m)(E + m)}{(V_0 - E - m)(E - m)}}$$

(A.31)

The argument for the validity of his Equations (A.30)- (A.31) is based on the hole theory (see also section 5.2 in Ref. [62]), saying that once $V_0 > E + m$, there would be an overlap between the occupied negative continuum for $z > 0$ and the empty positive continuum for $z < 0$, providing a mechanism for electron-positron pair creation if the "hole" at $z > 0$ can be identified with a positron. We doubt the "hole" theory seriously because there are only two electrons (with opposite spin orientations) staying at each energy level in the negative continuum. So it seems that there is no abundant source for electrons and "holes" to account for the huge value of $T > 1$ in Equation (A.30).

Fortunately, we learn from section 10.7 in Ref. [62] that if the Klein paradox in Dirac equation is treated at the level of QFT, their result turns out to be the same form as our Equations (A.24)-(A.28), rather than Equations (A.30) and (A.31).

NOTES

[1]The WF reads approximately as:

$$\Psi^{sym}_{K^0 K^0}\left(x_1, t_1; x_2, t_2\right) \sim e^{i(p_1 x_1 - E_1 t_1)} e^{i(p_2 x_2 - E_2 t_2)}$$

(2.14b)

which can be calculated from $\left\langle K^0 K^0 | \Psi(t_a, t_b) \right\rangle^{sym}$ with two terms. The squares of WF's amplitude reproduces the $I^{(sym)}_{like}\left(t_a, t_b\right)$ in Equation (2.12).

[2]Please see the derivation of Equations (2.20) and (2.21) from the quantum field theory (QFT) at the end of Section VI.

[3]Interestingly, if ignoring the coupling between ϕ and χ and $V = 0$ in Equation (25), they satisfy respectively the "two equations" written down by Schrödinger in his 6th paper in 1926, titled "Quantisation as a problem of proper values (Part IV)" (Annalen der Physik Vol. 81, No. 4, 1926, p. 104) when he invented NRQM in the form of wave mechanics.

[4]Here m always refers to the "rest mass" also the "inertial mass" for a particle or its antiparticle, see the excellent paper by Okun in Ref. [33].

[5]Some pictures of numerical calculation are shown in Ref. [35] and section 9.5C at Ref. [25], where an error in Equation (9.5.26) is corrected here.

[6]The reason why we use ψ_c' instead of ψ_c will be clear in Equations (5.12)-(5.15). Actually, we emphasize Dirac equation as a coupling equation of two two-component spinors, Equation (3), rather than merely a four-component spinor equation.

$$\Sigma = \begin{pmatrix} \sigma & 0 \\ 0 & \sigma \end{pmatrix}, \Sigma_c = \begin{pmatrix} \sigma_c & 0 \\ 0 & \sigma_c \end{pmatrix}$$

[7]

[8]The wonderful experiment by Wu et al. [6] reveals the decay configuration of a polarized neutron bearing a strong resemblance to a "comet" with its "head" oriented along neutron's spin parallel to z axis in space (note that a static neutron has no helicity h, see [45]) while its "tail" composed of emitted e^- and $\bar{v}_e$. So it was expected intuitively that [39] if one pushes the "comet" along its "head"'s direction, it (suddenly has a helicity $h = 1$ and) will be relatively more stable than it is pushed along its "tail" (when it has $h = -1$). That's what Equation (7.1) means and why the use of "spin state" fails to get it right.

[9]We had discarded the solution of $p' > 0$ in Equations (A.7)-(A.9) as a particle. However, if we consider $p' = -p_c > 0$ for an antiparticle, then similar to Equations (A.10) and (A.11), we would get $\rho_i^c > 0$ but both j_i^c and p_c are negative, meaning that the antiparticle is coming from $z = \infty$, not in accordance with our boundary condition. So the case of $p' > 0$ should be abandoned either as a particle or as an antiparticle.

[10]We find from the Google search that R. G. Winter in 1958 had written a paper titled "Klein paradox for the Klein-Gordon equation" and reached basically the same result as ours. So he was the first author dealing with this problem. Regrettably, it seems that his paper had never been published on some journal.

[11]Equation (A.23) means that the large (small) component of spinor is connected with large (small) component at both sides of $z = 0$. However, if instead of $\tilde{\psi}_i^c$, the ψ_i^c is used directly with its first (small) component being connected with the first (large) components of ψ_i and ψ_r, it would lead to a different expression of Equation (A.27): $\gamma \to \tilde{\gamma} = \sqrt{\dfrac{(E_c - m)(E - m)}{(E + m)(E_c + m)}}$, which is just the $1/\gamma$ (γ and $1/\gamma$ make no difference in the result of, say, Equations (A.24) and (A.25))

defined by Equation (8) on page 266 of Ref. [61] (see Equation (A31) below) or that by Equation (5.36) in Ref. [62]

REFERENCES

1.   A. Apostolakis, et al., (CPLEAR Collaboration) Physics Letters B, Vol. 422, 1998, pp. 339-348. doi:10.1016/S0370-2693(97)01545-1

2.   H. Feshbach and F. Villars, Review of Modern Physics, Vol. 30, 1958, pp. 24-45.doi:10.1103/RevModPhys.30.24

3.   T. D. Lee and C. N. Yang, Physical Review, Vol. 104, 1956, pp. 254-258.

4.   T. D. Lee and C. N. Yang, ibid, Vol. 105, 1957, pp. 1671-1675.

5.   T. D. Lee, R. Oehme and C. N. Yang, ibid, Vol. 106, 1957, pp. 340-345.

6.   C. S. Wu, E. Ambler, R. W. Hayward, D. D. Hoppes and R. P. Hudson, Physical Review, Vol. 105, 1957, pp. 1413-1415. doi:10.1103/PhysRev.105.1413

7.   J. H. Christensen, J. W. Cronin, V. L. Fitch and R. Turlay, Physical Review Letters, Vol. 13, 1964, pp. 138-140. doi:10.1103/PhysRevLett.13.138

8.   K. R. Schubert, B. Wolff, J.-M. Gaillard, M. R. Jane, T. J. Ratcliffe and J.-P. Repellin, Physics Letters B, Vol. 31, 1970, pp. 662-665. doi:10.1016/0370-2693(70)90029-8

9.   J. Beringer, et al., (Particle Data Group) Physical Review D, Vol. 86, 2012, Article ID: 010001. doi:10.1103/PhysRevD.86.010001

10. G. Lüders, Kgl. Danske Vidensk. Selsk. Mat.-Fys. Medd., Vol. 28, 1954.

11. G. Lüders, Annals of Physics (New York), Vol. 2, 1957, pp. 1-15.

12. W. Pauli, "Exclusion Principle, Lorentz Group and Reflection of Space-Time and Charge," In: W. Pauli, L. Rosenfeld and V. Weisskopf, Eds., Niels Bohr and the Development of Physics, McGraw-Hill, New York, 1955, pp. 30-51.

13. T. D. Lee and C. S. Wu, Annual Review of Nuclear Science, Vol. 15, 1965, pp. 381-476.doi:10.1146/annurev.ns.15.120165.002121

14. A. Einstein, B. Podolsky and N. Rosen, Physical Review, Vol. 47, 1935, pp. 777-780.doi:10.1103/PhysRev.47.777

15. D. Bohm, "Quantum Theory," Prentice Hall, Upper Saddle River, 1956.

16. J. S. Bell, Physics, Vol. 1, 1964, pp. 195-200.

17. H. Guan, "Basic Concepts in Quantum Mechanics," High Education Press, Beijing, 1990.

18. G. J. Ni, H. Guan, W. M. Zhou and J. Yan, Chinese Physics Letters, Vol. 17, 2000, pp. 393-395. doi:10.1088/0256-307X/17/6/002

19. O. Nachtmann, "Elementary Particle Physics: Concepts and Phenomena," Springer-Verlag, Berlin, 1990.

20. W. Greiner and B. Müller, "Gauge Theory of Weak Interactions," Springer-Verlag, Berlin, 1993.

21. E. J. Konopinski and H. M. Mahmaud, Physical Review, Vol. 92, 1953, pp. 1045-1049.doi:10.1103/PhysRev.92.1045

22. G. J. Ni, Journal of Fudan University (Natural Science), No. 3-4, 1974, pp. 125-134.

23. G. J. Ni and S. Q. Chen, Journal of Fudan University (Natural Science), Vol. 35, 1996, pp. 325-334.

24. G. J. Ni and S. Q. Chen, "Relation between Space-Time Inversion and Particle-Antiparticle Symmetry and the Microscopic Essence of Special Relativity," In: V. Dvoeglazov, Ed., Photon and Poincare Group, NOVA Science Publisher, New York, 1999, pp. 145-169.

25. G. J. Ni and S. Q. Chen, "Advanced Quantum Mechanics," Rinton Press, New Jersy, 2002.

26. G. J. Ni, Progress in Physics, Vol. 23, 2003, pp. 484-503.

27. G. J. Ni, "A New Insight into the Negative-Mass Paradox of Gravity and the Accelerating Universe," In: V. V. Dvoeglazov and A. A. Espinoza Garrido, Eds., Relativity, Gravitation, Cosmology, NOVA Science Publisher, New York, 2004, pp. 123-136.

28. G. J. Ni, J. J. Xu and S. Y. Lou, Chinese Physics B, Vol. 20, 2011, Article ID: 020302.

29. J. J. Sakurai, "Advanced Quantum Mechanics," Addison-Wesley Publishing Company, Boston, 1978.

30. J. J. Sakurai, "Modern Quantum Mechanics," John Wiley & Sons, Inc., NewYork, 1994.

31. J. D. Bjorken and S. D. Drell, "Relativistic Quantum Mechanics," McGraw-Hill, New York, 1964,

32. J. D. Bjorken and S. D. Drell, "Relativistic Quantum Fields," McGraw-Hill, New York, 1965.

33. L. B. Okun, Physics Today, Vol. 42, 1989, pp. 31-36.

34. G. Lochak, "De Broglie's Initial Conception of De Broglie Waves," In: S. Diner, D. Fargue, G. Lochak and F. Selleri, Eds., The Wave-Particle Dualism, D. Reidal Publishing Company, Dordrecht, 1984, pp. 1-25.

35. G. J. Ni, W. M. Zhou and J. Yan, "Comparison among Klein-Gordon Equation, Dirac Equation and Relativistic Schrödinger Equation," In: A.

E. Chubykalo, V. V. Dvoeglazov, D. J. Ernst, V. G. Kadyshevsky and Y. S. Kim, Eds., Lorentz Group, CPT and Neutrinos, World Scientific, London, 2000, pp. 68-81.

36. M. E. Peskin and D. V. Schroeder, "An Introdution to Quantum Field Theory," Addison-Wesley Publishing Company, Boston, 1995.

37. M. Jacob and G. C. Wicks, Annals of Physics (New York), Vol. 7, 1959, pp. 404-428.doi:10.1016/0003-4916(59)90051-X

38. S. Weinberg, Physical Review Letters, Vol. 19, 1967, pp. 1264-1266. doi:10.1103/PhysRevLett.19.1264

39. Z. Q. Shi and G. J. Ni, Chinese Physics Letters, Vol. 19, 2002, pp. 1427-1429.

40. Z. Q. Shi and G. J. Ni, Annales de la Fondation Louis de Bloglie, Vol. 29, 2004, pp. 1057-1066.

41. Z. Q. Shi and G. J. Ni, Handronic Journal, Vol. 29, 2006, pp. 401-407.

42. Z. Q. Shi and G. J. Ni, "Frontiers in Horizons in World Physics," Nova Science, Marselle, 2008, pp. 53-65.

43. Z. Q. Shi and G. J. Ni, Modern Physics Letters A, Vol. 26, 2011, pp. 987-998.doi:10.1142/S0217732311035250

44. A. Cho, Science, Vol. 326, 2009, pp. 1342-1343. doi:10.1126/science.326.5958.1342

45. L. H. Ryder, "Quantum Field Theory," Cambridge University Press, Cambridge, 1996.doi:10.1017/CBO9780511813900

46. T. Chang and G. J. Ni, "An Explanation of Possible Negative Mass-Square of Neutrinos," FIZIKA B (Zagreb), Vol. 11, 2002, pp. 49-56. arXiv.org:hep-ph/0009291

47. G. J. Ni and T. Chang, Journal of Shaanxi Normal University (Natural Science), Vol. 30, No. 3, 2002, pp. 32-39.

48. G. J. Ni, Journal of Shaanxi Normal University (Natural Science), Vol. 29, No. 1, 2001, pp. 1-5.

49. G. J. Ni, Journal of Shaanxi Normal University (Natural Science), Vol. 30, No. 4, 2002, pp. 1-6.

50. G. J. Ni, "A Minimal Three-Flavor Model for Neutrino Oscillation Based on Superluminal Property," In: V. V. Dvoeglazov and A. A. Espinoza, Eds., Relativity, Gravitation, Cosmology, NOVA Science Publisher, New York, 2004, pp. 137-148.

51. G. J. Ni, "Principle of Relativity in Physics and in Epistemology," In: V. Dvoeglazov, Ed., Relativity, Gravitation, Cosmology: New Development,

NOVA Science Publisher, New York, 2010, pp. 237-252.

52. G. J. Ni, "Cosmic Ray Spectrum and Tachyonic Neutrino," In: V. V. Dvoeglazov and A. A. Espinoza, Eds., Relativity, Gravitation, Cosmology: New Development, NOVA Science Publisher, New York, 2010, pp. 253-265.

53. M. Goldhaber, L. Grodgins and A. W. Sunyar, Physical Review, Vol. 109, 1958, pp. 1015-1017. doi:10.1103/PhysRev.109.1015

54. S. Weinberg, "Gravitation and Cosmology," John Wiley, New York, 1972.

55. Z. M. Xu and X. J. Wu, "General Relativity and Contemporary Cosmology," Press of Nanjing Normal University, Nanjing, 1999.

56. T. P. Cheng, "Relativity, Gravitation and Cosmology", 2nd Edition, Oxford University Press, Oxford, 2010.

57. K. Jagannathan and L. P. S. Singh, Physical Review D, Vol. 33, 1986, pp. 2475-2477.doi:10.1103/PhysRevD.33.2475

58. M. Villata, Europhysics Letters, Vol. 94, 2011, pp. 1-6. doi:10.1209/0295-5075/94/20001

59. A. Kellerbauer, et al., Nuclear Instruments and Methods in Physics Research Section B, Vol. 266, 2008, pp. 351- 356. doi:10.1016/j. nimb.2007.12.010

60. O. Klein, Zeitschrift für Physik, Vol. 53, 1929, pp. 157-165.doi:10.1007/ BF01339716

61. W. Greiner, "Relativistic Quantum Mechanics," SpringerVerlag, Berlin, 1990.

62. W. Greiner, B. Müller and J. Rafelski, "Quantum Electrodynamics of Strong Fields," Springer-Verlag, Berlin, 1985.

Chapter 8

MECHANICS OF STATIC SLIP AND ENERGY DISSIPATION IN SANDWICH STRUCTURES: CASE OF HOMOGENEOUS ELASTIC BEAMS IN TRANSVERSE MAGNETIC FIELDS

Charles A. Osheku

Centre for Space Transport and Propulsion, National Space Research and Development Agency, Federal Ministry of Science and Technology, FCT, Abuja, PMB 437, Nigeria

ABSTRACT

Mechanics of static slip and energy dissipation in sandwich structures with respect to two-layer homogeneous elastic beams in a transverse magnetic field is presented. The mathematical physics problem derives from nonuniform contact conditions of press sandwich layers or joints. On this theory, equations governing the stresses and the deflection profile are derived. By restricting analysis to the case of cantilever architecture, closed form polynomial expressions are computed for the deflection, interfacial slip, slip, and strain energies of the system. In particular, the effects of magnetoelasticity and interfacial pressure gradient on these properties are demonstrated for design analysis and engineering applications. In addition, explicit mathematical equations couched in magnetoelasticity and pressure gradient polynomial kernels with fractional coefficients for critical values of pressure for which no slip occurs at the tip and the optimum clamping pressure for optimal slip energy dissipation are derived. It is also shown for special cases that recent results in literature are recoverable from the theory reported in this paper.

INTRODUCTION

Investigation into the vibration and magnetoelastic stability of ferromagnetic flexible structures, beams, beam-plates, plates, and shells is abound in literature. Concerning these theoretical analyses or experimental studies, comprehensive reviews of trends are reported in [1–18]. For theoretical analyses, the effect of eddy current in the ferromagnetic material was neglected. Lee [1] in contrast to earlier investigators studied the dynamic stability with magnetic damping arising from eddy current and derived an explicit expression for the destabilizing effect. For experimental investigation, Moon and Pao [2] were credited with the pioneering work on magnetoelastic buckling of a ferromagnetic thin plate in transverse magnetic fields. Their results showed that a ferromagnetic plate buckles and loses its stability when the magnetic intensity approaches a critical value that is functionally related to the geometric ratio of length to plate thickness via a 3/2 power law. Based on these findings, a mathematical problem was contrived as the magnetic body coupled model to predict the experimental phenomenon of magnetoelastic instability and critical magnetic field. Additional experiment later showed that the natural frequency of a beam-plate decreased with increasing magnetic field intensity and becomes near to zero, as the field attains a critical value, which causes the same beam-plate to buckle statically. Following the emergence of large discrepancy between the theoretical predictions and experimental results of Moon and Pao in [3], research attentions were further devoted to the study of magnetoelastic stability and buckling problems.

Some investigators, Wallerstein and Peach [4], Miya et al. [5], Peach et al. [6], and so forth, directed their attentions to finding satisfactory explanations for these discrepancies. Notwithstanding the significance of previous findings, Lee [7] investigated the dynamic stability of electrically conducting beam-plates in transverse magnetic fields via a concise theory of flexural vibration of magnetoelastic plates immersed in transverse magnetic fields. Similarly, in the 1990s, additional theories were developed for the study of magnetoelastic buckling and bending of ferromagnetic plates in transverse and/or oblique magnetic fields via a generalized variational principle of magnetoelasticity by Zhou et al. [8], Zhou and Zheng [9], and Zhou and Miya [10]. In 2002, the experimental results of Wang et al. [11] confirmed the theoretical predictions in the 1990s.

Following renewed interest in magnetoelasticity and its applications in engineering systems, namely, magnetic storage elements, magnetic structural devices, geophysical physics, and plasma physics, attentions were directed on the study of magnetothermodynamic stress and perturbation of magnetic field vector in both solid and orthotropic thermoelastic cylinders. In this regard, Wang et al. [12] employed finite integral transforms to examine theoretically the magnetothermoelastic waves and perturbation of the magnetic field vector produced by thermal shock in a solid conducting cylinder. In this study, closed forms expressions were derived for magnetothermodynamic stress and perturbation response of an axial magnetic field vector in a solid cylinder. Comprehensively, Wang et al. [13] investigated the magnetothermoelastic responses and perturbation of the magnetic field vector in a conducting orthotropic thermoelastic cylinder subjected to thermal shock using finite Hankel integral transform, whilst Librescu et al. [14] and Wang et al. [15] studied the effect of magnetothermoelasticity of ferromagnetic conducting plates under excitations theoretically.

In related development, Wang and Dai [16] investigated the dynamic responses of piezoelectric hollow cylinders in axial magnetic field. This study led to the development of a concise analytical solution to reveal the interaction between mechanical and electromagnetoelastics responses of piezoelectric hollow cylinders subjected to arbitrary mechanical loadings and electric potential shock. An interpolation method was employed to solve the resulting Volterra integral equation of the second kind, arising from interaction between different physical fields. Furthermore, closed forms results were derived for dynamic stresses, electric-displacements and electric-potentials as well as perturbation responses using finite and Laplace integral transforms.

Meanwhile, the problem of stability loss and free vibration of electromagnetically conducting plate conveying an electric current in magnetic field environment was investigated by Hansanyan et al. [17]. Following the theoretical models in the 1990s, Wang and Lee [18] considered the magnetic damping effect induced by the eddy current and its effect on dynamic stability. Application of these structures is receiving significant attentions in magnetic propulsion devices for space transport and exploration. From experimental investigations, ferromagnetic flexible structures are usually subjected to magnetic forces arising from the coupling or mutual influence of the magnetization and magnetic fields.

Following recent advances in the mechanics of sandwich layered elastic structures, in an environment of nonuniform interface pressure by Damisa et al. [19, 20], Olunloyo et al. [21], Olunloyo et al. [22], and Osheku and Damisa [23] investigated the flexural vibration of a two-layer magnetoelastic beam in a transverse magnetic field. In their study, equations of mathematical physics governing the stresses and the structural vibration were derived via laminated beam theory employing Newtonian form of Cauchy's stress equations.

Although the study was restricted to the case of cantilever structure, the effects of magnetoelasticity, material conductivity, and interfacial pressure gradient on the system response were computed in the form of polynomial expression via Laplace and finite Fourier integral transforms.

The study also shows that each mode of vibration was governed by a two-dimensional family of natural frequencies. The natures of the closed forms expressions for the natural frequencies indicate that the oscillation ceases when the two become simultaneously zero. In both theory and experiment, this is the required condition for static or quasistatic buckling of any layered elastic structure in a transverse magnetic field.

In fact, it is an indication that with suitable geometric parameters and matching transverse magnetic field, critical damping can be enhanced. For special and limit cases, recent theoretical and experimental results were validated. Following the increasing significant of studying both theoretically and experimentally the required condition for static or quasistatic buckling of any layered elastic structure in a transverse magnetic field, this study is devoted to the comprehensive investigation of the characteristics of statically loaded homogenous two-layer sandwich magnetoelastic cantilever structure in a transverse magnetic field.

This paper is organized as follows. Section 1 introduces the problem under investigation within a general context. In Section 2, the essential analytical mechanics leading to the mathematical physics problem with additional specialized static boundary values ordinary differential equations are presented. In Section 3, formal analysis of the problem of interest using finite Fourier integral transform is discussed. Section 4 is concerned with the analysis of static slip, whilst in Section 5 the energy dissipation ability and damping capacity of the structure are analysed. In Section 6, simulated results are discussed. Finally, the paper ends with conclusion in Section 7.

FORMULATION OF THE GOVERNING DIFFERENTIAL EQUATION PROBLEM DEFINITION

As illustrated in Figure 1(a), the problem here is to examine analytically the effect of the pressure gradient on the damping properties of a statically loaded two-layer magnetoelastic beams clamped together in an environment of nonuniform pressure.

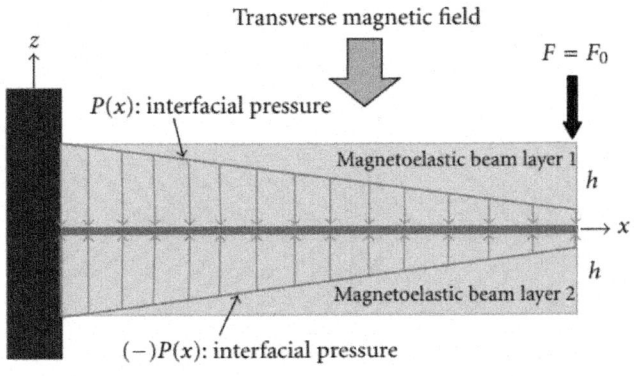

(a)

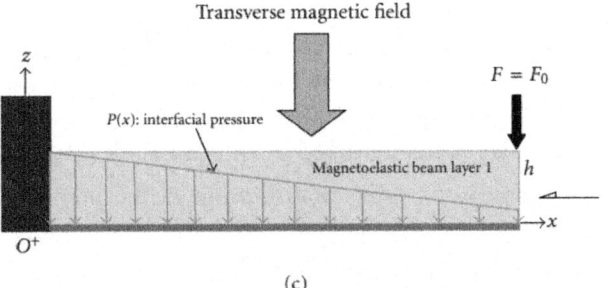

(b)

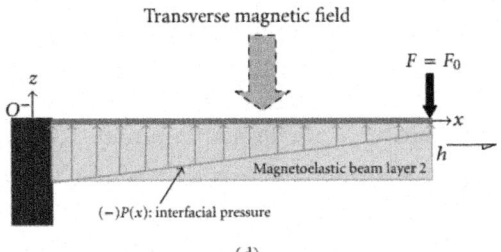

(c)

(d)

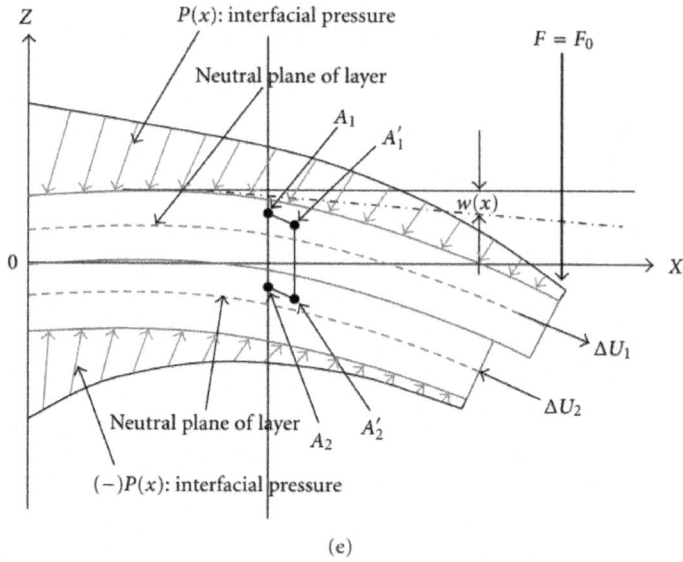

(e)

Figure 1: (a) Preslip geometry for the sandwich structure under static load. (b) conceptual description of the upper and lower layers neutral axes. (c) upper layer postslip geometry under static load. (d) lower layer postslip geometry under static load. (e) mechanism of interfacial slip geometry.

Underlying Assumptions

A two-layer elastic structure is subject to a transverse magnetic field. For the contrived structure, the upper and the lower layers are assumed to be perfectly press fit surfaces of homogenous magnetoelastic beams. The contact conditions between the mating layers as itemized in Damisa et al. [20] hold, namely;

- there is continuity of stress distributions at the interface to sufficiently hold the separate layers together both in the pre- and postslip conditions;
- the static deflection of each beam is small compared with the span;
- during bending, the magnetoelastic structure has (upper and lower) layers such that each has its neutral plane which may not necessarily coincide with its geometric mid plane of the resultant structure. These neutral planes are located at $z_1 = \psi(x)(h/2)$ and $z_2 = -\psi(x)(h/2)$, where $\psi(x)$ is a function of x as illustrated in Figure 1(b);
- the approximations involved in the forgoing beam theory are such that the field variables are linear and are expressible in terms of the derivatives of the transverse static deflection $W(x)$ and is taken to be same for both layers.

By defining u(x, z) and W(x) as displacements along x and z, respectively, the following relations hold from the classical theory of elasticity, namely,

$$\varepsilon_x = \frac{du}{dx}; \quad \gamma_{xz} = \left(\frac{dW}{dx} + \frac{du}{dz}\right) = 0,$$
(1)

$$\frac{du}{dz} = -\frac{dW}{dx} \text{ such that: } \int_{U_{01}}^{U_1} \frac{du}{dz} dz = \int_{U_{01}}^{U_1} du = -\int_{Z_1}^{Z_0} \frac{dW}{dx} dz.$$
(2)

Equation (2) can be evaluated as

$$U_1 - U_{01} = -\int_{Z_0}^{Z_1} \frac{dW}{dx} dz = -(z_0 - z_1)\frac{dW}{dx}.$$
(3)

Now $\forall z_0 = z$; $z_1 = \psi(x)(h/2)$, following assumption (iii) above, we can rewrite (3) as

$$U_1 - U_{01} = -\int_{Z_0}^{Z_1} \frac{dW}{dx} dz = -(z_0 - z_1)\frac{dW}{dx}$$

$$= -\left(z - \psi(x)\frac{h}{2}\right)\frac{dW}{dx},$$
(4)

to obtain the following expression:

$$U_1 = -\left(z - \psi(x)\frac{h}{2}\right)\frac{dW}{dx} + U_{01}(x),$$
(5)

where $U_{01}(x)$ is the point of initiation of interfacial slip in the upper layer.

Similarly, the following expression holds for the lower layer as

$$U_2 - U_{02} = -\int_{Z_0}^{Z_{12}} \frac{dW}{dx} dz = -(z_0 - z_2)\frac{dW}{dx} \quad \forall z = -\psi(x)\frac{h}{2}$$
(6)

and evaluated to obtain the following expression:

$$U_2 = -\left(z + \psi(x)\frac{h}{2}\right)\frac{dW}{dx} + U_{02}(x),$$
(7)

where $U_{02}(x)$ admits same definition in the lower layer.

From classical theory of elasticity, the in-plane bending stress for the upper layer takes the following form:

$$\sigma_x(x, z) = \sigma_x(x, z)_1 = E\varepsilon_{x1} = E\frac{dU_1}{dx}.$$
(8)

On substituting (5), the foregoing becomes

$$\sigma_x(x, z)_1$$

$$= E\frac{dU_1}{dx}$$

$$= -E\left(\frac{d}{dx}\left(\left(z - \psi(x)\frac{h}{2}\right)\frac{dW}{dx} + U_{01}(x)\right)\right)$$

$$= -E\left(\frac{d}{dx}\left(z - \psi(x)\frac{h}{2}\right)\frac{dW}{dx} + \left(z - \psi(x)\frac{h}{2}\right)\frac{d^2W}{dx^2} + \frac{dU_{01}(x)}{dx}\right)$$

$$= -E\left(\left(-\frac{h}{2}\left(\frac{d\psi(x)}{dx}\right)\frac{dW}{dx}\right) + \left(z - \psi(x)\frac{h}{2}\right)\frac{d^2W}{dx^2} + \frac{dU_{01}(x)}{dx}\right). \tag{9}$$

Now $-(h/2)(d\psi(x)/dx)(dW/dx)$ is a nonlinear term, whilst $dU_{01}(x)/dx$ is the strain at the point of initiation of static slip.

Following assumption (iv) (linear theory), $-(h/2) (d\psi(x)/dx)(dW/dx)$ is negligible while $dU_{01}(x)/dx = 0$ at the fixed end. Consequently,

$$\sigma_x(x, z)_1$$

$$= E\frac{dU_1}{dx}$$

$$= -E\left(\frac{d}{dx}\left(\left(z - \psi(x)\frac{h}{2}\right)\frac{dW}{dx} + U_{01}(x)\right)\right)$$

$$= -E\left(\frac{d}{dx}\left(z - \psi(x)\frac{h}{2}\right)\frac{dW}{dx}\right.$$

$$\left. + \left(z - \psi(x)\frac{h}{2}\right)\frac{d^2W}{dx^2} + \frac{dU_{01}(x)}{dx}\right)$$

$$= -E\left(z - \psi(x)\frac{h}{2}\right)\frac{d^2W}{dx^2}. \tag{10}$$

Similarly for layer (2), we have

$$\sigma_x(x, z) = \sigma_x(x, z)_2 = E\varepsilon_{x2} = E\frac{dU_2}{dx}. \tag{11}$$

On substituting $U_2 = -(z + \psi(x)(h/2))(\partial W/\partial_x) + U_{02}(x)$, (11) becomes

$\sigma_x(x,z)_2$

$$= E\frac{dU_2}{dx}$$

$$= -E\left(\frac{d}{dx}\left(\left(z+\psi(x)\frac{h}{2}\right)\frac{dW}{dx}+U_{02}(x)\right)\right)$$

$$= -E\left(\frac{d}{dx}\left(z+\psi(x)\frac{h}{2}\right)\frac{dW}{dx}+\left(z+\psi(x)\frac{h}{2}\right)\frac{d^2W}{dx^2}+\frac{dU_{02}(x)}{dx}\right)$$

$$= -E\left(\left(\frac{h}{2}\left(\frac{d\psi(x)}{dx}\right)\frac{dW}{dx}\right)+\left(z+\psi(x)\frac{h}{2}\right)\frac{d^2W}{dx^2}+\frac{dU_{02}(x)}{dx}\right). \tag{12}$$

Following assumption (iv) (linear theory), $(h/2)(d\psi(x)/dx) \times (dW/dx)$ is negligible while $dU_{02}(x)/dx = 0$ at the fixed end.

Consequently,

$\sigma_x(x,z)_2$

$$= E\frac{dU_2}{dx}$$

$$= -E\left(\frac{d}{dx}\left(\left(z+\psi(x)\frac{h}{2}\right)\frac{dW}{dx}+U_{02}(x)\right)\right)$$

$$= -E\left(\frac{d}{dx}\left(z+\psi(x)\frac{h}{2}\right)\frac{dW}{dx}\right.$$
$$\left.+\left(z+\psi(x)\frac{h}{2}\right)\frac{d^2W}{dx^2}+\frac{dU_{02}(x)}{dx}\right)$$

$$= -E\left(z+\psi(x)\frac{h}{2}\right)\frac{d^2W}{dx^2}. \tag{13}$$

Next, we invoke the static form of the generalized Cauchy stress equation in the absence of body forces, namely,

$$\nabla \cdot \vec{\tau} = 0, \tag{14}$$

where $\vec{\tau}$ is the stress tensor.

In the upper and lower halves, (14) admits the following forms:

$$\frac{\partial}{\partial x}\sigma_{(x)1} + \frac{\partial}{\partial z}\tau_{(xz)1} = 0,$$

$$\frac{\partial}{\partial z}\sigma_{(z)1} + \frac{\partial}{\partial x}\tau_{(xz)1} = 0, \tag{15a}$$

$$\frac{\partial}{\partial x}\sigma_{(x)2} + \frac{\partial}{\partial z}\tau_{(xz)2} = 0,$$

$$\frac{\partial}{\partial z}\sigma_{(z)2} + \frac{\partial}{\partial x}\tau_{(xz)2} = 0. \tag{15b}$$

On substitution of (10), (13), we rewrite the above as

$$-Ez\frac{d^3W}{dx^3} + \frac{Eh}{2}\frac{d}{dx}\left(\psi(x)\left(\frac{d^2W}{dx^2}\right)\right) + \frac{d\tau_{(xz)1}}{dz} = 0, \tag{16}$$

$$-Ez\frac{d^3W}{dx^3} - \frac{Eh}{2}\frac{d}{dx}\left(\psi(x)\left(\frac{d^2W}{dx^2}\right)\right) + \frac{d\tau_{(xz)2}}{dz} = 0. \tag{17}$$

Following Goodman and Klumpp [24], (16)-(17) must satisfy the following postslip boundary conditions along xz-plane, namely,

$$\sigma_{(z)1}(x,0) = -p(x,0); \quad \sigma_{(z)2}(x,-h) = p(x,0);$$

$$\tau_{(xz)1}(x,h) = 0; \quad \tau_{(xz)2}(x,-h) = 0;$$

$$\tau^2_{(xz)1}(x,0) = \mu^2\sigma^2_{(z)1}(x,0); \quad \tau^2_{(xz)2}(x,0) = \mu^2\sigma^2_{(z)2}(x,0);$$

$$\int_0^h \sigma_{(x)1}(x,z)dz = \int_0^h z\sigma_{(x)1}(x,z)dz = 0;$$

$$\int_{-h}^0 \sigma_{(x)2}(x,z)dz = \int_{-h}^0 \sigma_{(x)2}(x,z)dz = 0. \tag{18}$$

From Lee [7], the in-plane shear stress arising from electromagnetic surface traction follows from the generalized Maxwell's stress tensor τ^M defined as

$$\tau^M = \mu_M\vec{H} \otimes \vec{H} + \varepsilon_m\vec{E}_m \otimes \vec{E}_m$$

$$- \frac{1}{2}\left(\mu_m\vec{H} \cdot \vec{H} + \varepsilon_m\vec{E}_m \cdot \vec{E}_m\right)I_s. \tag{19}$$

By enforcing small perturbation on the primary bias field due to the field-structure interaction, following Lee [7], the following expression ensues

$$\vec{H}(x) = \vec{H}_0 + \vec{h}_1(x); \qquad \vec{E}_M(x) = 0 + \vec{e}(x). \tag{20}$$

Here the field quantities in lower-case letters are assumed to be small magnetic and electric perturbation variables. Consequently, their products can be neglected. Under this circumstance, the Maxwell's stress tensor in (19) reduces to the form

$$\tau^M = \left(\vec{B}_0 \otimes \vec{h}_1 + \vec{b}_m \otimes \vec{H}_0\right) - \frac{1}{2}\left(\vec{B}_0 \cdot \vec{h}_1 + \vec{b}_m \cdot \vec{H}_0\right)I_s. \tag{21}$$

Utilizing the relations in Lee [7], the in-plane and out-plane static frictional stresses are modified as

$$\tau_{(xz)1}(x,0) = \mu p(x,0) - \frac{2}{\mu_0}\left(1 - \frac{\mu_0}{\mu_m}\right)B_0{}^2\frac{dW}{dx},$$

$$\tau_{(xz)2}(x,0) = -\mu p(x,0) - \frac{2}{\mu_0}\left(1 - \frac{\mu_0}{\mu_m}\right)B_0{}^2\frac{dW}{dx}. \tag{22}$$

Consequently, (16) can be integrated to obtain

$$\tau_{(xz)1}(x,h)$$

$$= \left(\begin{array}{c}\dfrac{E(z^2 - zh)}{2}\dfrac{d^3W}{dx^3} + \dfrac{2}{\mu_0}\left(1 - \dfrac{\mu_0}{\mu_m}\right)B_0{}^2\dfrac{(z-h)}{h}\dfrac{dW}{dx} \\ -\dfrac{\mu p(x,0)(z-h)}{h}\end{array}\right). \tag{23}$$

Similar expression can be derived for the lower layer as

$$\tau_{(xz)2}(x,h)$$

$$= \left(\begin{array}{c}\dfrac{E(z^2 + zh)}{2}\dfrac{d^3W}{dx^3} + \dfrac{2}{\mu_0}\left(1 - \dfrac{\mu_0}{\mu_m}\right)B_0{}^2\dfrac{(z+h)}{h}\dfrac{dW}{dx} \\ +\dfrac{\mu p(x,0)(z+h)}{h}\end{array}\right). \tag{24}$$

On substituting (23)-(24) into the second parts of (15a)-(15b) the generalized ordinary differential equation governing the static deflection of the sandwich magnetoelastic structure is.

$$EI\frac{d^4W}{dx^4} + \frac{bh}{\mu_0}\left(1 - \frac{\mu_0}{\mu_m}\right)B_0{}^2\frac{d^2W}{dx^2} = \frac{1}{2}bh\mu\frac{dP}{dx}. \tag{25}$$

The following specialized ordinary differential equations can be formulated from the foregoing as follows.

Case 1. For the case of a beam-plate of thickness , (25) becomes

$$EI^*\frac{d^4W}{dx^4} + \frac{bh}{\mu_0}\left(1 - \frac{\mu_0}{\mu_m}\right)B_0{}^2\frac{d^2W}{dx^2}$$

$$= \frac{1}{2}bh\mu\frac{dP}{dx} \quad \forall I^* = \frac{bh^3}{12(1 - v^2)}. \tag{26a}$$

By dropping the variable b, the formulated equation governing the static deflection takes the form

$$D\frac{d^4 W}{dx^4} + \frac{h}{\mu_0}\left(1 - \frac{\mu_0}{\mu_m}\right)B_0{}^2\frac{d^2 W}{dx^2}$$

$$= \frac{1}{2}h\mu\frac{dP}{dx} \quad \forall D = \frac{Eh^3}{12(1 - v^2)}.$$

(26b)

Case 2: For the case of a beam-plate of thickness 2h, (25) becomes

$$EI^{**}\frac{d^4 W}{dx^4} + \frac{bh}{\mu_0}\left(1 - \frac{\mu_0}{\mu_m}\right)B_0{}^2\frac{d^2 W}{dx^2}$$

$$= \frac{1}{2}bh\mu\frac{dP}{dx} \quad \forall I^{**} = \frac{b(2h)^3}{12(1 - v^2)} = \frac{2bh^3}{3(1 - v^2)}.$$

(27a)

By dropping the variable b, the formulated equation governing the static deflection takes the form

$$D^*\frac{d^4 W}{dx^4} + \frac{2h}{\mu_0}\left(1 - \frac{\mu_0}{\mu_m}\right)B_0{}^2\frac{d^2 W}{dx^2}$$

$$= h\mu\frac{dP}{dx} \quad \forall D = \frac{2Eh^3}{3(1 - v^2)}.$$

(27b)

On the other hand, we can rewrite (26b) to obtain the following specialized ordinary differential equations.

Case 3: A two-layer sandwich homogenous magnetoelastic beam-plate of thickness h with non-uniform pressure at the interface. For such a problem, the formulated equation governing the static deflection takes the form

$$EI\frac{d^4 W}{dx^4} + \frac{bh}{\mu_0}\left(1 - \frac{\mu_0}{\mu_m}\right)(1 - v^2)B_0{}^2\frac{d^2 W}{dx^2}$$

$$= \frac{1}{2}bh\mu(1 - v^2)\frac{dP}{dx} \quad \forall I = \frac{bh^3}{12}.$$

(28)

Case 4: A two-layer sandwich homogenous magnetoelastic beam-plate of thickness 2h with non-uniform pressure at the interface. For such a problem, the formulated equation governing the static deflection takes the form

$$EI^*\frac{d^4 W}{dx^4} + \frac{2bh}{\mu_0}\left(1 - \frac{\mu_0}{\mu_m}\right)(1 - v^2)B_0{}^2\frac{d^2 W}{dx^2}$$

$$= bh\mu(1 - v^2)\frac{dP}{dx} \quad \forall I^* = \frac{2Eh^3}{3(1 - v^2)}.$$

(29)

ANALYSIS OF STATIC DEFLECTION

The generalized governing differential equation for the static deflection of each layer takes the following form:

$$\frac{d^4 W}{dx^4} + \frac{12}{\mu_0}\left(1 - \frac{\mu_0}{\mu_m}\right)\frac{B_0^2}{Eh^2}\frac{d^2 W}{dx^2} = \alpha\frac{dP}{dx}; \quad \forall \alpha = \frac{6\mu}{Eh^2}. \tag{30}$$

For the cantilever architecture under investigation, the usual boundary conditions hold as follows:

$$W(0) = \frac{dW(0)}{dx} = \frac{d^2 W(L)}{dx^2} = 0, \tag{31}$$

in conjunction with the generalized end condition reported in Damisa et al. [19], namely,

$$\int_0^h \tau_{(xz)_1}(x)dz = \frac{F}{2b} \quad \text{at } x = L. \tag{32a}$$

By limiting our investigation to linear interface pressure profile, we obtain

$$p(x) = p_0\left(1 + \frac{\varepsilon}{L}x\right). \tag{32b}$$

Equation (30) takes the form

$$\frac{d^4 W}{dx^4} + \frac{12}{\mu_0}\left(1 - \frac{\mu_0}{\mu_m}\right)\frac{B_0^2}{Eh^2}\frac{d^2 W}{dx^2} = \alpha\frac{\varepsilon}{L}p_0. \tag{33}$$

The solution to the above is sorted via the Fourier finite sine transform namely,

$$[\cdot]^F = \int_0^L [\cdot]\sin\left(\frac{n\pi x}{L}\right)dx; \qquad [\cdot] = \frac{2}{L}\sum_{n=0}^{\infty}[\cdot]^F \sin\left(\frac{n\pi x}{L}\right). \tag{34}$$

Equation (33) in the Fourier transform plane takes the form

$$\frac{n^4\pi^4}{L^4}W^F(\lambda_n) - \frac{12B_0^2}{Eh^2}\left(\frac{1}{\mu_0}\left(1 - \frac{\mu_0}{\mu_m}\right)\right)\frac{n^2\pi^2}{L^2}W^F(\lambda_n)$$

$$= \left(\begin{array}{c}\left(\frac{n^3\pi^3}{L^3}(-1)^{n+1} - \frac{12B_0^2}{Eh^2}\left(\frac{1}{\mu_0}\left(1 - \frac{\mu_0}{\mu_m}\right)\right)\frac{n\pi}{L}(-1)^{n+1}\right)W(L) - \frac{n\pi}{L}W_{xx}(0) \\ \alpha P_0\varepsilon\left(\frac{1+(-1)^{n+1}}{n\pi}\right)\end{array}\right). \tag{35}$$

Following Damisa et al. [19], the bending moment and static deflection in the Fourier transform plane are computed as

$$W_{xx}(0) = \left\{ \left(\frac{6F}{Ebh^3} - \frac{6\mu P_0}{Eh^2}\left(1 + \frac{\varepsilon}{2}\right) \right)L + \frac{12B_0^2}{Eh^2}\frac{1}{\mu_0}\left(1 - \frac{\mu_0}{\mu_m}\right)W(L) \right\},$$

$$W^F(\lambda_n) = \frac{\left(\begin{array}{c} \left(\frac{n^3\pi^3}{L^3}(-1)^{n+1} - \frac{n\pi}{L}(-1)^{n+1}\frac{12B_0^2}{Eh^2}\frac{1}{\mu_0}\left(1 - \frac{\mu_0}{\mu_m}\right) \right)W(L) \\ -n\pi\left(\frac{6F}{Ebh^3} - \frac{6\mu P_0\left(1 + \frac{\varepsilon}{2}\right)}{Eh^2} \right) - \frac{n\pi}{L}\frac{12B_0^2}{Eh^2}\frac{1}{\mu_0}\left(1 - \frac{\mu_0}{\mu_m}\right)W(L) + \frac{6\mu P_0\varepsilon}{Eh^2}\left(\frac{1 + (-1)^{n+1}}{n\pi} \right) \end{array} \right)}{\left(\frac{n^4\pi^4}{L^4} - \frac{12B_0^2}{Eh^2}\left(\frac{1}{\mu_0}\left(1 - \frac{\mu_0}{\mu_m}\right) \right)\frac{n^2\pi^2}{L^2} \right)}.$$

(36)

The Fourier inversion of the above yields

$$W(x) = \frac{\left\{ \begin{array}{c} 2W(L)\sum_{n=1}^{\infty}(-1)^{n+1}\frac{\sin n\pi\bar{x}}{n\pi} - 24W(L)\chi^2\left(1 - \frac{\mu_0}{\mu_m}\right)\sum_{n=1}^{\infty}(-1)^{n+1}\frac{\sin n\pi\bar{x}}{n^3\pi^3} \\ -12\left(\begin{array}{c} L^3\left(\frac{F}{Ebh^3} - \frac{\mu P_0\left(1 + \frac{\varepsilon}{2}\right)}{Eh^2} \right)\sum_{n=1}^{\infty}\frac{\sin n\pi\bar{x}}{n^3\pi^3} \\ +\left(2\chi^2\left(1 - \frac{\mu_0}{\mu_m}\right)W(L) \right)\sum_{n=1}^{\infty}\left(\frac{(-1)^{n+1}}{n^3\pi^3} \right)\sin n\pi\bar{x} \end{array} \right) + \frac{3L^3\mu}{8Eh^2}P_0\varepsilon\sum_{n=1}^{\infty}\frac{\sin 2n\pi\bar{x}}{n^5\pi^5} \end{array} \right\}}{\left(1 - 12\left(1 - \frac{\mu_0}{\mu_m}\right)\frac{\chi^2}{n^2\pi^2} \right)} \qquad \forall \chi = \frac{B_0^2 L^2}{\mu_0 Eh^2}.$$

(37)

Utilizing binomial expansion, (37) is rewritten as

$$W(x) = \left(1 + 12\left(1 - \frac{\mu_0}{\mu_m}\right)\frac{\chi^2}{n^2\pi^2}\right)$$

$$\times \left\{ \begin{array}{c} 2W(L)\sum_{n=1}^{\infty}(-1)^{n+1}\frac{\sin n\pi\bar{x}}{n\pi} - 24W(L)\chi^2\left(1 - \frac{\mu_0}{\mu_m}\right)\sum_{n=1}^{\infty}(-1)^{n+1}\frac{\sin n\pi\bar{x}}{n^3\pi^3} \\ -12\left(\begin{array}{c} L^3\left(\frac{F}{Ebh^3} - \frac{\mu P_0\left(1 + \frac{\varepsilon}{2}\right)}{Eh^2} \right)\sum_{n=1}^{\infty}\frac{\sin n\pi\bar{x}}{n^3\pi^3} \\ +2\chi^2\left(1 - \frac{\mu_0}{\mu_m}\right)W(L)\sum_{n=1}^{\infty}\frac{\sin n\pi\bar{x}}{n^3\pi^3} \end{array} \right) + \frac{3L^3\mu}{8Eh^2}P_0\varepsilon\sum_{n=1}^{\infty}\frac{\sin 2n\pi\bar{x}}{n^5\pi^5} \end{array} \right\}.$$

(38)

The semiinfinite series in (38) can be converted to spatial polynomials via the following closed form Fourier series representations:

$$\bar{x} = \frac{1}{\pi}\sum_{n=1}^{\infty}\frac{(-1)^{n+1}}{n}\sin n\pi\bar{x},$$

$$\forall 0 < \bar{x} < 1,$$

$$\sum_{n=1}^{\infty}\frac{\sin n\bar{x}}{n^3} = \frac{\pi^2\bar{x}}{6} - \frac{\pi\bar{x}^2}{4} + \frac{\bar{x}^3}{12},$$

$$\forall 0 < \bar{x} < 2,$$

$$\sum_{n=1}^{\infty} \frac{\sin n\bar{x}}{n^5} = \frac{\pi^4 \bar{x}}{90} - \frac{\pi^2 \bar{x}^3}{36} + \frac{\pi \bar{x}^4}{48} - \frac{\bar{x}^5}{240},$$

$$\forall 0 < \bar{x} < 2,$$

$$\sum_{n=1}^{\infty} \frac{\sin n\bar{x}}{n^7} = \frac{2\pi^6 \bar{x}}{405} - \frac{\pi^4 \bar{x}^3}{540} + \frac{\pi^2 \bar{x}^5}{720} - \frac{\pi \bar{x}^6}{1440} + \frac{\bar{x}^7}{980},$$

$$\forall 0 < \bar{x} < 2. \tag{39}$$

Consequently, we can write (39) in the form

$$W(x) = \left\{ \begin{array}{l} 2W(L)\bar{x} - 12L^3 \left(\left(\frac{F}{Ebh^3} - \frac{\mu P_0(1+\varepsilon/2)}{Eh^2} \right)\Lambda_1 \right) + \frac{3L^3 \mu}{8Eh^2} P_0 \varepsilon \Lambda_2 - W(L) \left(24\chi^2 \left(1 - \frac{\mu_0}{\mu_m} \right) + 15\chi^4 \left(1 - \frac{\mu_0}{\mu_m} \right)^2 \right)\Lambda_2 \\[2mm] -144L^5 \chi^2 \left(1 - \frac{\mu_0}{\mu_m} \right) \left(\frac{F}{Ebh^3} - \frac{\mu P_0(1+\varepsilon/2)}{Eh^2} \right)\Lambda_3 + \frac{9L^3 \mu}{2Eh^2} \chi^2 \left(1 - \frac{\mu_0}{\mu_m} \right) P_0 \varepsilon \, \Lambda_4 \end{array} \right\}, \tag{40}$$

Where

$$\Lambda_1 = \left(\frac{\bar{x}}{6} - \frac{\bar{x}^2}{4} + \frac{\bar{x}^3}{12} \right);$$

$$\Lambda_2 = \left(\frac{\bar{x}}{45} - \frac{2\bar{x}^3}{9} + \frac{\bar{x}^4}{3} - \frac{2\bar{x}^5}{5} \right);$$

$$\Lambda_3 = \left(\frac{\bar{x}}{90} - \frac{\bar{x}^3}{36} + \frac{\bar{x}^4}{48} - \frac{\bar{x}^5}{240} \right);$$

$$\Lambda_4 = \frac{4\bar{x}}{405} - \frac{2\bar{x}^3}{135} + \frac{2\bar{x}^5}{45} - \frac{2\bar{x}^6}{45} + \frac{64\bar{x}^7}{245}. \tag{41}$$

Imposing the condition $d\overline{W}/dx = 0$ in (40), the deflection at the end of the sandwich magnetoelastic cantilever structure is

$$W(L) = \left(\begin{array}{l} \left(\frac{F}{Ebh^3} - \frac{\mu P_0(1+\varepsilon/2)}{Eh^2} - \frac{\mu P_0 \varepsilon}{240Eh^2} \right) + \chi^2 \left(1 - \frac{\mu_0}{\mu_m} \right) \left(\frac{F}{Ebh^3} - \frac{\mu P_0}{Eh^2} \right) \\[2mm] + \frac{13}{50}\chi^4 \left(1 - \frac{\mu_0}{\mu_m} \right)^2 \frac{F}{Ebh^3} - \frac{\mu P_0(1+\varepsilon/2)}{Eh^2} - \frac{5\mu P_0 \varepsilon}{26Eh^2} + \frac{2}{25}\chi^6 \left(1 - \frac{\mu_0}{\mu_m} \right)^3 \left(\frac{F}{Ebh^3} - \frac{\mu P_0}{Eh^2} \right) \end{array} \right) \tag{42}$$

Substitution of the above with rearrangement gives

$$\overline{W}(\bar{x}) = \left\{ \begin{array}{l} (1 - \mu\overline{P}_0)(3\bar{x}^2 - \bar{x}^3) + \mu\overline{P}_0 \varepsilon \left(-3\bar{x}^2 + \frac{11}{12}\bar{x}^3 + \frac{1}{8}\bar{x}^4 - \frac{1}{20}\bar{x}^5 \right) \\[2mm] +\chi^2 \left(1 - \frac{\mu_0}{\mu_m} \right) \left((1 - \mu\overline{P}_0) \left(-\frac{28}{13}\bar{x}^3 + 11\bar{x}^4 + \frac{51}{5}\bar{x}^5 \right) + \mu\overline{P}_0 \varepsilon \left(-\frac{214}{45}\bar{x}^3 - \frac{83}{15}\bar{x}^4 + \frac{267}{50}\bar{x}^5 - \frac{1}{5}\bar{x}^6 + \frac{288}{245}\bar{x}^8 \right) \right) \\[2mm] +\chi^4 \left(1 - \frac{\mu_0}{\mu_m} \right)^2 \left((1 - \mu\overline{P}_0) \left(\frac{26}{3}\bar{x}^3 - 13\bar{x}^4 + \frac{78}{5}\bar{x}^5 \right) + \mu\overline{P}_0 \varepsilon \left(-\frac{121}{72}\bar{x}^3 + \frac{121}{48}\bar{x}^4 + \frac{121}{40}\bar{x}^5 \right) \right) \\[2mm] +\chi^6 \left(1 - \frac{\mu_0}{\mu_m} \right)^3 \left((1 - \mu\overline{P}_0) \left(\frac{-354}{75}\bar{x}^3 - \frac{177}{25}\bar{x}^4 + \frac{1107}{125}\bar{x}^5 \right) + \mu\overline{P}_0 \varepsilon \left(-\frac{1}{20}\bar{x}^3 + \frac{20}{13}\bar{x}^4 - \frac{24}{13}\bar{x}^5 \right) \right) \\[2mm] +\chi^8 \left(1 - \frac{\mu_0}{\mu_m} \right)^4 \left((1 - \mu\overline{P}_0) \left(\frac{97}{75}\bar{x}^3 - \frac{97}{50}\bar{x}^4 + \frac{296}{125}\bar{x}^5 \right) + \mu\overline{P}_0 \varepsilon \left(-\frac{1}{6}\bar{x}^3 - \frac{20}{13}\bar{x}^4 + \frac{33}{50}\bar{x}^5 \right) \right) \\[2mm] +\chi^{10} \left(1 - \frac{\mu_0}{\mu_m} \right)^5 \left((1 - \mu\overline{P}_0) \left(\frac{4}{15}\bar{x}^3 - \frac{2}{5}\bar{x}^4 + \frac{12}{25}\bar{x}^5 \right) \right) \end{array} \right\}, \tag{43}$$

where the following nondimensionalized parameters have been introduced as

$$\overline{W}(\bar{x}) = \frac{W(\bar{x})Ebh^3}{L^3 F}; \qquad \overline{P}_0 = \frac{P_0}{(F/bh)}; \qquad \bar{x} = \frac{x}{L}.$$

(44)

For the special case $\varepsilon \to 0$ in (43), the transverse deflection at uniform pressure is

$\overline{W}(\bar{x})$

$$= \left\{ \begin{array}{l} (1 - \mu \overline{P}_0)(3\bar{x}^2 - \bar{x}^3) \\ +\chi^2\left(1 - \frac{\mu_0}{\mu_m}\right)\left((1 - \mu \overline{P}_0)\left(-\frac{28}{13}\bar{x}^3 + 11\bar{x}^4 + \frac{51}{5}\bar{x}^5\right)\right) \\ +\chi^4\left(1 - \frac{\mu_0}{\mu_m}\right)^2\left((1 - \mu \overline{P}_0)\left(\frac{26}{3}\bar{x}^3 - 13\bar{x}^4 + \frac{78}{5}\bar{x}^5\right)\right) \\ +\chi^6\left(1 - \frac{\mu_0}{\mu_m}\right)^3\left((1 - \mu \overline{P}_0)\left(\frac{-354}{75}\bar{x}^3 - \frac{177}{25}\bar{x}^4 + \frac{1107}{125}\bar{x}^5\right)\right) \\ +\chi^8\left(1 - \frac{\mu_0}{\mu_m}\right)^4\left((1 - \mu \overline{P}_0)\left(\frac{97}{75}\bar{x}^3 - \frac{97}{50}\bar{x}^4 + \frac{296}{125}\bar{x}^5\right)\right) \\ +\chi^{10}\left(1 - \frac{\mu_0}{\mu_m}\right)^5\left((1 - \mu \overline{P}_0)\left(\frac{4}{15}\bar{x}^3 - \frac{2}{5}\bar{x}^4 + \frac{12}{25}\bar{x}^5\right)\right) \end{array} \right\},$$

(45)

which for the case of $\chi = 0$ agrees with the results in Damisa et al. [19].

ANALYSIS OF STATIC SLIP

As shown in Figure 1(e), during bending, each half of the layered elastic structure has its own neutral plane that does not necessarily coincide with the geometric mid plane through the interface because of the frictional stresses. For the sandwich structure, the geometrical description of the gross interfacial slip is defined in Figure 1(e). In view of the foregoing, the expressions for the displacements of the two adjacent opposite points follow from Taylor series approximation as:

$\Delta U_1(x,z)$

$$= \left(\Delta U_1(0,0+) + \left(z - \psi(x)\frac{h}{2}\right)\frac{dW(x)}{dx} + \frac{1}{2}\left(z - \psi(x)\frac{h}{2}\right)^2 \right.$$
$$\left. \times \frac{d^2W(x)}{dx^2} + \frac{1}{6}\left(z - \psi(x)\frac{h}{2}\right)^3\frac{d^3W(x)}{dx^3}\right),$$

(46)

$\Delta U_2(x,z)$

$$= \left(\Delta U_2(0,0-) + \left(z + \psi(x)\frac{h}{2}\right)\frac{dW(x)}{dx} + \frac{1}{2}\left(z + \psi(x)\frac{h}{2}\right)^2 \right.$$
$$\left. \times \frac{d^2W(x)}{dx^2} + \frac{1}{6}\left(z + \psi(x)\frac{h}{2}\right)^3\frac{d^3W(x)}{dx^3}\right),$$

(47)

which for the case of first order theory reduce to the forms

$$\Delta U_1(x,z) = \Delta U_1(0,0+) + \left(z - \psi(x)\frac{h}{2}\right)\frac{dW(x)}{dx},$$

$$\Delta U_2(x,z) = \Delta U_2(0,0-) + \left(z + \psi(x)\frac{h}{2}\right)\frac{dW(x)}{dx}.$$

$$(48)$$

For this problem, $\Delta u_1(0, 0+)$ and $\Delta u_2(0, 0-)$ must be zero at the fixed end. Hence, the relative static slip at the interface of the elastic structure is given by

$$\Delta U(x,0) = \Delta U_1(x,0+) - \Delta U_2(x,0-).$$

$$(49)$$

Following Goodman and Klumpp [24], (49) becomes

$$\Delta U(x,0) = E^{-1}\int_0^x \{(\sigma_x)_1(\xi,0+) - (\sigma_x)_2(\xi,0-)\}d\xi,$$

$$(50)$$

where ξ is a dummy axial spatial variable of integration across the interface, and 0+, 0− denote the origin of the transverse spatial variable for each layer; where subscripts 1 and 2 refer to the upper and lower laminates.

On substituting (23)-(24) into the first parts of (15a)-(15b), the derived corresponding spatial bending stresses are, namely,

$(\sigma_x)_1(x,z)$

$$= -\frac{E}{2}(2z - h)\frac{d^2W(x)}{dx^2} + \frac{B_0^2}{\mu_0}\left(1 - \frac{\mu_0}{\mu_m}\right)$$

$$\times (W(x) - W(L)) + \mu P_0(1 + \varepsilon\bar{x})\frac{(x - L)}{h},$$

$(\sigma_x)_2(x,z)$

$$= -\frac{E}{2}(2z + h)\frac{d^2W(x)}{dx^2} - \frac{B_0^2}{\mu_0}\left(1 - \frac{\mu_0}{\mu_m}\right)$$

$$\times (W(x) - W(L)) - \mu P_0(1 + \varepsilon\bar{x})\frac{(x - L)}{h}.$$

$$(51)$$

This gives (50) as

$\Delta U(\bar{x})$

$$= \int_0^{\bar{x}}\left\{\frac{d^2\overline{W}(\bar{x})}{d\bar{\xi}^2} + 2\overline{B}_0^2\left(1 - \frac{\mu_0}{\mu_m}\right)\right.$$

$$\left.\times\left(\overline{W}(\bar{\xi}) - \overline{W}(1)\right) + 2\mu\overline{P}_0\left(1 + \varepsilon\bar{\xi}\right)\left(\bar{\xi} - 1\right)\right\}d\bar{\xi},$$

$$(52)$$

and on introducing the following nondimensionalized parameters:

$$\Delta \bar{U} = \frac{\Delta U(\bar{x})\, Ebh^2}{L^2 F}; \qquad \bar{P}_0 = \frac{P_0}{(F/bh)}.$$

(53)

Equation (52) simplifies to the form

$$\Delta \bar{U} = \left\{ \begin{array}{l} \left(4\mu\bar{P}_0 - 3\right)\left(\bar{x}^2 - 2\bar{x}\right) + \mu\bar{P}_0\varepsilon\left(-6\bar{x} + \dfrac{7\bar{x}^2}{4} + \dfrac{7\bar{x}^3}{6} - \dfrac{\bar{x}^4}{4}\right) \\[2mm] +\chi^2\left(1 - \dfrac{\mu_0}{\mu_m}\right)\left(\left(1 - \mu\bar{P}_0\right)\Pi_1(\bar{x}) + \mu\bar{P}_0\varepsilon\Pi_2(\bar{x})\right) + \chi^4\left(1 - \dfrac{\mu_0}{\mu_m}\right)^2\left(\left(1 - \mu\bar{P}_0\right)\Pi_3(\bar{x}) + \mu\bar{P}_0\varepsilon\Pi_4(\bar{x})\right) \\[2mm] +\chi^6\left(1 - \dfrac{\mu_0}{\mu_m}\right)^3\left(\left(1 - \mu\bar{P}_0\right)\Pi_5(\bar{x}) + \mu\bar{P}_0\varepsilon\Pi_6(\bar{x})\right) + \chi^8\left(1 - \dfrac{\mu_0}{\mu_m}\right)^4\left(\left(1 - \mu\bar{P}_0\right)\Pi_7(\bar{x}) + \mu\bar{P}_0\varepsilon\Pi_8(\bar{x})\right) \\[2mm] +\chi^{10}\left(1 - \dfrac{\mu_0}{\mu_m}\right)^5\left(\left(1 - \mu\bar{P}_0\right)\Pi_9(\bar{x}) + \mu\bar{P}_0\varepsilon\Pi_{10}(\bar{x})\right) + \chi^{12}\left(1 - \dfrac{\mu_0}{\mu_m}\right)^6\left(1 - \mu\bar{P}_0\right)\Pi_{11}(\bar{x}) \end{array} \right\},$$

(54)

Where

$$\Pi_1(\bar{x}) = \left(2\bar{x} - \frac{84}{13}\bar{x}^2 + 44\bar{x}^3 + 51\bar{x}^4\right),$$

$$\Pi_2(\bar{x}) = \left(-\frac{241}{120}\bar{x} - \frac{214}{15}\bar{x}^2 - \frac{332}{15}\bar{x}^3 + \frac{6463}{240}\bar{x}^4\right.$$

$$\left. -\frac{47}{40}\bar{x}^5 - \frac{1}{120}\bar{x}^6 - \frac{2304}{245}\bar{x}^7\right),$$

$$\Pi_3(\bar{x}) = \left(\frac{1283}{65}\bar{x} + \frac{1}{26}\bar{x}^2 - 52\bar{x}^3 - \frac{1021}{13}\bar{x}^4 + \frac{11}{5}\bar{x}^5 + \frac{17}{10}\bar{x}^6\right),$$

$$\Pi_4(\bar{x}) = \left(4\bar{x} - \frac{121}{24}\bar{x}^2 + \frac{121}{12}\bar{x}^3 + \frac{5017}{360}\bar{x}^4\right.$$

$$\left. -\frac{83}{75}\bar{x}^5 + \frac{267}{300}\bar{x}^6 - \frac{1}{35}\bar{x}^7 + \frac{32}{245}\bar{x}^9\right),$$

$$\Pi_5(\bar{x}) = \left(\frac{169}{5}\bar{x} - \frac{354}{25}\bar{x}^2 - \frac{708}{25}\bar{x}^3 + \frac{6967}{150}\bar{x}^4 - \frac{13}{5}\bar{x}^5 + \frac{39}{15}\bar{x}^6\right),$$

$$\Pi_6(\bar{x}) = \left(\frac{2783}{720}\bar{x} - \frac{3}{20}\bar{x}^2 + \frac{80}{13}\bar{x}^3\right.$$

$$\left. -\frac{36133}{3744}\bar{x}^4 + \frac{121}{240}\bar{x}^5 + \frac{121}{240}\bar{x}^6\right),$$

$$\Pi_7(\bar{x}) = \left(-\frac{368}{125}\bar{x} - \frac{97}{25}\bar{x}^2 - \frac{194}{25}\bar{x}^3\right.$$

$$\left. +\frac{533}{50}\bar{x}^4 - \frac{177}{125}\bar{x}^5 + \frac{1107}{750}\bar{x}^6\right),$$

$$\Pi_8(\bar{x}) = \left(\frac{-93}{260}\bar{x} - \frac{1}{2}\bar{x}^2 + \frac{80}{13}\bar{x}^3 - \frac{53}{16}\bar{x}^4 + \frac{4}{13}\bar{x}^5 - \frac{4}{13}\bar{x}^6 \right),$$

$$\Pi_9(\bar{x}) = \left(\frac{1291}{750}\bar{x} - \frac{97}{25}\bar{x}^2 - \frac{194}{25}\bar{x}^3 \right.$$

$$\left. + \frac{3649}{300}\bar{x}^4 + \frac{97}{250}\bar{x}^5 - \frac{148}{375}\bar{x}^6 \right),$$

$$\Pi_{10}(\bar{x}) = \left(\frac{-1019}{975}\bar{x} - \frac{1}{2}\bar{x}^2 + \frac{80}{13}\bar{x}^3 - \frac{401}{120}\bar{x}^4 - \frac{4}{13}\bar{x}^5 + \frac{11}{100}\bar{x}^6 \right),$$

$$\Pi_{11}(\bar{x}) = \left(\frac{26}{75}\bar{x} + \frac{1}{15}\bar{x}^4 - \frac{1}{5}\bar{x}^5 + \frac{2}{25}\bar{x}^6 \right). \tag{55}$$

On setting $\varepsilon \rightarrow 0$ in (54), the static slip at uniform pressure is

$$\Delta\bar{U} = \left\{ \begin{matrix} (4\mu\bar{P}_0 - 3)(\bar{x}^2 - 2\bar{x}) + \chi^2\left(1 - \frac{\mu_0}{\mu_m}\right)(1 - \mu\bar{P}_0)\Pi_1(\bar{x}) + \chi^4\left(1 - \frac{\mu_0}{\mu_m}\right)^2(1 - \mu\bar{P}_0)\Pi_3(\bar{x}) \\ +\chi^6\left(1 - \frac{\mu_0}{\mu_m}\right)^3(1 - \mu\bar{P}_0)\Pi_5(\bar{x}) + \chi^8\left(1 - \frac{\mu_0}{\mu_m}\right)^4(1 - \mu\bar{P}_0)\Pi_7(\bar{x}) \\ +\chi^{10}\left(1 - \frac{\mu_0}{\mu_m}\right)^5(1 - \mu\bar{P}_0)\Pi_9(\bar{x}) + \chi^{12}\left(1 - \frac{\mu_0}{\mu_m}\right)^6(1 - \mu\bar{P}_0)\Pi_{11}(\bar{x}) \end{matrix} \right\}. \tag{56}$$

The case $\chi = 0$ agrees with the result in [19]. The foregoing expression allows us to compute the maximum slip at the tip of the magnetoelastic structure in the context of the pressure gradient as

$$\Delta\bar{U} = \left\{ \begin{matrix} (3 - 4\mu\bar{P}_0) - 3.33\mu\bar{P}_0\varepsilon + \chi^2\left(1 - \frac{\mu_0}{\mu_m}\right)(91(1 - \mu\bar{P}_0) - 24\mu\bar{P}_0\varepsilon) \\ +\chi^4\left(1 - \frac{\mu_0}{\mu_m}\right)^2(-107(1 - \mu\bar{P}_0) - 22.9\mu\bar{P}_0\varepsilon) + \chi^6\left(1 - \frac{\mu_0}{\mu_m}\right)^3(38(1 - \mu\bar{P}_0) + 1.23\mu\bar{P}_0\varepsilon) \\ +\chi^8\left(1 - \frac{\mu_0}{\mu_m}\right)^4(-3.86(1 - \mu\bar{P}_0) + 1.98\mu\bar{P}_0\varepsilon) + \chi^{10}\left(1 - \frac{\mu_0}{\mu_m}\right)^5(2.24(1 - \mu\bar{P}_0) + 1.07\mu\bar{P}_0\varepsilon) \\ -0.3\chi^{12}\left(1 - \frac{\mu_0}{\mu_m}\right)^6(1 - \mu\bar{P}_0) \end{matrix} \right\}. \tag{57}$$

On setting $\varepsilon \rightarrow 0$ in (57), the maximum static slip at the tip for the case of uniform pressure is

$$\Delta\bar{U}_{max(tip)} = \left\{ \begin{matrix} (3 - 4\mu\bar{P}_0) + 91\chi^2\left(1 - \frac{\mu_0}{\mu_m}\right)(1 - \mu\bar{P}_0) - 107\chi^4\left(1 - \frac{\mu_0}{\mu_m}\right)^2(1 - \mu\bar{P}_0) \\ +38\chi^6\left(1 - \frac{\mu_0}{\mu_m}\right)^3(1 - \mu\bar{P}_0) - 3.86\chi^8\left(1 - \frac{\mu_0}{\mu_m}\right)^4(1 - \mu\bar{P}_0) \\ +2.24\chi^{10}\left(1 - \frac{\mu_0}{\mu_m}\right)^5(1 - \mu\bar{P}_0) - 0.3\chi^{12}\left(1 - \frac{\mu_0}{\mu_m}\right)^6(1 - \mu\bar{P}_0) \end{matrix} \right\}. \tag{58}$$

This suggests that in the presence of coulomb friction and interfacial pressure $\bar{P}_0$, there are critical values of pressure for which no slip occurs at the tip such as

$$\bar{P}_{0(critical)} = 0.75\mu^{-1}\frac{Y_1\left(\chi,\bar{B}_0\right)}{Y_2\left(\chi,\bar{B}_0\right)},$$

(59)

Where

$$Y_1\left(\chi,\bar{B}_0\right) = \left(\begin{array}{l} 1 + \frac{91}{3}\chi^2\left(1 - \frac{\mu_0}{\mu_m}\right) - \frac{107}{3}\bar{B}_0^4\left(1 - \frac{\mu_0}{\mu_m}\right)^2 + \frac{38}{3}\chi^6\left(1 - \frac{\mu_0}{\mu_m}\right)^3 - \frac{3.86}{3}\chi^8\left(1 - \frac{\mu_0}{\mu_m}\right)^4 \\ + \frac{2.24}{3}\chi^{10}\left(1 - \frac{\mu_0}{\mu_m}\right)^5 - \frac{1}{10}\chi^{12}\left(1 - \frac{\mu_0}{\mu_m}\right)^6 \end{array}\right),$$

$$Y_2\left(\chi,\bar{B}_0\right) = \left(\begin{array}{l} 1 + \frac{3.33}{4} + \chi^2\left(1 - \frac{\mu_0}{\mu_m}\right)\left(\frac{91}{4} + 6\varepsilon\right) - \chi^4\left(1 - \frac{\mu_0}{\mu_m}\right)^2\left(\frac{107}{4} - \frac{22.9}{4}\varepsilon\right) \\ + \chi^6\left(1 - \frac{\mu_0}{\mu_m}\right)^3\left(\frac{38}{4} - \frac{1.23}{4}\varepsilon\right) - \chi^8\left(1 - \frac{\mu_0}{\mu_m}\right)^4\left(\frac{3.86}{4} + \frac{1.23}{4}\varepsilon\right) \\ + \chi^{10}\left(1 - \frac{\mu_0}{\mu_m}\right)^5\left(\frac{2.24}{4} - \frac{1.07}{4}\varepsilon\right) - \frac{3}{10}\chi^{12}\left(1 - \frac{\mu_0}{\mu_m}\right)^5 \end{array}\right).$$

(60)

The case $\varepsilon \to 0$ gives the expression for critical values of pressure for which no slip occurs at the tip as

$$\bar{P}_{0(critical)} = 0.75\mu^{-1}\frac{Y_1\left(\chi,\bar{B}_0\right)}{Y_3\left(\chi,\bar{B}_0\right)},$$

(61)

Where

$$Y_3\left(\chi,\bar{B}_0\right) = \left(\begin{array}{l} 1 + \frac{3.33}{4} + \frac{91}{4}\chi^2\left(1 - \frac{\mu_0}{\mu_m}\right) - \frac{107}{4}\chi^4\left(1 - \frac{\mu_0}{\mu_m}\right)^2 + \frac{38}{4}\chi^6\left(1 - \frac{\mu_0}{\mu_m}\right)^3 - \frac{3.86}{4}\chi^8\left(1 - \frac{\mu_0}{\mu_m}\right)^4 \\ + \frac{2.24}{4}\chi^{10}\left(1 - \frac{\mu_0}{\mu_m}\right)^5 - \frac{3}{10}\chi^{12}\left(1 - \frac{\mu_0}{\mu_m}\right)^5 \end{array}\right).$$

(62)

which for the case of $\bar{B}_0 = 0$ agrees with the result in [19].

ENERGY DISSIPATION

The energy dissipated per static slip, following Damisa [25] is given by the relation

$$D = 4\mu b \int_0^L p(x)\Delta u(x)dx,$$

(63)

which can also be expressed as

$$\bar{D} = 4\mu\bar{p}_0 \int_0^1 \left(1 + \frac{\varepsilon}{2}\right)\Delta\bar{u}\,d\bar{x} \quad \forall\bar{p}_{av} = \bar{p}_0 \int_0^1 (1 + \varepsilon x)d\bar{x},$$

(64)

on substituting for $\Delta\bar{u}$ gives

$$\bar{D} = \begin{pmatrix} \left(8\mu\bar{p}_0 - \frac{32}{3}\mu^2\bar{p}_0^2\right) + 4\mu\bar{p}_0\varepsilon - \frac{421}{30}\mu^2\bar{p}_0^2\varepsilon - \frac{261}{60}\mu^2\bar{p}_0^2\varepsilon^2 \\[2mm] +\chi^2\left(1 - \frac{\mu_0}{\mu_m}\right)\left(80\mu\bar{p}_0 - 80\mu^2\bar{p}_0^2 + 20\mu\bar{p}_0\varepsilon - 69.2\mu^2\bar{p}_0^2\varepsilon - \frac{29.2}{2}\mu^2\bar{p}_0^2\varepsilon^2\right) \\[2mm] +\chi^4\left(1 - \frac{\mu_0}{\mu_m}\right)^2\left(-72.8\mu\bar{p}_0 + 72.8\mu^2\bar{p}_0^2 - 36.4\mu\bar{p}_0\varepsilon + 14\mu^2\bar{p}_0^2\varepsilon - 11.2\mu^2\bar{p}_0^2\varepsilon^2\right) \\[2mm] +\chi^6\left(1 - \frac{\mu_0}{\mu_m}\right)^3\left(57.2\mu\bar{p}_0 - 57.2\mu^2\bar{p}_0^2 + 28.6\mu\bar{p}_0\varepsilon - 21.8\mu^2\bar{p}_0^2\varepsilon + 3.4\mu^2\bar{p}_0^2\varepsilon^2\right) \\[2mm] +\chi^8\left(1 - \frac{\mu_0}{\mu_m}\right)^4\left(-10.4\mu\bar{p}_0 + 10.4\mu^2\bar{p}_0^2 - 5.2\mu\bar{p}_0\varepsilon + 7.4\mu^2\bar{p}_0^2\varepsilon + 1.1\mu^2\bar{p}_0^2\varepsilon^2\right) \\[2mm] +\chi^{10}\left(1 - \frac{\mu_0}{\mu_m}\right)^5\left(0.28\mu\bar{p}_0 - 0.28\mu^2\bar{p}_0^2 + 0.14\mu\bar{p}_0\varepsilon - 0.46\mu^2\bar{p}_0^2\varepsilon + 0.3\mu^2\bar{p}_0^2\varepsilon^2\right) \\[2mm] +\chi^{12}\left(1 - \frac{\mu_0}{\mu_m}\right)^6\left(0.68\mu\bar{p}_0 - 0.68\mu^2\bar{p}_0^2 + 0.32\mu\bar{p}_0\varepsilon - 0.32\mu^2\bar{p}_0^2\varepsilon^2\right) \end{pmatrix},$$

(65)

where $\bar{D} = DEbh^3/(L^3F_0^2)$ is the dimensionless static energy dissipated.

Analysis of Optimum Clamping Pressure

The optimum clamping pressure can be found from the partial derivative of the energy dissipated if we set

$$\frac{\partial\bar{D}}{\partial\bar{p}_0} = 0.$$

(66)

Thus we can derive the general expression for the optimum clamping pressure as

$$\bar{p}_{opt} = 0.375\mu^{-1}\frac{\Lambda_1}{\Lambda_2},$$

(67)

Where

$$\Lambda_1 = \begin{pmatrix} 1 + \frac{1}{2}\varepsilon + \frac{5}{2}\chi^2\left(1 - \frac{\mu_0}{\mu_m}\right)(1 + 4\varepsilon) + \frac{91}{20}\chi^4\left(1 - \frac{\mu_0}{\mu_m}\right)^2(2 - \varepsilon) + \frac{143}{40}\chi^6\left(1 - \frac{\mu_0}{\mu_m}\right)^3(2 + \varepsilon) \\[2mm] +\frac{13}{20}\chi^8\left(1 - \frac{\mu_0}{\mu_m}\right)^4(-2 - \varepsilon) + \frac{7}{400}\chi^{10}\left(1 - \frac{\mu_0}{\mu_m}\right)^5(2 + \varepsilon) + \frac{17}{12}\chi^{12}\left(1 - \frac{\mu_0}{\mu_m}\right)^6(1 + \varepsilon) \end{pmatrix},$$

$$\Lambda_2 = \begin{pmatrix} 1 + \frac{421}{320}\varepsilon + \frac{261}{640}\varepsilon^2 + \chi^2\left(1 - \frac{\mu_0}{\mu_m}\right)\left(\frac{15}{2} + \frac{81}{25}\varepsilon + \frac{137}{100}\varepsilon^2\right) \\ +\chi^4\left(1 - \frac{\mu_0}{\mu_m}\right)^2\left(-\frac{683}{100} - \frac{131}{100}\varepsilon + \frac{21}{20}\varepsilon^2\right) + \chi^6\left(1 - \frac{\mu_0}{\mu_m}\right)^3\left(\frac{107}{20} + \frac{51}{20}\varepsilon - \frac{8}{25}\varepsilon^2\right) \\ +\chi^8\left(1 - \frac{\mu_0}{\mu_m}\right)^4\left(-\frac{49}{50} - \frac{69}{100}\varepsilon - \frac{103}{1000}\varepsilon^2\right) + \chi^{10}\left(1 - \frac{\mu_0}{\mu_m}\right)^5\left(\frac{3}{100} + \frac{43}{1000}\varepsilon - \frac{3}{100}\varepsilon^2\right) \\ +\chi^{12}\left(1 - \frac{\mu_0}{\mu_m}\right)^6\left(\frac{16}{25} + \frac{3}{100}\varepsilon\right) \end{pmatrix}.$$

$$(68)$$

In the limit as $\chi \to 0$, we recover the optimum clamping pressure in [19] as

$$\hat{\overline{P}}_0 = 0.375\mu^{-1}\left\{\left(1 + \frac{1}{2}\varepsilon\right)\left(1 + \frac{421}{320}\varepsilon + \frac{261}{640}\varepsilon^2\right)^{-1}\right\}.$$

$$(69)$$

By letting $\varepsilon \to 0$, the optimum clamping pressure for the magnetoelastic structure computed is

$$\overline{P}_{opt} = 0.375\mu^{-1}\frac{\Lambda_3}{\Lambda_4},$$

$$(70)$$

Where

Λ_3

$$= \begin{pmatrix} 1 + \frac{5}{2}\chi^2\left(1 - \frac{\mu_0}{\mu_m}\right) + \frac{91}{10}\chi^4\left(1 - \frac{\mu_0}{\mu_m}\right)^2 + \frac{143}{20}\chi^6\left(1 - \frac{\mu_0}{\mu_m}\right)^3 \\ -\frac{13}{10}\chi^8\left(1 - \frac{\mu_0}{\mu_m}\right)^4 + \frac{7}{200}\chi^{10}\left(1 - \frac{\mu_0}{\mu_m}\right)^5 + \frac{17}{12}\chi^{12}\left(1 - \frac{\mu_0}{\mu_m}\right)^6 \end{pmatrix},$$

Λ_4

$$= \begin{pmatrix} 1 + \frac{15}{2}\chi^2\left(1 - \frac{\mu_0}{\mu_m}\right) - \frac{683}{100}\chi^4\left(1 - \frac{\mu_0}{\mu_m}\right)^2 + \frac{107}{20}\chi^6\left(1 - \frac{\mu_0}{\mu_m}\right)^3 \\ -\frac{49}{50}\chi^8\left(1 - \frac{\mu_0}{\mu_m}\right)^4 + \frac{3}{100}\chi^{10}\left(1 - \frac{\mu_0}{\mu_m}\right)^5 + \frac{16}{25}\chi^{12}\left(1 - \frac{\mu_0}{\mu_m}\right)^6 \end{pmatrix},$$

$$(71)$$

and for the special case $\chi \to 0$, the optimum clamping pressure reduce to the form $\hat{\overline{P}}_0 = 0.375\mu^{-1} = 3/8\mu$ in Damisa [25]. At this optimal pressure, the corresponding energy dissipation is

$$\bar{D}_{max} = \frac{3}{2}\Psi$$

$$\times \begin{pmatrix} (2+\varepsilon) - \Psi\left(1 + \frac{421}{320}\varepsilon\right) - \Psi\frac{261}{640}\varepsilon^2 \\ +\chi^2\left(1 - \frac{\mu_0}{\mu_m}\right)\left(20 - 5\varepsilon - \Psi\left(\frac{15}{2} + \frac{519}{80}\varepsilon + \frac{219}{160}\varepsilon^2\right)\right) \\ +\chi^4\left(1 - \frac{\mu_0}{\mu_m}\right)^2\left(-\frac{91}{10}(2+\varepsilon) + \Psi\left(\frac{273}{40} + \frac{21}{16}\varepsilon - \frac{21}{20}\varepsilon^2\right)\right) \\ +\chi^6\left(1 - \frac{\mu_0}{\mu_m}\right)^3\left(\frac{143}{20}(2+\varepsilon) - \Psi\left(\frac{429}{80} + \frac{327}{160}\varepsilon - \frac{153}{320}\varepsilon^2\right)\right) \\ +\chi^8\left(1 - \frac{\mu_0}{\mu_m}\right)^4\left(-\frac{13}{10}(2+\varepsilon) - \Psi\left(\frac{39}{40} - \frac{13}{10}\varepsilon - \frac{33}{320}\varepsilon^2\right)\right) \\ +\chi^{10}\left(1 - \frac{\mu_0}{\mu_m}\right)^5\left(\frac{7}{100}(1+5\varepsilon) - \Psi\left(\frac{21}{800} + \frac{69}{1600}\varepsilon - \frac{9}{320}\varepsilon^2\right)\right) \\ +\chi^{12}\left(1 - \frac{\mu_0}{\mu_m}\right)^6\left(\frac{17}{100} + \frac{2}{25}\varepsilon - \Psi\left(\frac{51}{800} + \frac{3}{100}\varepsilon^2\right)\right) \end{pmatrix}, \tag{72}$$

Where

$$\Psi = \frac{M_1}{M_2}$$

M_1

$$= \begin{pmatrix} 1 + \frac{5}{2}\chi^2\left(1 - \frac{\mu_0}{\mu_m}\right) + \frac{91}{10}\chi^4\left(1 - \frac{\mu_0}{\mu_m}\right)^2 + \frac{143}{20}\chi^6\left(1 - \frac{\mu_0}{\mu_m}\right)^3 \\ -\frac{13}{10}\chi^8\left(1 - \frac{\mu_0}{\mu_m}\right)^4 + \frac{7}{200}\chi^{10}\left(1 - \frac{\mu_0}{\mu_m}\right)^5 + \frac{17}{12}\chi^{12}\left(1 - \frac{\mu_0}{\mu_m}\right)^6 \end{pmatrix},$$

M_2

$$= \begin{pmatrix} 1 + \frac{15}{2}\chi^2\left(1 - \frac{\mu_0}{\mu_m}\right) - \frac{683}{100}\chi^4\left(1 - \frac{\mu_0}{\mu_m}\right)^2 + \frac{107}{20}\chi^6\left(1 - \frac{\mu_0}{\mu_m}\right)^3 \\ -\frac{49}{50}\chi^8\left(1 - \frac{\mu_0}{\mu_m}\right)^4 + \frac{3}{100}\chi^{10}\left(1 - \frac{\mu_0}{\mu_m}\right)^5 + \frac{16}{25}\chi^{12}\left(1 - \frac{\mu_0}{\mu_m}\right)^6 \end{pmatrix} M, \tag{73}$$

which indicates that even when linear pressure variation is admitted, the maximum dissipated energy still remains independent of the coefficient of friction as reported in [19].

Analysis of Damping Capacity of the Magnetoelastic Structure

The damping capacity of any structure is a measure of the ratio of its slip energy dissipation to total strain energy under any conditions. For this case, we shall first derive the expression for the total strain energy following Damisa et al. [19]. It is a combination of the energy introduced by the bending moment as well as that stored from the deflection of the free end. While the former can be evaluated from the theorem of Castigliano, the later can be computed from the free end deflection theory. For this problem, we can derive the total strain energy expression as

$$\overline{X} = \overline{X}_1 + \overline{X}_2,$$

(74)

Where

$$\overline{X}_1 = \left(\begin{array}{l} 1 - \mu\overline{P}_0(2+\varepsilon) + \mu^2\overline{P}_0^{\,2}\left(1+\varepsilon+\dfrac{\varepsilon^2}{4}\right) + \chi^2\left(1-\dfrac{\mu_0}{\mu_m}\right)\left(1-\mu\overline{P}_0(2+\varepsilon)\overline{W}_{\max}\right) \\ +\chi^4\left(1-\dfrac{\mu_0}{\mu_m}\right)^2\overline{W}_{\max}^{\,2} \end{array} \right),$$

$$\forall \overline{W}_{\max} = \left(\begin{array}{l} 1 + \chi^2\left(1-\dfrac{\mu_0}{\mu_m}\right) \\ +\dfrac{3}{20}\chi^4\left(1-\dfrac{\mu_0}{\mu_m}\right)^2 \end{array} \right)\left(\begin{array}{l} \left(1-\mu\overline{P}_0\left(1+\dfrac{\varepsilon}{2}\right) - \dfrac{\mu\overline{P}_0\varepsilon}{240}\right) \\ +\dfrac{4}{5}\chi^2\left(1-\dfrac{\mu_0}{\mu_m}\right)\left(1-\mu\overline{P}_0\left(1+\dfrac{\varepsilon}{2}\right)+\dfrac{\mu\overline{P}_0\varepsilon}{576}\right) \end{array} \right),$$

$$\overline{W}_{\max}^{\,2} = \Gamma(\chi)\left(\begin{array}{l} \left(\begin{array}{l} \left(1-\mu\overline{P}_0(2+\varepsilon) + \mu^2\overline{P}_0^{\,2}\left(1+\varepsilon+\dfrac{\varepsilon^2}{4}\right)\right) \\ +\left(\dfrac{\mu\overline{P}_0\varepsilon}{120} - \dfrac{\mu^2\overline{P}_0^{\,2}}{120}\left(1+\dfrac{\varepsilon}{2}\right) + \dfrac{\mu^2\overline{P}_0^{\,2}\varepsilon^2}{240^2}\right) \end{array} \right) \\ +\dfrac{8}{5}\chi^2\left(1-\dfrac{\mu_0}{\mu_m}\right)\left(\begin{array}{l} \left(1-\mu\overline{P}_0\left(1+\dfrac{\varepsilon}{2}\right)+\dfrac{\mu\overline{P}_0\varepsilon}{576}\right) \\ +\mu\overline{P}_0\left(1+\dfrac{\varepsilon}{2}\right)-\mu^2\overline{P}_0^{\,2}\left(1+\varepsilon+\dfrac{\varepsilon^2}{4}\right)+\dfrac{\mu^2\overline{P}_0^{\,2}}{576}\left(\varepsilon+\dfrac{\varepsilon^2}{2}\right) \\ +\left(-\dfrac{\mu\overline{P}_0\varepsilon}{240}+\dfrac{\mu^2\overline{P}_0^{\,2}}{240}\left(1+\dfrac{\varepsilon}{2}\right)-\dfrac{\mu^2\overline{P}_0^{\,2}\varepsilon^2}{(240)(576)}\right) \end{array} \right) \\ +\dfrac{16}{25}\chi^4\left(1-\dfrac{\mu_0}{\mu_m}\right)^2\left(\begin{array}{l} \left(1-\mu\overline{P}_0(2+\varepsilon)+\mu^2\overline{P}_0^{\,2}\left(1+\varepsilon+\dfrac{\varepsilon^2}{4}\right)\right) \\ +\left(\dfrac{\mu\overline{P}_0\varepsilon}{288}-\dfrac{\mu^2\overline{P}_0^{\,2}}{288}\left(1+\dfrac{\varepsilon}{2}\right)\right)+\dfrac{\mu^2\overline{P}_0^{\,2}\varepsilon^2}{(16^2)(36^2)} \end{array} \right) \end{array} \right),$$

(75)

Where

$$\Gamma(\chi) = \left(1+\chi^2\left(1-\dfrac{\mu_0}{\mu_m}\right)+\dfrac{3}{20}\chi^4\left(1-\dfrac{\mu_0}{\mu_m}\right)^2\right);$$

(76)

While

$$\overline{X}_2 = \overline{W}_{\max}^{\,2} = \Gamma(\chi)\left(\begin{array}{l} \left(\begin{array}{l} \left(1-\mu\overline{P}_0(2+\varepsilon)+\mu^2\overline{P}_0^{\,2}\left(1+\varepsilon+\dfrac{\varepsilon^2}{4}\right)\right) \\ +\left(\dfrac{\mu\overline{P}_0\varepsilon}{120}-\dfrac{\mu^2\overline{P}_0^{\,2}}{120}\left(1+\dfrac{\varepsilon}{2}\right)+\dfrac{\mu^2\overline{P}_0^{\,2}\varepsilon^2}{240^2}\right) \end{array} \right) \\ +\dfrac{8}{5}\chi^2\left(1-\dfrac{\mu_0}{\mu_m}\right)\left(\begin{array}{l} \left(1-\mu\overline{P}_0\left(1+\dfrac{\varepsilon}{2}\right)+\dfrac{\mu\overline{P}_0\varepsilon}{576}\right) \\ +\mu\overline{P}_0\left(1+\dfrac{\varepsilon}{2}\right)-\mu^2\overline{P}_0^{\,2}\left(1+\varepsilon+\dfrac{\varepsilon^2}{4}\right)+\dfrac{\mu^2\overline{P}_0^{\,2}}{576}\left(\varepsilon+\dfrac{\varepsilon^2}{2}\right) \\ +\left(-\dfrac{\mu\overline{P}_0\varepsilon}{240}+\dfrac{\mu^2\overline{P}_0^{\,2}}{240}\left(1+\dfrac{\varepsilon}{2}\right)-\dfrac{\mu^2\overline{P}_0^{\,2}\varepsilon^2}{(240)(576)}\right) \end{array} \right) \\ +\dfrac{16}{25}\chi^4\left(1-\dfrac{\mu_0}{\mu_m}\right)^2\left(\begin{array}{l} \left(1-\mu\overline{P}_0(2+\varepsilon)+\mu^2\overline{P}_0^{\,2}\left(1+\varepsilon+\dfrac{\varepsilon^2}{4}\right)\right)+ \\ \left(\dfrac{\mu\overline{P}_0\varepsilon}{288}-\dfrac{\mu^2\overline{P}_0^{\,2}}{288}\left(1+\dfrac{\varepsilon}{2}\right)\right)+\dfrac{\mu^2\overline{P}_0^{\,2}\varepsilon^2}{(16^2)(36^2)} \end{array} \right) \end{array} \right).$$

(77)

For this problem, the damping capacity following Damisa et al. [19] is

$$\Phi = \frac{\overline{D}}{\overline{X}},$$

(78)

whilst the maximum damping capacity can be computed as

$$\Phi_{max} = \frac{\overline{D}_{max}}{\overline{X}_{max}},$$

(79)

Where

$$\overline{X}_{max} = \widehat{X}_1 + \widehat{X}_2,$$

$$\widehat{X}_1 = \left(\begin{array}{l} 1 - \frac{3}{8}\Psi(2+\varepsilon) + \frac{9}{64}\Psi^2\left(1+\varepsilon+\frac{\varepsilon^2}{4}\right) + \chi^2\left(1-\frac{\mu_0}{\mu_m}\right)\left(1-\frac{3}{8}\Psi(2+\varepsilon)\widehat{W}_{max}\right) \\ +\chi^4\left(1-\frac{\mu_0}{\mu_m}\right)^2\widehat{W}_{max}^2 \end{array} \right),$$

$$\forall \widehat{W}_{max} = \left(\begin{array}{l} 1 + \chi^2\left(1-\frac{\mu_0}{\mu_m}\right) \\ +\frac{3}{20}\chi^4\left(1-\frac{\mu_0}{\mu_m}\right)^2 \end{array} \right)\left(\begin{array}{l} \left(1-\frac{3}{8}\Psi\left(1+\frac{\varepsilon}{2}\right)-\frac{\Psi\varepsilon}{640}\right) \\ +\frac{4}{5}\chi^2\left(1-\frac{\mu_0}{\mu_m}\right)\left(1-\frac{3}{8}\Psi\left(1+\frac{\varepsilon}{2}\right)+\frac{\Psi\varepsilon}{(8)(192)}\right) \end{array} \right)$$

$$\widehat{W}_{max}^2 = \Gamma(\chi)\left(\begin{array}{l} \left(\begin{array}{l} \left(1-\frac{3}{8}\Psi(2+\varepsilon)+\frac{9}{64}\Psi^2\left(1+\varepsilon+\frac{\varepsilon^2}{4}\right)\right) \\ +\left(\frac{\Psi\varepsilon}{320}-\frac{3\Psi^2}{(40)(64)}\left(1+\frac{\varepsilon}{2}\right)+\frac{\Psi^2\varepsilon^2}{(10^2)(8^4)}\right) \end{array} \right) \\ +\frac{8}{5}\chi^2\left(1-\frac{\mu_0}{\mu_m}\right)\left(\begin{array}{l} \left(1-\frac{3}{8}\Psi\left(1+\frac{\varepsilon}{2}\right)+\frac{\Psi\varepsilon}{(8)(192)}\right) \\ +\frac{3}{8}\Psi\left(1+\frac{\varepsilon}{2}\right)-\frac{9}{64}\Psi^2\left(1+\varepsilon+\frac{\varepsilon^2}{4}\right)+\frac{\Psi^2}{64^2}\left(\varepsilon+\frac{\varepsilon^2}{2}\right) \\ +\left(-\frac{\Psi\varepsilon}{640}+\frac{\Psi^2}{(10^2)(8^4)}\left(1+\frac{\varepsilon}{2}\right)-\frac{\Psi^2\varepsilon^2}{(64^2)(240)}\right) \end{array} \right) \\ +\frac{16}{25}\chi^4\left(1-\frac{\mu_0}{\mu_m}\right)^2\left(\begin{array}{l} \left(1-\frac{3}{8}\Psi(2+\varepsilon)+\frac{9}{64}\Psi^2\left(1+\varepsilon+\frac{\varepsilon^2}{4}\right)\right) \\ +\left(\frac{\Psi\varepsilon}{768}-\frac{\Psi^2}{2^{11}}\left(1+\frac{\varepsilon}{2}\right)\right)+\frac{\Psi^2\varepsilon^2}{(2^{23})(3^2)} \end{array} \right) \end{array} \right),$$

(80)

While

$$\widehat{X}_2 = \widehat{W}_{max}^2.$$

(81)

ANALYSIS OF RESULTS

In this paper, static slip and energy dissipation in two-layer sandwich homogeneous elastic beams in a transverse magnetic field is studied. The problem physics derives from energy dissipation via contact frictional stresses in press fit joints. In the contrived problem, a constant tip force prompts the two-layer cantilever elastic beams in an environment of uniform transverse magnetic field. Simulation studies employed the characteristic values listed in Table 1. (Figures 2, 3, 4, 5, 6, 7, 8, 9, 10, 11, 12, 13, and 14) illustrate the effect of magnetoelasticity, relative permeability, and pressure gradient at optimum interface clamping pressure. In Figures 2–7, a tip force in an environment of weak magnetic field ($\chi = 0.05$) prompts the structure. We displayed in the Figure 2 result for homogeneous laminates of relative permeability $\mu_r = 0.6$. As can be seen, the deflection in the absence of pressure gradient ($\varepsilon = 0$) is lower and higher than the cases of negative and positive pressure gradients. Such a response pattern is expected ideally.

Table 1: Material and geometric parameters

Definition	Symbol	Value
Magnetic permeability	μ_0	$4\pi \times 10^{-7}$ mho/m
Interfacial pressure	P	1×10^9 Nm^{-2}
Geometry		
Length	L	0.1 m–5 m
Width	b	0.3 m
Thickness	h	0.001 m–0.4 m
Modulus of rigidity of materials	E	1.2×10^{11} Nm^{-2}
Density of material	ρ	2810 kg m^{-3}

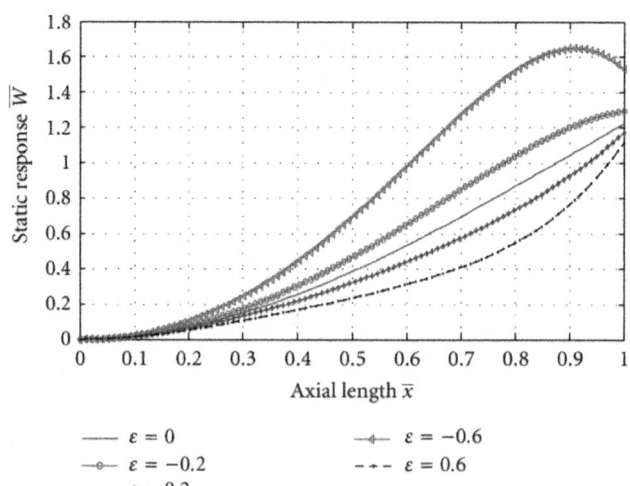

$$\text{---} \quad \varepsilon = 0 \qquad\qquad \text{--+--} \quad \varepsilon = -0.6$$
$$\text{--o--} \quad \varepsilon = -0.2 \qquad \text{--·--} \quad \varepsilon = 0.6$$
$$\text{--+--} \quad \varepsilon = 0.2$$

Figure 2: $\overline{W}$ versus $\overline{x}$ with different values of ε for the following case: $\chi = 0.05$; $\mu_r = 0.6$; $\mu\overline{P}_0 = \mu\overline{P}_{\text{opt}}$.

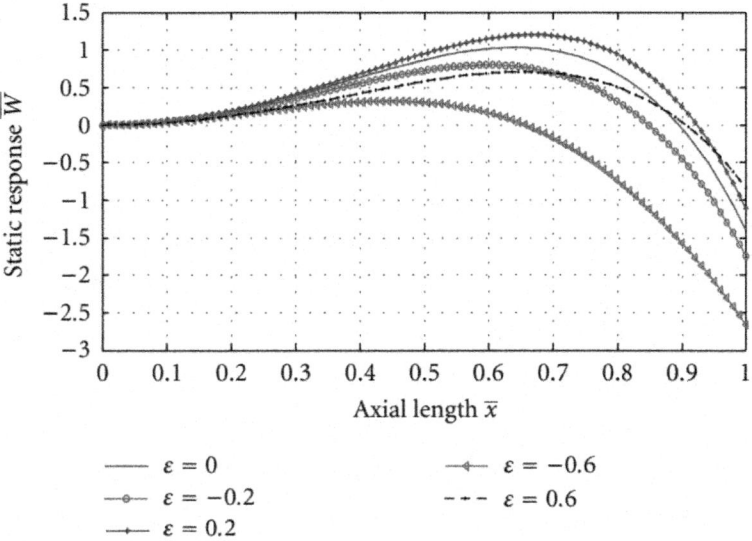

Figure 3: $\overline{W}$ versus $\overline{x}$ with different values of ε for the following case: $\chi = 0.05$; $\mu_r = 0.6$; $\mu\overline{P}_0 = \mu\overline{P}_{\text{opt}}$.

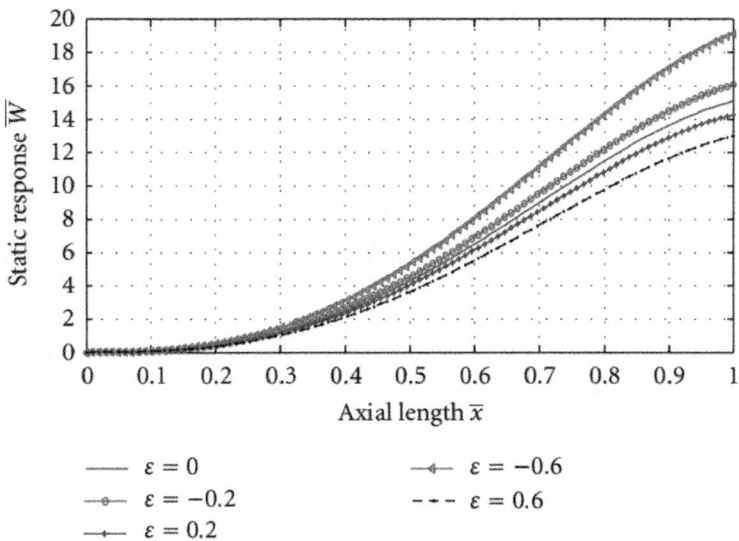

Figure 4: $\overline{W}$ versus $\overline{x}$ with different values of ε for the following case: $\chi = 1.5$; $\mu_r = 0.6$; $\mu\overline{P}_0 = \mu\overline{P}_{\text{opt}}$.

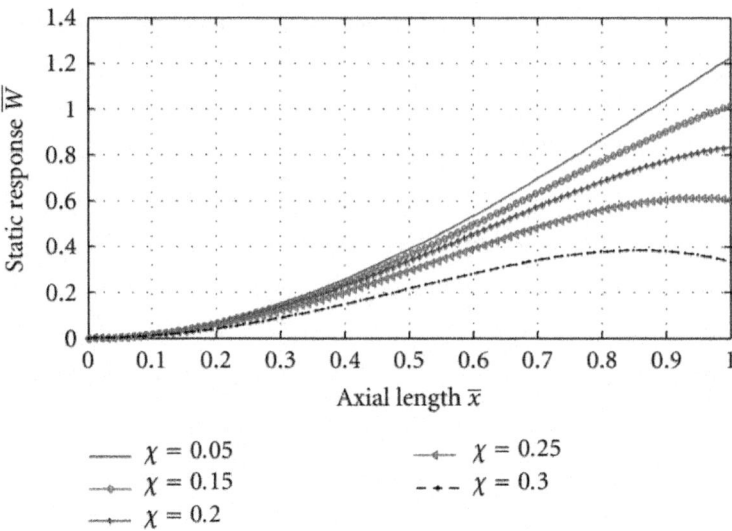

Figure 5: $\overline{W}$ versus $\overline{x}$ with different values of χ for the following case: $\varepsilon = 0$; $\mu_r = 0.6$; $\mu\overline{P}_0 = \mu\overline{P}_{opt}$.

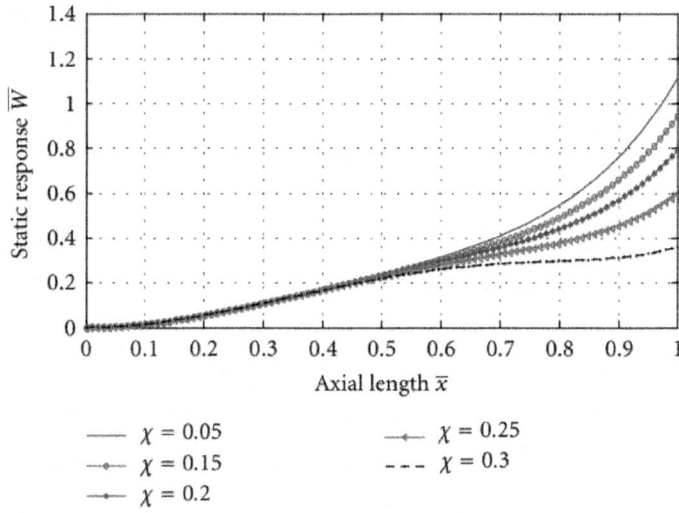

Figure 6: $\overline{W}$ versus $\overline{x}$ with different values of χ for the following case: $\varepsilon = 0$; $\mu_r = 0.6$; $\mu\overline{P}_0 = \mu\overline{P}_{opt}$.

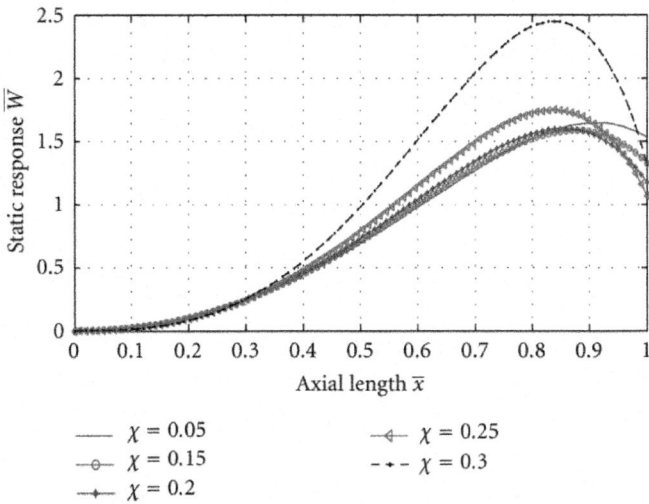

Figure 7: $\overline{W}$ versus $\overline{x}$ with different values of χ for the following case: $\varepsilon = -0.6$; $\mu_r = 0.6$; $\mu\overline{P}_0 = \mu\overline{P}_{opt}$.

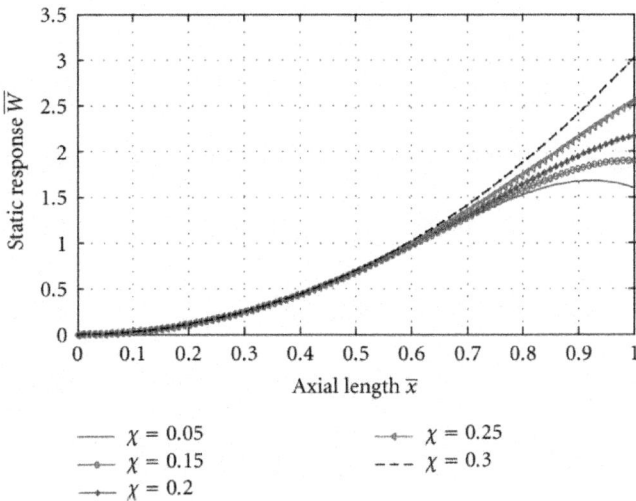

Figure 8: $\overline{W}$ versus $\overline{x}$ with different values of χ for the following case: $\varepsilon = -0.6$; $\mu_r = 6$; $\mu\overline{P}_0 = \mu\overline{P}_{opt}$.

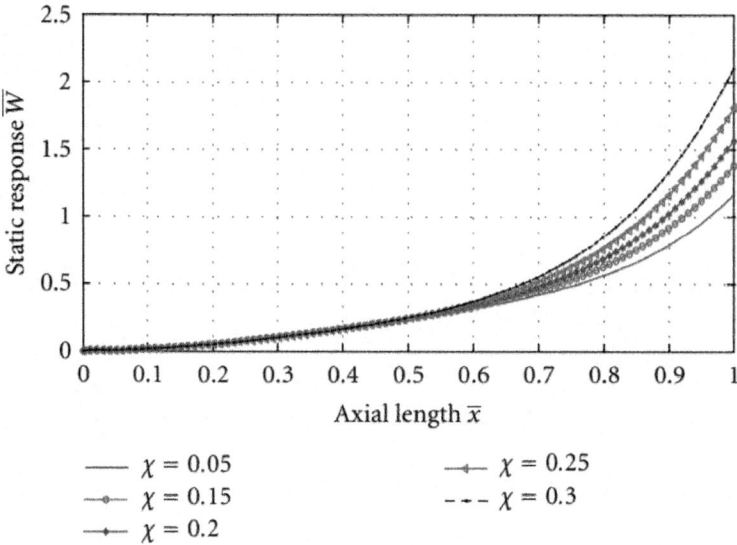

Figure 9: $\overline{W}$ versus $\overline{x}$ with different values of χ for the following case: $\varepsilon = -0.6$; $\mu_r = 6$; $\mu \overline{P}_0 = \mu \overline{P}_{opt}$.

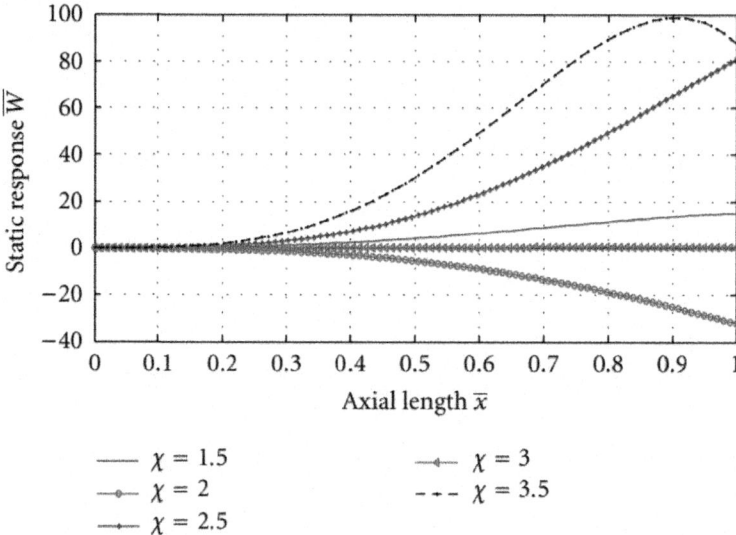

Figure 10: $\overline{W}$ versus $\overline{x}$ with different values of χ for the following case: $\varepsilon = 0$; $\mu_r = 6$; $\mu \overline{P}_0 = \mu \overline{P}_{opt}$.

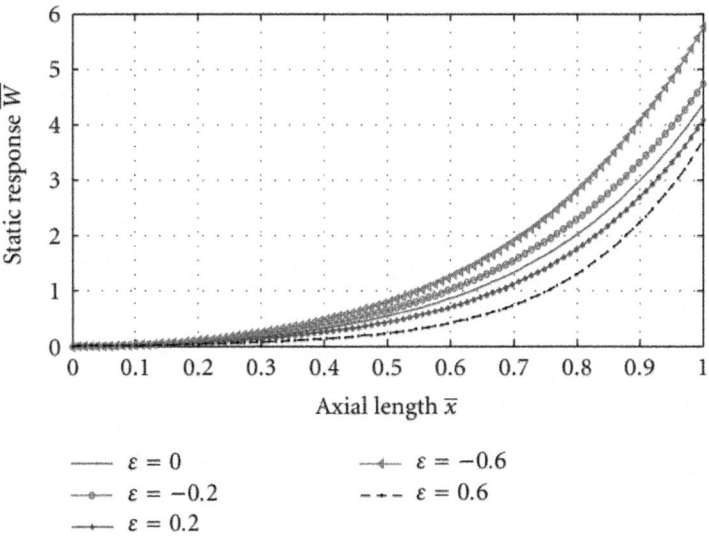

Figure 11: $\overline{W}$ versus $\overline{x}$ with different values of ε for the following case: $\chi = 0.5$; $\mu_r = 6$; $\mu\overline{P}_0 = \mu\overline{P}_{opt}$.

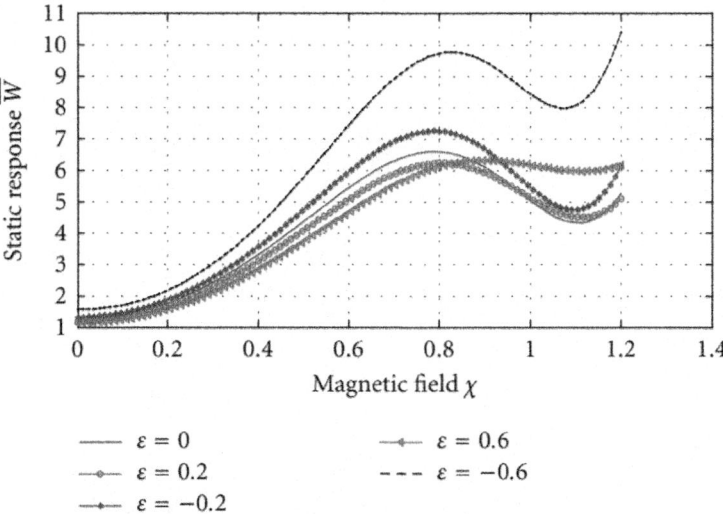

Figure 12: $\overline{W}$ versus χ with different values of ε for the following case: $\overline{x} = 1$; $\mu_r = 6$; $\mu\overline{P}_0 = \mu\overline{P}_{opt}$.

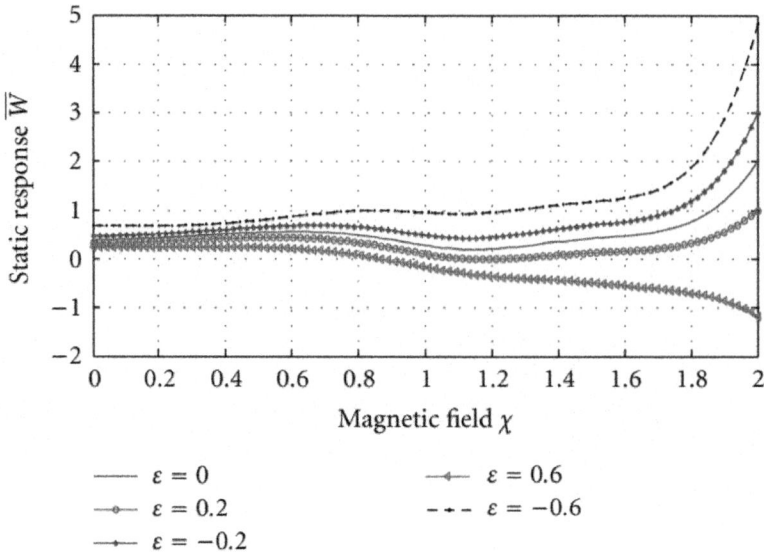

Figure 13: $\overline{W}$ versus χ with different values of ε for the following case: $\overline{x} = 0.5$; $\mu_r = 6$; $\mu\overline{P}_0 = \mu\overline{P}_{\text{opt}}$.

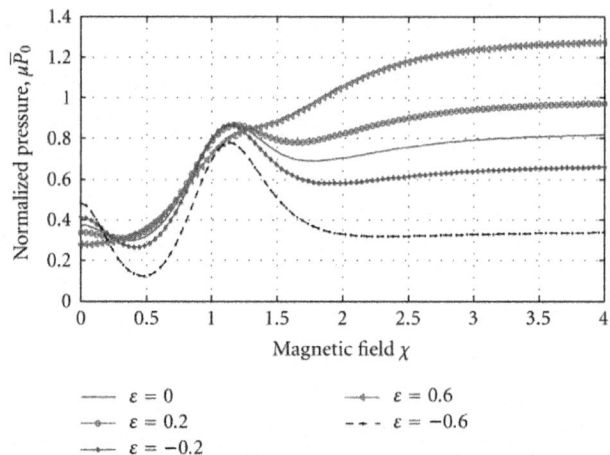

Figure 14: $\mu\overline{P}_0$ versus χ with different values of ε for the following case: $\mu_r = 6$.

The rigidity or stiffness of the structure from the fixed end should increase with progressive increment in the tightening torque axially. It is conceivable to expect higher deflection for the case $\varepsilon = -0.6$ than for $\varepsilon = -0.2$. Intuitively, a converse pattern should be expected in the reversed order as correctly noted in the figure. Furthermore, for $\varepsilon = -0.6$, there is a tendency for buckling to occur with axial progressive decrement in the tightening torque close to the free end, whilst for $\varepsilon = -0.2$, buckling is expected to occur at the free end. In Figure 3, the magnetic field ($\chi = 0.5$) intensity is relatively higher. Under the same parametric variables, we note a reversion in the pattern of deflection and tendency for buckling in the neighborhood of the free end irrespective of the value of ε.

The effect of much higher magnetic field intensity ($\chi = 1.5$) on the deflection response is illustrated in Figure4. Nature of curves indicated identical trajectories with higher deflection and symmetric ordering with respect to the role of the pressure gradient. In Figure 5, we illustrate the effect of varying the magnetic field intensity for the case of uniform clamping pressure on the static deflection profile. In the environment of lower magnetic field intensity χ, the trajectory is uniform but as field intensity increases, proportional decrements to respective optimum points are noted. Next, we display in Figures 6 and 7 the effects of pressure gradients on the deflection pattern in the same magnetic field environment. For the case $\varepsilon = 0.6$, the pattern is the same from the fixed end to the middle of the structure. Beyond this point, the effect of the field intensity becomes noticeable with higher deflection in the weak magnetic environment. We also note identical pattern from the fixed end to the middle of the structure for the case $\varepsilon = -0.6$. Beyond this point, the effect of the field intensity becomes noticeable. Natures of curves indicate higher deflection with respective optimum points in the strong magnetic environment. Next, we show in Figures 8 and 9 the effect of increasing the relative permeability on the static deflection.

As shown in Figures 8 and 9, the effect of increasing the field intensity is noticeable beyond the middle of the structure. Nevertheless, for the case $\varepsilon = -0.6$, a bisegmented asymmetrical deflection appeared in the neighborhood of the free end of the structure compared to the case $\varepsilon = 0.6$ where the pattern of deflection is progressively monotonic. Profiles for the case $\varepsilon = 0$ (uniform pressure) in much higher field intensity are illustrated in Figure 10. In contrast to results in Figures 8 and 9, we note a very visible bisegmented asymmetrical deflection slightly away from the fixed point to the free end of the structure. Next, we study the static response of the structure at optimum clamping pressure for the case displayed in Figure 11. Nature of curves indicate that the deflections for negative pressure gradient are higher than for the case of uniform pressure and positive gradient as reported in [19].

The cumulative effect of pressure gradient and field intensity on the mid structure and tip deflections are demonstrated in Figures 12 and 13. The picture in Figure 12 indicates a three-region segmentation except for the case $\varepsilon = $

0.6 as modulated by the field intensity. In the first region, the field intensity lies within the range $0 < \chi \leq 0.2$. Here, the deflections are ordered in consonant with the forms and values of pressure gradient with each profile, indicating a monotonic variable that is approaching a local maximum.

In the second region, the field intensity lies within the range $0.5 \leq \chi \leq 1$. Here, each profile is also a monotonic variable approaching a local minimum, and in the last region, the field intensity lies within the range $1 \leq \chi \leq 1.2$ and the respective tip deflection is progressively monotonic. Next, the deflection pattern in the middle of the structure is displayed in Figure 13. In contrast to tip deflection, a relatively stable structure is noted in the range $0 \leq \chi \leq 0.4$ and beyond, and a segmented distortional response characterized the structure. To what extent is the optimum clamping pressure influenced by the pressure gradient and the magnetic field strength? Such modulating effect is demonstrated in Figure 14. In the range $0 \leq \chi \leq 1.5$, a cyclic propagation ensued and beyond, and the interface pressure becomes stable and higher for positive pressure gradient. The profiles of interfacial static slip are shown in Figures 15–21. Effect of pressure gradient on the tip static slip as a function of the magnetic field intensity is displayed in Figure 15. Family of curves indicate a three-region profile. In the first region, the slip for positive pressure gradient is higher than the cases for uniform pressure and negative pressure gradient. As the magnetic intensity increases to $\chi = 0.2$, the interfacial slip becomes constant irrespective of the nature of the pressure gradient. Beyond this point, the second region begins and a reversal in the pressure gradient becomes apparent.

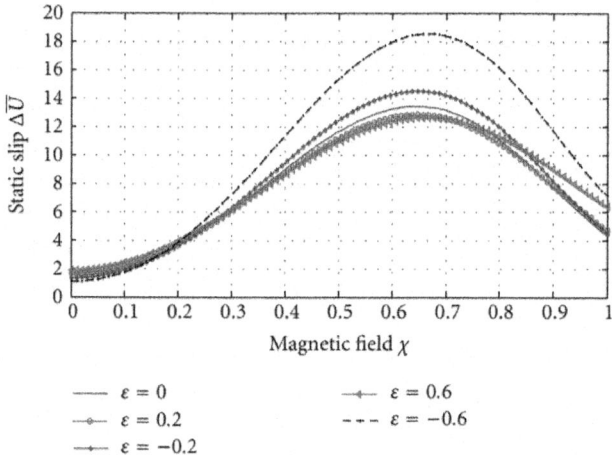

Figure 15: $\Delta \overline{U}$ versus χ with different values of ε for the following case: $\overline{x} = 1$; $\mu_r = 6$

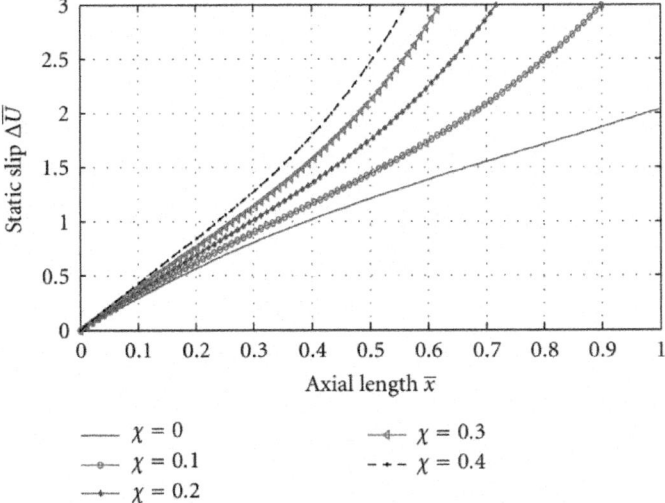

Figure 16: $\Delta\overline{U}$ versus $\overline{x}$ with different values of χ for the following case; $\varepsilon = 0$; $\mu_r = 6$; $\mu\overline{P}_0 = \mu\overline{P}_{opt}$.

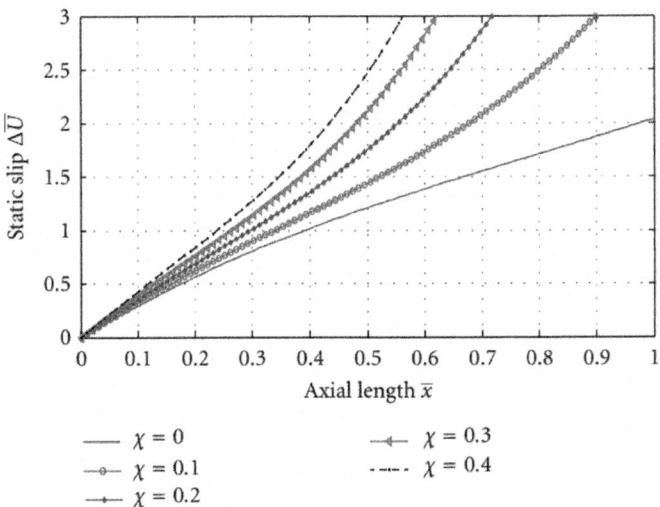

Figure 17: $\Delta\overline{U}$ versus $\overline{x}$ with different values of χ for the following case: $\varepsilon = 0.6$; $\mu_r = 6$; $\mu\overline{P}_0 = \mu\overline{P}_{opt}$.

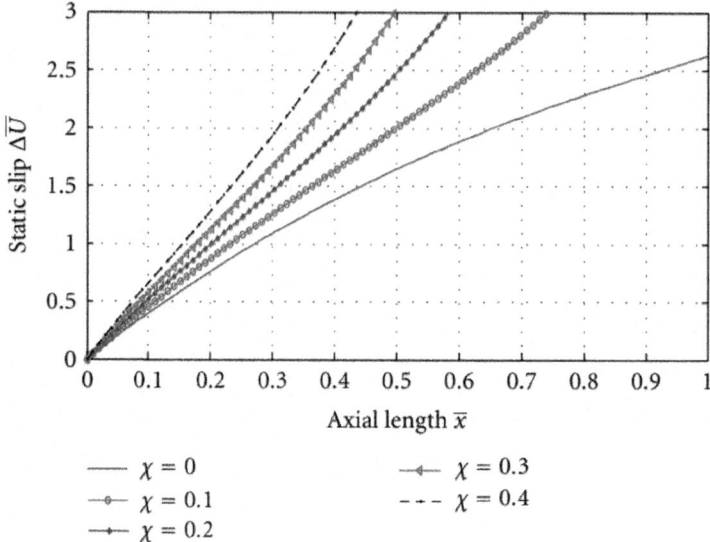

Figure 18: $\Delta\overline{U}$ versus $\overline{x}$ with different values of χ for the following case: $\varepsilon = -0.6$; $\mu_r = 6$; $\mu\overline{P}_0 = \mu\overline{P}_{\text{opt}}$.

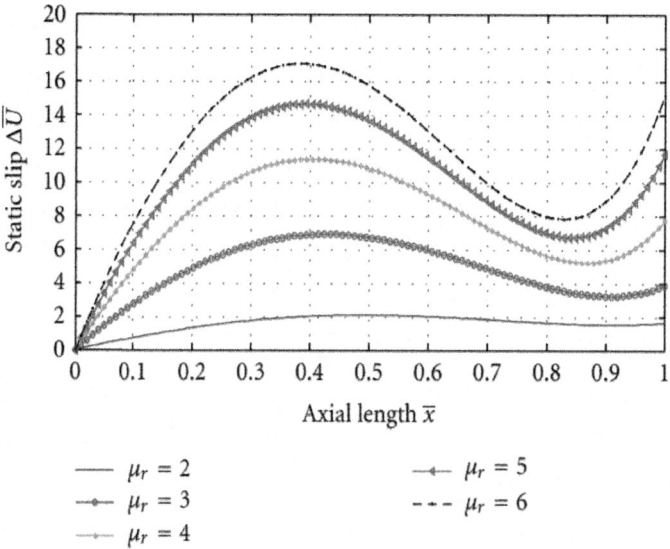

Figure 19: $\Delta\overline{U}$ versus $\overline{x}$ with different values of μ_r for the following case: $\varepsilon = 0$; $\chi = 1.5$; $\mu\overline{P}_0 = \mu\overline{P}_{\text{opt}}$.

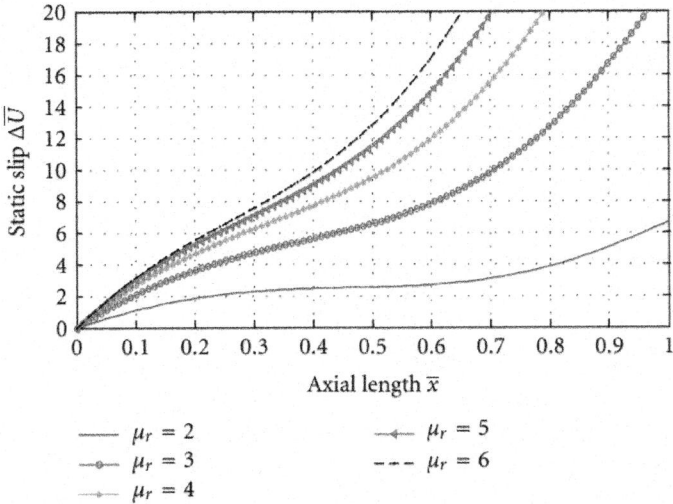

Figure 20: $\Delta\overline{U}$ versus $\overline{x}$ with different values of μ_r for the following case: $\varepsilon = 0.6$; $\chi = 1.5$; $\mu\overline{P}_0 = \mu\overline{P}_{\text{opt}}$.

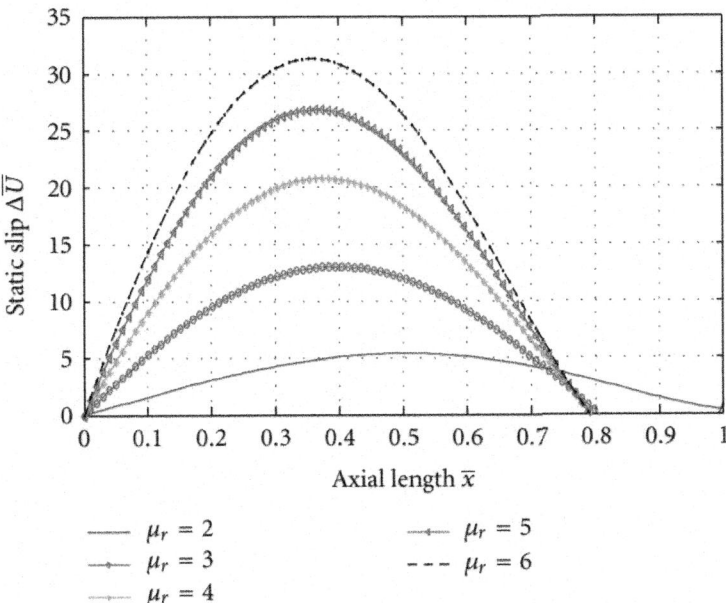

Figure 21: $\Delta\overline{U}$ versus $\overline{x}$ with different values of μ_r for the following case: $\varepsilon = -0.6$; $\chi = 1.5$; $\mu\overline{P}_0 = \mu\overline{P}_{\text{opt}}$.

In this domain, each curve indicates a monotonic progressing slip that attains a maximum value subject to common critical magnetic field intensity. Next, we display in Figures 16, 17, and 18 gross interfacial slip profiles for the cases of uniform pressure, positive- and negative-pressure gradients, respectively. As noted in [19], gross slip for negative pressure gradients is higher compared to the cases of uniform pressure and positive pressure gradients.

For the special case, $\chi = 0$, the profiles replicated the results in [19] for the same values of ε. Irrespective of the form of ε, the gross slip is proportional to the magnetic intensity and admits approximate linear profile when compared to the case $\chi = 0$. The gross slip for uniform optimum pressure as influenced by different relative permeability values in an environment of strong magnetic field defined by $\chi = 1.5$ is shown in Figure 19. In general, interfacial slip increases as the permeability increases. Apart from the curve typified by $\mu_r = 2$, which rises progressively to an optimum value, the rest profiles admit cyclic variation with local maxima and minima values.

To what extent are the profiles modified for the cases of positive and negative pressure gradients? Such modification or effects are displayed in Figures 20 and 21 respectively. In Figure 20, profiles admit the same noncyclic configuration pattern that are proportional to the permeability values across the structures, and in Figure 21 slip profiles are proportionally parabolic curves. Energy dissipation ability as influenced by the magnetic field intensity and interfacial pressure are shown in Figures 22–24. Natures of profiles indicate a three-region regime. As shown in the plotted curves, the profiles are parabolic in the first region with local maxima as reported in [19, 20] and here, the interfacial pressure values are restricted in consonance with the form of pressure gradient. The same observations are noted in Damisa [25], where energy dissipation is maximum at respective optimum clamping pressure defined by corresponding critical transition into the second region. Such is anticipated via (72). In the second region, the values of the energy dissipation quanta are reversed in accordance with the respective magnetic field intensity prior to entering the third region. In this region, the dissipation quanta are proportional to the magnetic field intensities. The same trends are noted for energy dissipation profiles for the cases of positive and negative pressure gradients as shown in Figures 23 and 24.

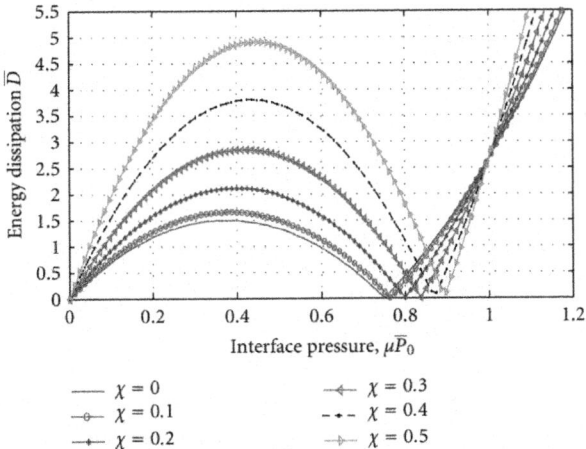

Figure 22: $\overline{D}$ versus $\mu\overline{P}_0$ with different values of χ for the following case: $\varepsilon = 0$; μ_r = 6.

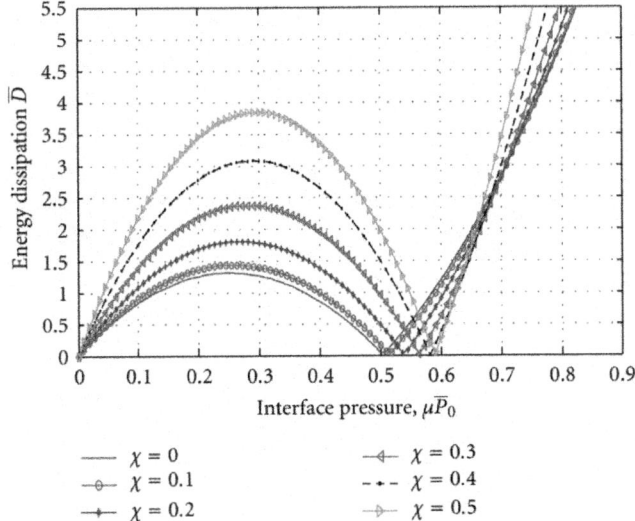

Figure 23: $\overline{D}$ versus $\mu\overline{P}_0$ with different values of χ for the following case: ε = 0.6; μ_r = 6.

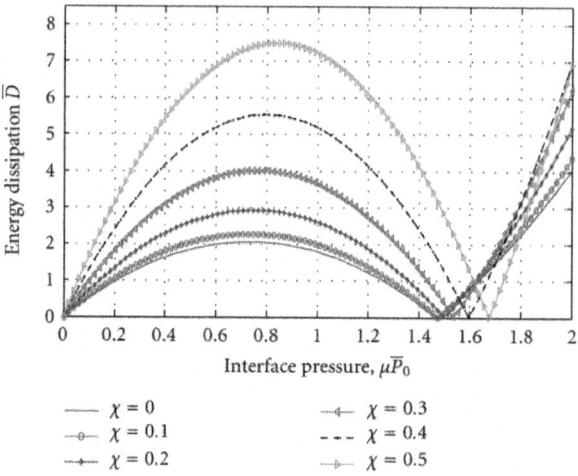

Figure 24: $\overline{X}$ versus $\mu\overline{P}_0$ with different values of χ for the following case: $\varepsilon = 0.6$; $\mu_r = 6$.

As expected, the magnetic environment has an increasing effect on the energy dissipation mechanism as shown in Figures 22–24. Nevertheless, as correctly noted in [19, 20], dissipation is more with negative pressure gradient than for the uniform pressure and positive pressure gradient. The strain energy profiles as influenced by the magnetic field intensity and pressure gradient parameter are illustrated in Figures 25, 26, and 27. For the case of uniform pressure, symmetric parabolic curves proportional to the magnetic intensity ensued as shown in Figure 25, whilst skewed asymmetric parabolas are displayed for the positive and negative pressure gradients.

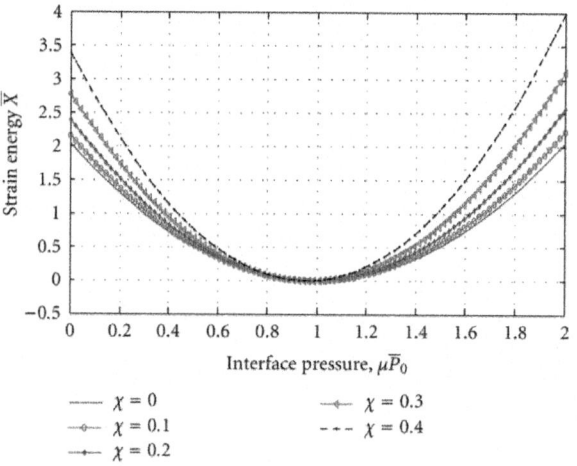

Figure 25: $\overline{X}$ versus $\mu\overline{P}_0$ with different values of χ for the following case: $\varepsilon = 0$; $\mu_r = 6$.

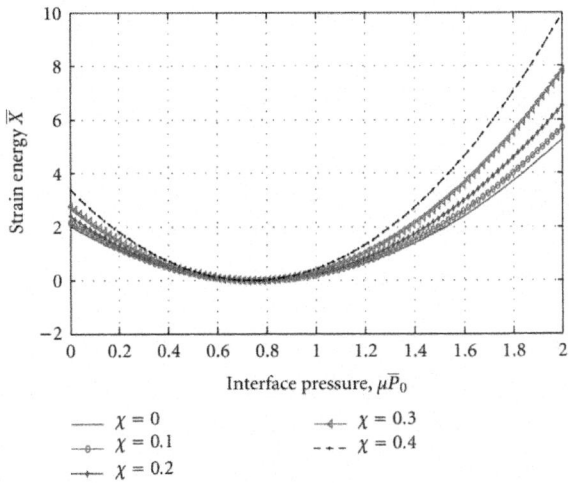

Figure 26: $\overline{D}$ versus $\mu\overline{P}_0$ with different values of χ for the following case: $\varepsilon = 0.6$; $\mu_r = 6$.

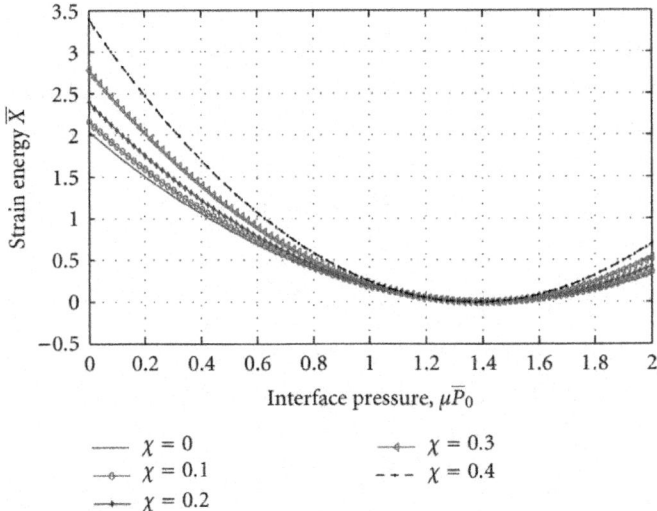

Figure 27: $\overline{X}$ versus $\mu\overline{P}_0$ with different values of χ for the following case: $\varepsilon = -0.6$; $\mu_r = 6$.

CONCLUSION

A well-posed mathematical physics problem on the mechanics of interfacial slip and energy dissipations mechanism with a two-layer sandwich homogenous elastic beam in a transverse magnetic field is presented. By employing operational methods, closed form polynomial expressions are derived for the responses defined in the body of the paper. In particular, the effects of magnetoelasticity and interfacial pressure gradient are demonstrated for design analysis and engineering applications. For special and limit cases, recent theoretical and experimental results are validated from the theory reported in the paper.

REFERENCES

1. J. S. Lee, "Destabilizing effects of magnetic dampimg in plate strip," Journal of Engineering Mechanics, vol. 118, no. 1, pp. 161–173, 1992.

2. F. C. Moon and Y. H. Pao, "Magnetoelastic buckling of a thin plate," Journal of Applied Mechanics, vol. 35, pp. 53–58, 1968.

3. F. C. Moon and Y. H. Pao, "Vibration and dynamic instability of a beam-plate in a transverse magnetic field," Journal of Applied Mechanics, vol. 36, pp. 92–100, 1969.

4. D. V. Wallerstein and M. O. Peach, "Magnetoelastic buckling of beams and thin plates of magnetically soft material," Journal of Applied Mechanics, vol. 39, pp. 451–455, 1972.

5. K. Miya, T. Takagi, and Y. Ando, "Finite element analysis of magnetoelastic buckling of a ferromagnectic beam-plate," Journal of Applied Mechanics, vol. 47, no. 2, pp. 377–382, 1980.

6. M. O. Peach, N. S. Christopherson, J. M. Dalrymple, and G. L. Viegelahn, "Magnetoelastic buckling: why theory and experiment disagree," Experimental Mechanics, vol. 28, no. 1, pp. 65–69, 1988.

7. J. S. Lee, "Dynamic stability of conducting beam-plates in transverse magnetic fields," Journal of Engineering Mechanics, vol. 122, no. 2, pp. 89–94, 1996.

8. Y. H. Zhou, X. J. Zheng, and K. Miya, "Magnetoelastic bending and snapping of ferromagnetic plates in oblique magnetic fields," Fusion Engineering and Design, vol. 30, no. 4, pp. 325–337, 1995.

9. Y. H. Zhou and X. Zheng, "A general expression of magnetic force for soft ferromagnetic plates in complex magnetic fields," International Journal of Engineering Science, vol. 35, no. 15, pp. 1405–1417, 1997.

10. Y. H. Zhou and K. Miya, "A theoretical prediction of natural frequency of a ferromagnetic beam plate with low susceptibility in an in-plane magnetic field," Journal of Applied Mechanics, vol. 65, no. 1, pp. 121–126, 1998.

11. X. Wang, G. Lu, and S. R. Guillow, "Magnetothermodynamic stress and perturbation of magnetic field vector in a solid cylinder," Journal of Thermal Stresses, vol. 25, no. 10, pp. 909–926, 2002.

12. X. Wang, J. S. Lee, and X. Zheng, "Magneto-thermo-elastic instability of ferromagnetic plates in thermal and magnetic fields," International Journal of Solids and Structures, vol. 40, no. 22, pp. 6125–6142, 2003.

13. X. Wang and H. L. Dai, "Magnetothermodynamic stress and perturbation of magnetic field vector in an orthotropic thermoelastic cylinder," International Journal of Engineering Science, vol. 42, no. 5-6, pp. 539–556, 2004.

14. L. Librescu, D. Hasanyan, Z. Qin, and D. R. Ambur, "Nonlinear magnetothermoelasticity of anisotropic plates immersed in a magnetic field," Journal of Thermal Stresses, vol. 26, no. 11-12, pp. 1277–1304, 2003.

15. X. Wang, J. S. Lee, and X. Zheng, "Magneto-thermo-elastic instability of ferromagnetic plates in thermal and magnetic fields," International Journal of Solids and Structures, vol. 40, no. 22, pp. 6125–6142, 2003.

16. X. Wang and H. L. Dai, "Magnetothermodynamic stress and perturbation of magnetic field vector in an orthotropic thermoelastic cylinder," International Journal of Engineering Science, vol. 42, no. 5-6, pp. 539–556, 2004.

17. D. J. Hasanyan, L. Librescu, and D. R. Ambur, "A few results on the foundation of the theory and behavior of nonlinear magnetoelastic plates carrying an electrical current," International Journal of Engineering Science, vol. 42, no. 15-16, pp. 1547–1572, 2004.

18. X. Wang and J. S. Lee, "Dynamic stability of ferromagnetic beam-plates with magnetoelastic interaction and magnetic damping in transverse magnetic fields," Journal of Engineering Mechanics, vol. 132, no. 4, pp. 422–428, 2006.

19. O. Damisa, V. O. S. Olunloyo, C. A. Osheku, and A. A. Oyediran, "Static analysis of slip damping with clamped laminated beams," European Journal of Scientific Research, vol. 17, no. 4, pp. 455–475, 2007.

20. O. Damisa, V. O. S. Olunloyo, C. A. Osheku, and A. A. Oyediran, "Dynamic analysis of slip damping in clamped layered beams with non-uniform pressure distribution at the interface," Journal of Sound and

Vibration, vol. 309, no. 3–5, pp. 349–374, 2008.

21. V. O. S. Olunloyo, C. A. Osheku, and O. Damisa, "Vibration damping in structures with layered viscoelastic beam-plate," Journal of Vibration and Acoustics, Transactions of the ASME, vol. 130, no. 6, Article ID 061002, 2008.

22. V. O. S. Olunloyo, O. Damisa, C. A. Osheku, and A. A. Oyediran, "Analysis of the effects of laminate depth and material properties on the damping associated with layered structures in a pressurized environment," Transactions of the Canadian Society for Mechanical Engineering, vol. 34, no. 2, pp. 165–196, 2010.

23. C. A. Osheku and O. Damisa, "Vibration of conducting two-layer sandwich homogeneous elastic beams in transverse magnetic fields part I," ASCE Journal of Aerospace Engineering ASENG-245. In press.

24. L. E. Goodman and J. H. Klumpp, "Analysis of slip damping with reference to turbine blade vibration,"Journal of Applied Mechanics, vol. 23, pp. 421–429, 1956.

25. O. Damisa, "Slip damping of Timoshenko beam close to resonance," Journal of Engineering Research, vol. 11, no. 1-2, pp. 13–26, 2003.

Chapter 9

DYNAMICS AND STATICS OF FLEXIBLE AXIALLY SYMMETRIC SHALLOW SHELLS

J. Awrejcewicz[1], V. A. Krysko[2], and I. V. Kravtsova[3]

[1]Department of Automatics and Biomechanics, Technical University of Lodz, 1/15 Stefanowskiego Street, 90924 Lodz, Poland

[2]Department of Mathematics, Saratov State University, 410054 Saratov, Russia

[3]I. V. Kravtsova: Department of Mathematics, Saratov State University, 410054 Saratov, Russia

ABSTRACT

In this work, we propose the method for the investigation of stochastic vibrations of deterministic mechanical systems represented by axially symmetric spherical shells. These structure members are widely used as sensitive elements of pressure measuring devices in various branches of measuring and control industry, machine design, and so forth. The proposed method can be easily extended for the investigation of shallow spherical shells, goffer-type membranes, and so on. The so-called charts of control parameters for a shell subjected to a transversal uniformly distributed and local harmonic loading force and resistance moment are constructed. The scenarios of the transition of vibration of shallowtype system into chaotic state are investigated with the use of the theory of differential equations and the theory of nonlinear dynamics. The method of the control of chaotic vibrations of flexible spherical shells subjected to a transversal harmonic load through a synchronized action of either harmonic resistance moment or force is proposed, illustrated, and discussed.

INTRODUCTION

Stochastic vibrations of flexible shallow shells are rather rarely investigated. It is mainly static problems as well as those of dynamical stability loss subjected to impulse time loads constant in time that is the subject of investigation, and hence the similar problems but subjected to a harmonic transversal load are not often studied. A reason for that is the fact that the general theory of nonlinear dissipative mechanical systems is not yet fully developed. We focus our attention on filling the gap occurring in this field. It is worth noticing that the investigation of vibrations of flexible plates and shells has been recently initiated [1–8, 15, 17, 18].

On the other hand, the control of chaos in deterministic systems belongs to diffi- cult and challenging tasks of nonlinear dynamical systems investigation, since chaotic vibrations are relatively often exhibited by various engineering, economic, and biological systems.

Firstly, the problem of chaos control has been stated in [12–14] and in the classical work of Ott et al. [22]. In [26] one may find even earlier sources of this idea. Problems related to the control of chaos have been considered in hydrodynamics [27], chemistry [23], and biology and medicine [25]. However, this problem has not been investigated in the theory of shells, at least to the authors' knowledge.

PROBLEM FORMULATION AND COMPUTATIONAL ALGORITHM

Consider a spherical axially symmetric shallow shell creating a closed two-dimensional subspace of space R^2 in the polar coordinate system introduced in the following manner: $\Omega = \{(r,z) \mid r \in [0,b], -h/2 \leq z \leq h/2\}$. Equations governing the dynamics of shallow axially symmetric shells are cast in the form [29]

$$w'' + \varepsilon w' = -\frac{\partial^4 w}{\partial r^4} - \frac{2}{r}\frac{\partial^3 w}{\partial r^3} + \frac{1}{r^2}\frac{\partial^2 w}{\partial r^2} - \frac{1}{r^3}\frac{\partial w}{\partial r} - \frac{\phi}{r}\left(1 - \frac{\partial^2 w}{\partial r^2}\right) - \frac{\partial \phi}{\partial r}\left(1 - \frac{1}{r}\frac{\partial w}{\partial r}\right) + 4q,$$

$$\frac{\partial^2 \phi}{\partial r^2} + \frac{1}{r}\frac{\partial \phi}{\partial r} - \frac{1}{r^2}\frac{\partial \phi}{\partial r} = \frac{\partial w}{\partial r}\left(1 - \frac{1}{2r}\frac{\partial w}{\partial r}\right),$$

(2.1)

where $\varphi = \partial F/\partial r$. The nondimensional quantities follow: $\omega_0 = \sqrt{Eg/\gamma R^2}$; $\bar{\varepsilon} = \sqrt{(g/\gamma E)}(R/h)\varepsilon$; $\bar{F} = \eta(F/Eh^3)$; $\bar{w} = \sqrt{\eta}(w/h)$; $\bar{r} = b(r/c)$; $\bar{q} = \bar{q}_3 = (\sqrt{\eta}/4)(q_3/E)(R/h)^2$; $\eta = 12(1 - \nu^2)$; $b = \sqrt[4]{\eta}(c/\sqrt{Rh})$; where t is the time; ε is the damping coefficient of the shell surrounding medium, F is the stress function, w is the function of displacements; R, c are the main and secondary radii of the curvature of supported contour, respectively; h is the shell thickness; b is the sloping parameter; ν is Poisson's coefficient; r is the distance from a rotation axis to a point lying in the shell's mean surface; q is the parameter of external load; ω_0 is the frequency of linear vibrations. To simplify, bars over nondimensional quantities are omitted in (2.1). Derivatives with respect to time are denoted by ('). One has to add boundary and initial conditions to the system (2.1), as well as conditions in the shell vertex. For a ball-type movable support in the meridian direction, the boundary conditions read as

(1) (a) homogeneous boundary conditions

$$\phi = w = 0, \quad \frac{\partial^2 w}{\partial r^2} + \frac{\nu}{r} \frac{\partial w}{\partial r}, \quad \text{for } r = b;$$

(2.2)

(b) nonhomogeneous boundary conditions

$$\phi = w = 0, \quad \frac{\partial^2 w}{\partial r^2} = M_0 \sin(\omega_p t), \quad \text{for } r = b;$$

(2.3)

(2) (a) homogeneous boundary conditions for a ball-type unmovable support

$$\frac{\partial \phi}{\partial r} - \nu \frac{\phi}{b} = 0, \quad w = 0, \quad \frac{\partial^2 w}{\partial r^2} + \frac{\nu}{r} \frac{\partial w}{\partial r} = 0, \quad \text{for } r = b;$$

(2.4)

(b) nonhomogeneous boundary conditions

$$\frac{\partial \phi}{\partial r} - \nu \frac{\phi}{b} = 0, \quad w = 0, \quad \frac{\partial^2 w}{\partial r^2} + \frac{\nu}{r} \frac{\partial w}{\partial r} = M_0 \sin(\omega_p t), \quad \text{for } r = b;$$

(2.5)

(3) a movable clamped (resistance) contour

$$\phi = w = 0, \quad \frac{\partial w}{\partial r} = 0, \quad \text{for } r = b;$$

(2.6)

(4) an unmovable clamped (resistance) contour

$$\frac{\partial \phi}{\partial r} - \nu \frac{\phi}{b} = 0, \quad w = 0, \quad \frac{\partial w}{\partial r} = 0, \quad \text{for } r = b.$$

(2.7)

The initial conditions read as

$$w = f_1(r,0) = 0, \quad w' = f_2(r,0) = 0, \quad 0 \le t < \infty.$$

(2.8)

In the close vicinity of the shell vertex, the following estimations hold:

$$\phi \approx Ar, \quad \phi' \approx A, \quad w \approx B + Cr^2, \quad w' \approx 2Cr, \quad w'' \approx 2C, \quad w''' \approx 0.$$

(2.9)

In order to reduce the distributed system (2.1)–(2.9) to a lumped one, the method of finite differences with approximation $O(\Delta^2)$ is applied. Let us express the system (2.1)– (2.9) via finite-difference relations with respect to the spatial variable r:

$$w_i'' + \varepsilon w_i' = -\frac{w_{i+1} - w_{i-1}}{2\Delta}\left(\frac{1}{r_i^3} - \frac{\phi_{i+1} - \phi_{i-1}}{2r_i\Delta}\right) + \frac{w_{i+1} - 2w_i + w_{i-1}}{r_i\Delta^2}\left(\phi_i + \frac{1}{r_i}\right)$$

$$-\frac{\phi_{i+1} - \phi_{i-1}}{2\Delta} - \frac{\phi_i}{r_i} - \frac{w_{i+2} - 4w_{i+1} + 6w_i - 4w_{i-1} + w_{i-2}}{\Delta^4}$$

$$-\frac{w_{i+2} - 2w_{i+1} + 2w_{i-1} - w_{i-2}}{r_i\Delta^3} + 4q_i,$$

$$\phi_{i+1}\left(-\frac{1}{\Delta^2} - \frac{1}{2r_i\Delta}\right) + \phi_i\left(\frac{2}{\Delta^2} + \frac{1}{r_i^2}\right) + \phi_{i-1}\left(-\frac{1}{\Delta^2} + \frac{1}{2r_i\Delta}\right)$$

$$= -\frac{w_{i+1} - w_{i-1}}{2\Delta}\left(1 - \frac{w_{i+1} - w_{i-1}}{4r_i\Delta}\right),$$

$$\tag{2.10}$$

where $\Delta = b/n$, n is a member of distributing parts of two shell radii. The boundary conditions have the following form now:

(1) (a) homogeneous boundary conditions

$$\phi_n = 0, \quad w_{i+1} = \frac{v\Delta - 2b}{2b + v\Delta}w_{i-1}, \quad w_n = 0 \quad \text{for } r_n = b;$$

$$\tag{2.11}$$

(b) nonhomogeneous boundary conditions

$$\phi_n = 0, \quad w_{i+1} = \frac{M_0\sin(\omega_p t) - (1/\Delta^2 - v/2b\Delta)w_{i-1}}{(1/\Delta^2 + v/2b\Delta)}, \quad w_n = 0 \quad \text{for } r_n = b;$$

$$\tag{2.12}$$

(2) (a) homogenous boundary conditions for a ball-type unmovable support

$$\phi_{i+1} = \phi_{i-1} + \frac{2\Delta v}{b}\phi_i, \quad w_{i+1} = \frac{v\Delta - 2b}{2b + v\Delta}w_{i-1}, \quad w_n = 0 \quad \text{for } r_n = b;$$

$$\tag{2.13}$$

(b) nonhomogeneous boundary conditions

$$\phi_{i+1} = \phi_{i-1} + \frac{2\Delta v}{b}\phi_i,$$

$$w_{i+1} = \frac{M_0\sin(\omega_p t) - (1/\Delta^2 - v/2b\Delta)w_{i-1}}{(1/\Delta^2 + v/2b\Delta)}, \quad w_n = 0 \quad \text{for } r_n = b;$$

$$\tag{2.14}$$

(3) a movable clamped (resistance) contour

$$\phi_n = 0, \quad w_{n+1} = w_{n-1}, \quad w_n = 0 \quad \text{for } r_n = b;$$

$$\tag{2.15}$$

(4) an unmovable clamped (resistance) contour

$$\phi_{i+1} = \phi_{i-1} + \frac{2\Delta v}{b}\phi_i, \quad w_{n+1} = w_{n-1}, \quad w_n = 0 \quad \text{for } r_n = b.$$

$$\tag{2.16}$$

The initial conditions have the form

$$w_n = f_1(r_k,0) = 0, \quad w'_n = f_2(r_k,0) = 0 \quad (0 \le k \le n), \quad 0 \le t < \infty.$$

$$(2.17)$$

If one neglects some small terms, after the change of differential operators by central finite-difference ones for $r = \Delta$, the following conditions hold in the shell vertex:

$$\phi_0 = \phi_2 - 2\phi_1, \quad w_0 = \frac{4}{3}w_1 - \frac{1}{3}w_2, \quad w_{-1} = \frac{8}{3}w_1 - \frac{8}{3}w_2 + w_3.$$

$$(2.18)$$

Although the applied load can be changed in an arbitrary manner, in this work only either uniformly distributed load or sinusoidal ($q = q_0 \sin(\omega p t)$, where $\omega_p = 2\pi/T$, T is the period of vibration) force is analysed for boundary conditions (2.11), (2.13), (2.15), and (2.16). If sinusoidal resistance moment (2.12), (2.14) occurs, then, on the shell surface we take $q = 0$.

After reduction of the problem (2.1)–(2.9) into a normal form, the associated Cauchy problem is solved using the 4th-order Runge-Kutta method. Time step is defined by the conditions of solution stability ($\Delta t = 3.90625 \cdot 10^{-3}$).

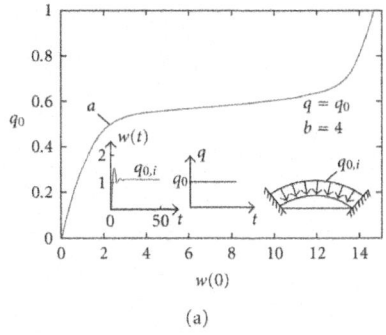

(a)

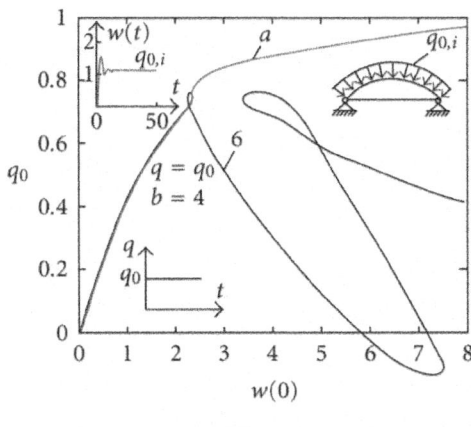

(b)

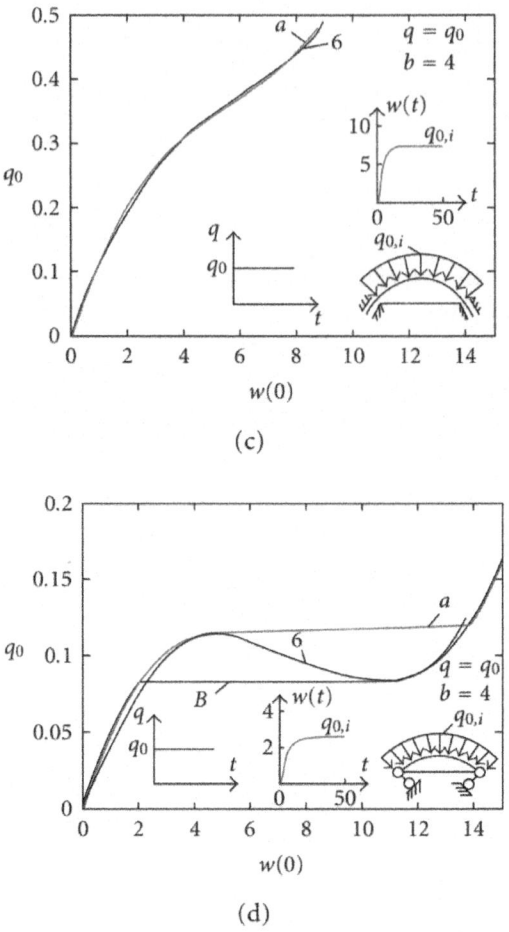

Figure 3.1. (a) Shell with a stiff clamped contour ($q = q_0$, $\varepsilon = 1$). (b) Shell with a ball-type unmovable contour ($q = q_0$, $\varepsilon = 1$). (c) Shell with a movable contour ($q = q_0$, $\varepsilon = 1$). (d) Shell with a ball-type movable contour ($q = q_0$, $\varepsilon = 1$).

RELIABILITY OF RESULTS: STATIC SOLUTION OF SHELLS STABILITY

The developed algorithm and program package allows to solve both static and dynamic problems. In order to solve static problems, the so-called "setup" method proposed by Fedos'ev [9, 10] is used. The main idea of this method follows. For $\varepsilon = \varepsilon cr$, the dependency $\{q_m, w_m(t)\}$, m = 1,2,..., denotes numbers of loading values for which a solution has been found with the use of the setup method. Applying this method, one may compute the characteristics q(w) and investigate the stress-strain state of shells. The analysis of the applied iterational setup method is given in [3]. In Figures 3.1(a)–3.1(d), the dependencies q(w) obtained via both our algorithm and the Valishvili [29] method are reported for the boundary conditions shown in the bottom-right corner of each figure and for the given parameter b. Curves denoted by (a) correspond to our method, whereas those denoted by (b) have been obtained by Valishvili [29]. It is worth noticing that the solution obtained by Valishvili and shown in Figure 3.1(b) is not unique with respect to both load

Table 3.1

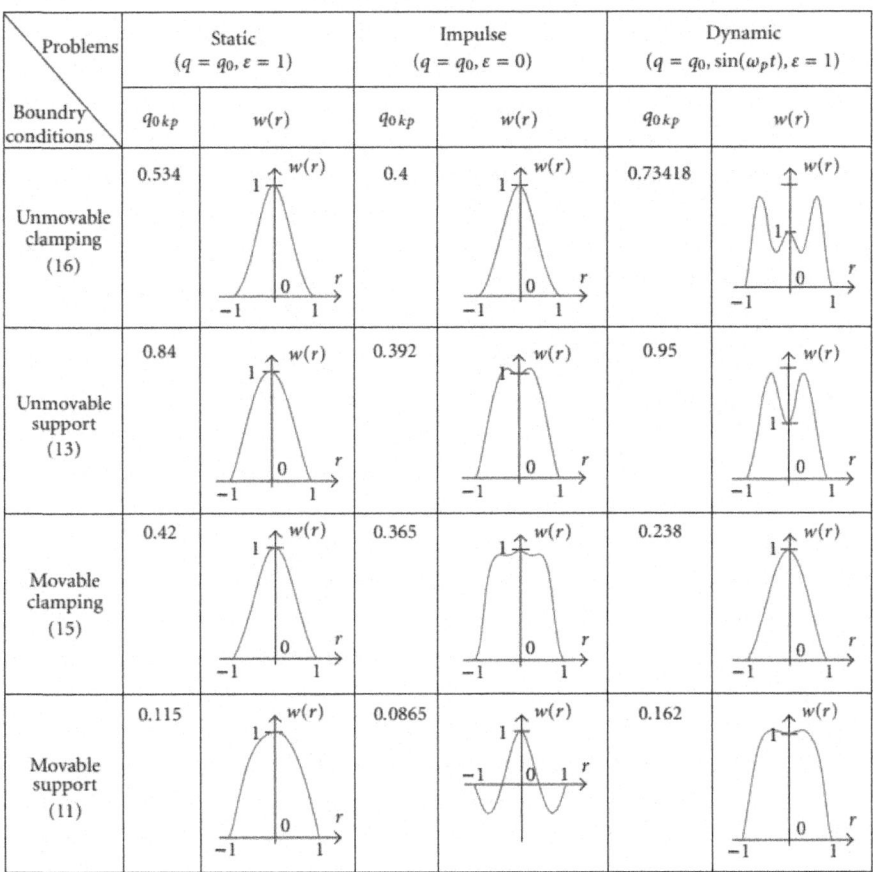

Problems	Static $(q = q_0, \varepsilon = 1)$		Impulse $(q = q_0, \varepsilon = 0)$		Dynamic $(q = q_0, \sin(\omega_p t), \varepsilon = 1)$	
Boundry conditions	q_{0kp}	$w(r)$	q_{0kp}	$w(r)$	q_{0kp}	$w(r)$
Unmovable clamping (16)	0.534		0.4		0.73418	
Unmovable support (13)	0.84		0.392		0.95	
Movable clamping (15)	0.42		0.365		0.238	
Movable support (11)	0.115		0.0865		0.162	

and deflection. Therefore, it is impossible to estimate either upper or lower critical loads. Contrarily, a dynamical approach proposed in this paper allows for a highly accurate prediction of critical loads, for which the so-called shell concavity and/or convexity occurs. Notice that loops occur, since the given characteristics $q_0(w)$ are constructed for the shell's centre, whereas other shell's points behave in independent manner, that is, the stability loss takes place not in the shell centre but in its quadrants. Therefore, the shell centre creates those loops depending on the acting load (see the graphs of deflections reported in Table 3.1). The analysis of the obtained results indicates high accuracy of our method and its high efficiency, as well as it points out the potentially wide spectrum of possible applications during solutions to various static problems.

CONVERGENCE OF THE METHOD WITH RESPECT TO SPATIAL COORDINATES

In this work, the solution to the problem for $n = 20$ is studied. Initially, the problem of convergence of the solution depending on the number n of shell radius partition [17]

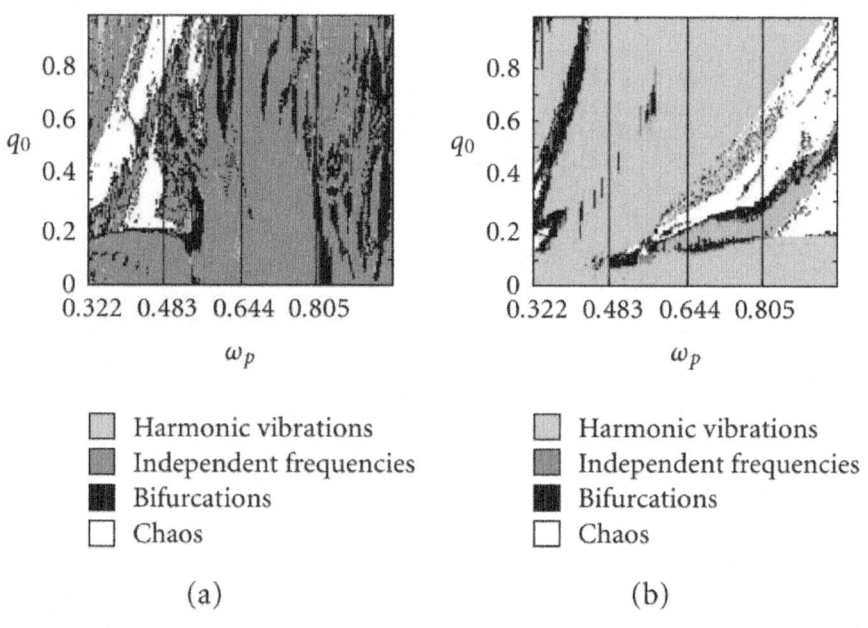

(a)   (b)

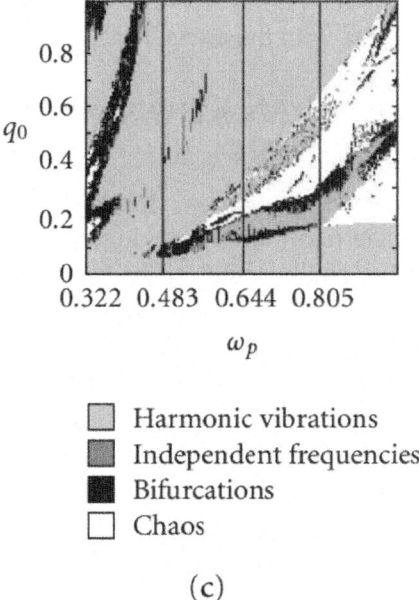

q_0

0.322 0.483 0.644 0.805

ω_p

☐ Harmonic vibrations
■ Independent frequencies
■ Bifurcations
☐ Chaos

(c)

Figure 4.1. Charts of control parameters $\{q_0, \omega_p\}$.

is analysed. It has been found that n = 20 is optimal for harmonic and chaotic areas. Power spectra for 2n and n of vibration character charts for control parameters $\{q_0, \omega_p\}$ coincide.

In Figure 4.1, charts serving for the identification of vibration character for control parameters $\{q_0, \omega_p\}$ with different steps of shell radius partition n are reported. Namely, for Figure 4.1(a), $\{q_0, \omega_p\}$ is obtained for n = 10; for Figure 4.1(b), the chart $\{q_0, \omega_p\}$ is obtained for $0 \leq q_0 \leq 1$ for n = 20; for Figure 4.1(c), the chart of control parameters is obtained for n = 30. The comparison of charts shown in Figures 4.1(b) and 4.1(c) exhibits their coincidence, that is, the process of partition is convergent (further in this work, the notation shown in Figure 4.1 is applied).

CONVERGENCE OF THE METHOD WITH RESPECT TO THE ELEMENTS OF THE SET OF CONTROL PARAMETERS $\{Q_0, \Omega_p\}$

Initially, the problem related to the identification of the charts of control parameters versus the step of the parameters q_0 and ω_p has been studied. In Figure 5.1, the charts of identification of vibration characters are analysed for three cases: (a) 100 × 100, (b) 200 × 200, and (c) 400 × 400. The comparison of Figures 5.1(a) and 5.1(c) exhibits their full coincidence, and hence further results have been computed with the mesh 200 × 200.

SCENARIOS OF THE TRANSITION FROM HARMONIC TO CHAOTIC VIBRATIONS

In what follows, we consider vibrations for a class of problems of shells theory with respect to the harmonic excitation $q = q_0 \sin(\omega_p t)$ uniformly or locally distributed on the shell surface. Possible scenarios of the transition from harmonic to chaotic states are studied, and various hypotheses of the mechanism of transition from regular flow to turbulence are also illustrated and discussed. In spite of Landau's [19] hypothesis that

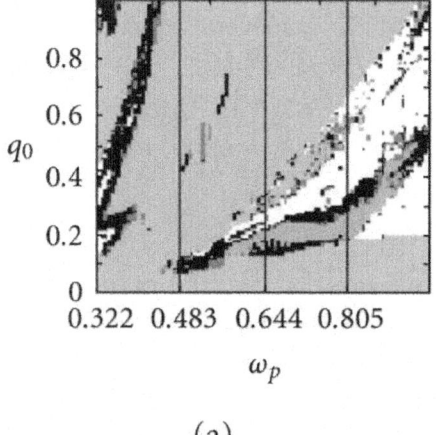

(a)

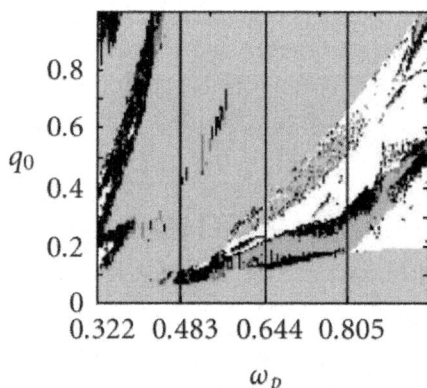

(b)

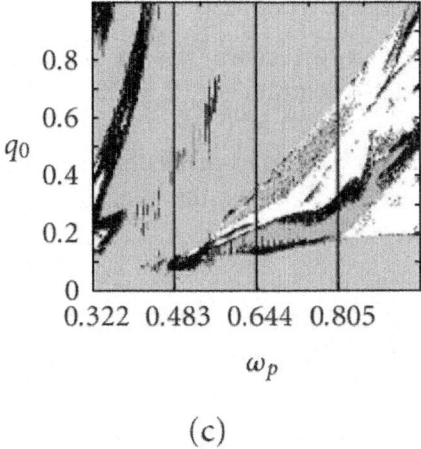

(c)

Figure 5.1: Charts of control parameters $\{q_0, \omega_p\}$.

was formulated earlier than any other, all other mechanisms are associated with finitedimensional models: Ruelle-Takens-Newhouse [24], Feigenbaum [11], and PomeauManneville [21]. It is worth noticing that so far there is no unique mechanism of transition into turbulence. In what follows, we investigate the mechanism of the occurrence of turbulence for transversal vibrations of flexible axially symmetric shells in more detail.

Feigenbaum Scenario

The transition of our mechanical system from harmonic to chaotic vibrations is carried out in accordance with the Feigenbaum scenario [11]. The Feigenbaum model has been verified numerically in the case of simple mathematical models. It is well known that period-doubling bifurcations are exhibited in Rossler at- ¨ tractor, and others. The sequence of period doubling bifurcation is also found in our system. In Table 6.1, for central shell point with ball-type movable contour (2.11) and for b = 4, some dependencies for boundary values of q0 are reported, that is, phase portrait w(w'), power spectrum S, db(ω_p), and Poincare sections ´ wt(w_{t+T}), where T is the period of excitation. Considering the results given in Table 6.1 yields the convergent series

$$d_n = \frac{q_{0,n} - q_{0,n-1}}{q_{0,n+1} - q_{0,n}} = 4.65608466, \quad n = 5.$$

(6.1)

Notice that theoretically predicted d = 4.66916224. It is worth noticing that the difference between theoretical and numerical experiment is 0.28%. The values of series q_0,n and dn are shown in Table 6.2.

Ruelle-Takens-Newhouse Scenario

The Ruelle-Takens-Newhouse scenario has an important scientific meaning, since it yields the example contrary to the so far existing so-called Landau-Hopf scenario [19]. In 1971, Ruelle and Takens [24] showed that a finite sequence of bifurcation is required to reach chaotic state. Initially, they found three Hopf bifurcations, and T3 torus then created can loose its stability and chaotic attractor may occur.

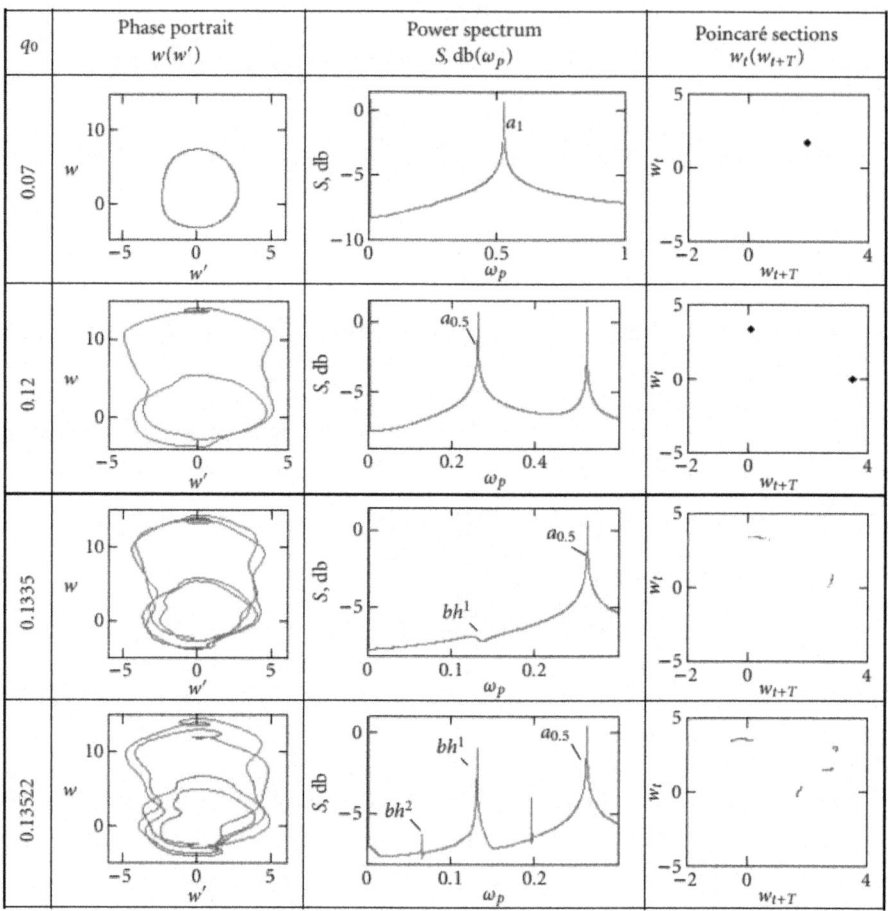

Next, in 1978, Ruelle, Takens, and Newhouse proved the theorem according to which a strange chaotic attractor is generated only through two successive Hopf bifurcations [24].

Indeed, Ruelle and Takens have shown that after Hopf bifurcations, the motion is bounded by nonsmooth manifolds with the complex topology. Such manifolds have been called strange attractors. The strange attractors do not have integer dimensions, that is, it is something between a surface and space. The fractal dimension has been studied in particular by Mandelbrot [20]. Notice that attractor obtained through the Ruelle-TakensNewhous scenario should satisfy some defined conditions, that is, if it satisfies the socalled "axiom A," then it is a chaotic one (in practice, such class of attractors rarely appears in nature). Also, a chaotic attractor is very sensitive to initial conditions. Feigenbaum et al., and Rand et al., in 1982 independently considered the following problem: How quasiperiodic motion with two independent frequencies ω_1 and ω_2 on tori can be phase-locked after the substitution of small excitation.

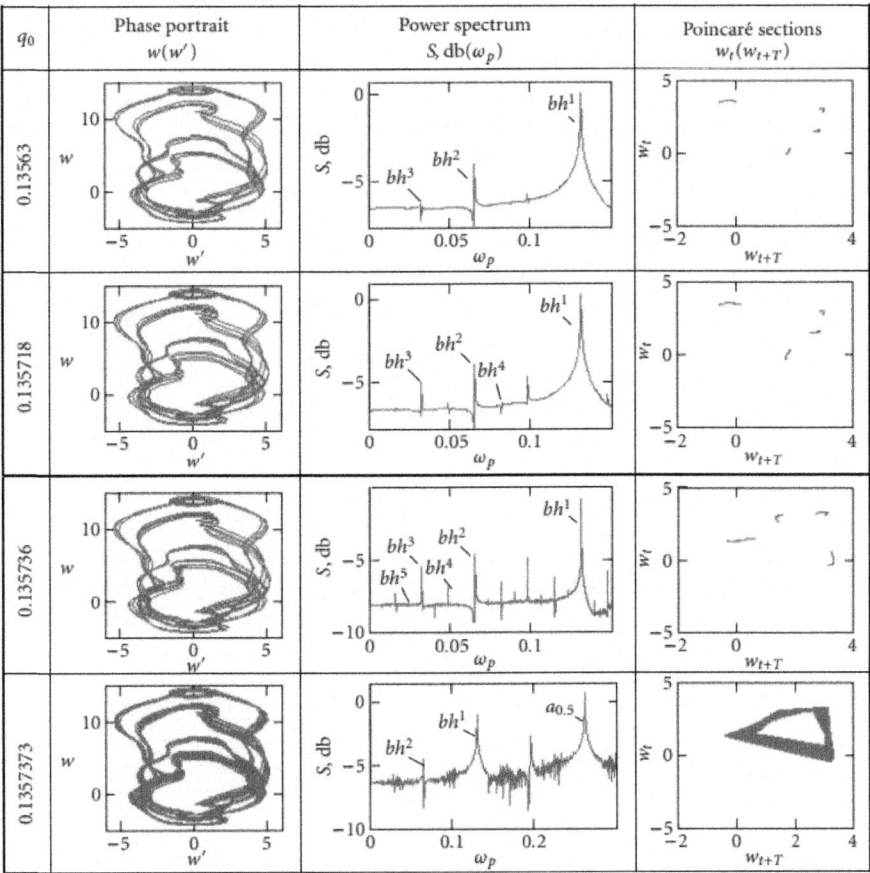

Table 6.2

n	1st bifurcation	2nd bifurcation	3rd bifurcation	4th bifurcation	5th bifurcation
$q_{0,n}$	0.1335	0.13522	0.13563	0.135718	0.1357369
d_n	—	4.19512	4.659091	4.656084	—

Recall that due to rational values $\omega_1/\omega_2 = p/q$, a trajectory closes after q cycles (synchronization). However, when ω_1/ω_2 irrational, one gets the quasiperiodic motion, that is, the trajectory is never closed and occupies the whole torus surface.

Modified Ruelle-Takens-Newhouse Scenario

For a spherical shell with ball-type movable face and contour (2.13), with sliding (2.15) and stiff (2.16) clamping and harmonic excitation of uniformly distributed load, the new mechanism of transition from harmonic to chaotic vibrations has been obtained. Recall that in the Ruelle-Takens-

Table 6.3

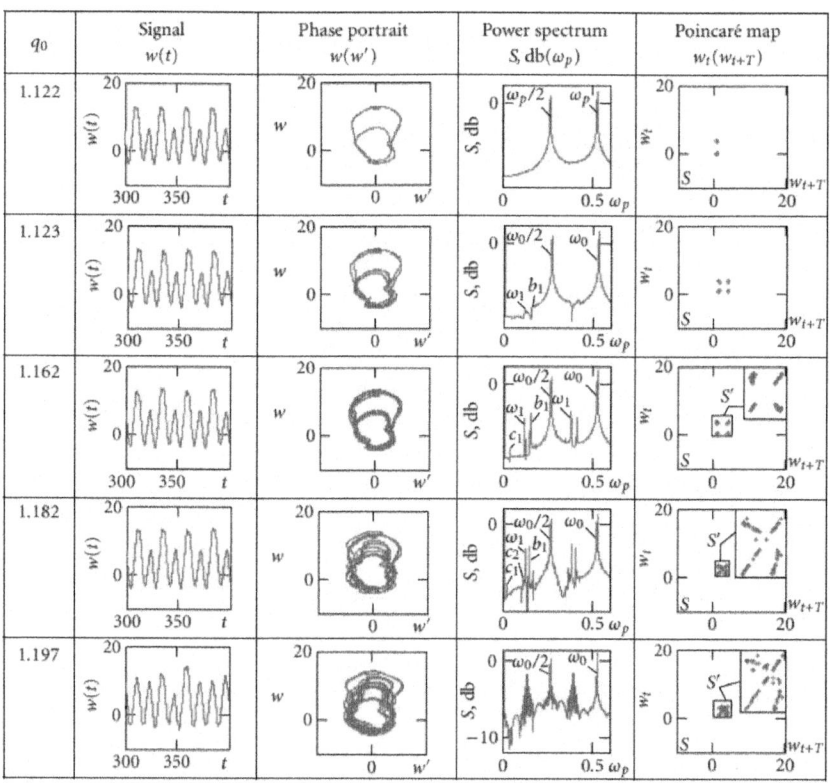

Newhouse scenario, three independent frequencies are required to reach chaos. Contrarily, in the modified scenario, the transition is realized through one independent frequency and its linear combinations with the frequency of excitation. Let us consider this novel scenario in more detail.

The applied fundamental characteristics follow: signal $w(0,t)$, phase portrait $w(w\dot{})$, power spectrum S, $db(\omega_p)$, and Poincare map $´$ $w_t(w_{t+T})$ in the dependency of the limiting values q0 are given in Table 6.3. The values q_0 are called limiting because between the shown limits of q_0, the dynamical picture remains unchanged. Let us consider this scenario using example of the problem with stiff clamped resistance contour.

(1)   Vibrations consist of the fundamental excitation frequency a_1, and they are harmonic. The phase portrait is composed of a limited set of one circled cycle ($_{q0}$ = 0.68).

(2)   Further increase of the parameter q_0 up to the value of q_0 = 0.699 yields a new independent frequency b1, that is, there is two-frequency motion (with frequencies a_1 and b_1). Notice that the motion is not synchronized, that is, $a_1/b_1 = m_1/n_1 = 3.169$ is the irrational value.

3)   Further increase up to q_0 = 0.7 results in the activation of the series of linearly dependent frequencies $b_n = n \cdot b_1$ and $a_n = a_1 - (n - 1)b_1$. Due to the continuation of this process, two frequencies a_k and $b_k \in [b_1, a_1]$ (q_0 = 0.706) approach each other. Then, one more series of linear combinations appears, that is, $c_2 = b_2 - a_2$, $c_n = x \pm c_2$ $(x = a_n, b_n)$ $(q_0 = 0.73418)$

The increase of q_0 u_p to q_0 = 0.73418 puts the considered system into chaos. Variation of q_0 on the amount of $2 \cdot 10{-}5$, that is, up to the value of q_0 = 0.7342, causes the occurrence of the shell stiff stability loss, and the system starts to vibrate again harmonically with the excitation frequency $_{a1}$. Here we may treat this process as the dynamical stability loss of spherical shells subjected to periodic harmonic excitations.

THE INVESTIGATION OF CHAOTIC VIBRATIONS OF THE HARMONICALLY EXCITED SHELL

Let us consider vibrations of the harmonically excited shell. The system behaviour is investigated for three different values of the sloping parameter b = 3,4,5, and for the following three types of clamping: ball-type movable contour (2.13), sliding clamping (2.15), and stiff clamping (2.16). In order to investigate the shell behaviour in different cases, the following characteristics are monitored: charts of control $\{q_0, \omega_p\}$ for various values of the shell sloping parameters (Figure 7.1), and the graphs of the dependency $q_0(w_{max})$ (Figure 7.2).

The charts of control parameters $\{q_0, \omega_p\}$ for all the problems associated with uniform harmonic excitations have been constructed (see Figure 7.1). Three vertical lines correspond to the following frequencies: the middle line corresponds to $\omega_p = \omega_0$ (frequency of linear vibrations), whereas the left and right lines correspond to $\omega_p - \omega_0/4$ and $\omega_p + \omega_0/4$, respectively. Notice that the analysis of three charts exhibits all possible manifolds of complex vibrations of spherical shells. According to the investigation of the shell behaviour in various conditions, the increase of the sloping parameter b = 3,4,5 yields the increase of chaotic zone for any type of clamping (see the charts of control parameters). However, the increase of the sloping parameter does not change the scenario of transition from harmonic to chaotic vibrations. Comparing the shell behaviour depending on the clamping type, the following observations are made. Zones of chaotic vibrations are larger for a ball-type unmovable contour and stiff clamping of the shell, and in the majority of the analysed cases, a route to chaos is realized through the modified Ruelle-Takens-Newhouse scenario [24], proposed in this work. In the case of the ball-type movable contour (2.11) and movable clamping (2.16), the Feigenbaum scenario [11] dominates.

For all types of clamping and the shell sloping parameter values, the dependencies $q_0(w_{max})$ (Figure 7.2) are constructed. Under the obtained graphs, the scales characterizing various signal types are shown.

In Figure 7.2(a), in the graph of dependency $q_0(w_{max})$ (for the shell with the ball-type movable contour and with sloping parameter b = 3), one inflection point as well as firstorder discontinuity are observed. In these points, the change of signal types occurs, which is manifested by the change of color in the scale of vibrations type. In Figure 7.2(b), the curve associated with the same type of clamping and with sloping parameter b = 4 is reported. Observe the inflection point and the series of stiff bifurcations. In addition, the

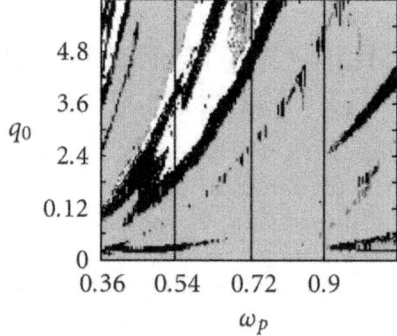

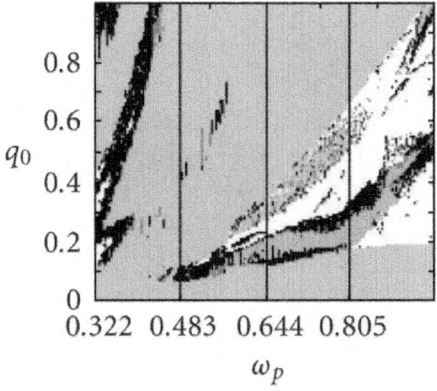

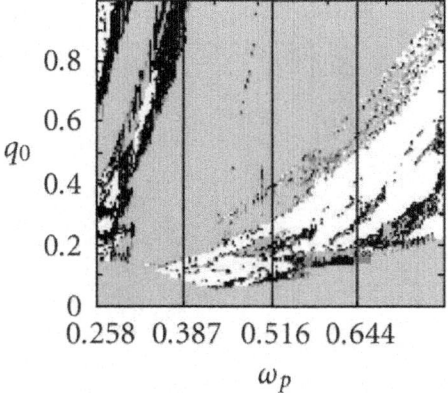

(c) $b = 5$.

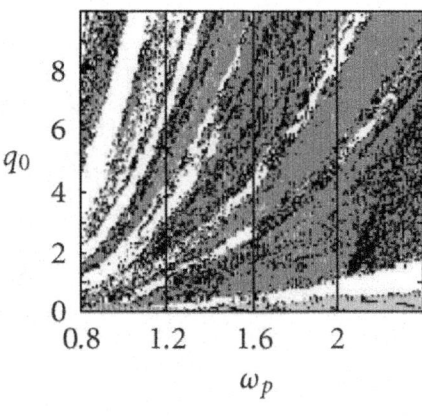

(d) $b = 3$.

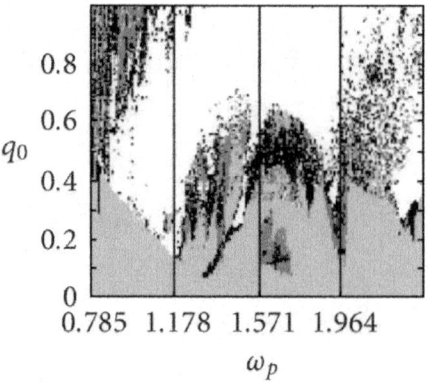

(e) $b = 4$.

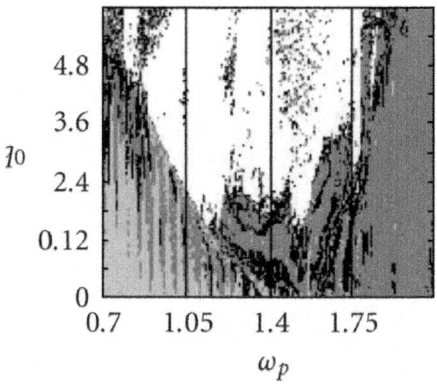

(f) $b = 5$.

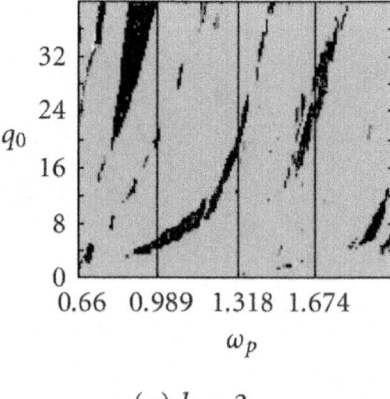

(g) $b = 3$.

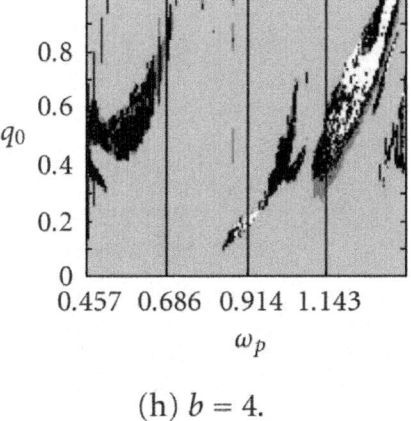

(h) $b = 4$.

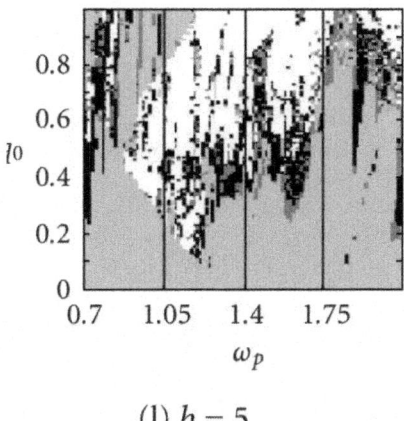

(l) $b = 5$.

Figure 7.1: Charts of control parameters $\{q_0, \omega_p\}$. (a)–(c) Ball-type movable contour. (d)–(f) Balltype unmovable contour. (g)–(i) Sliding clamping. (j)–(l) Stiff clamping.

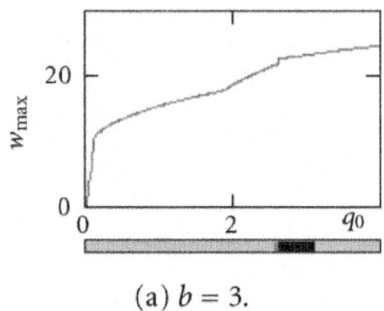

(a) $b = 3$.

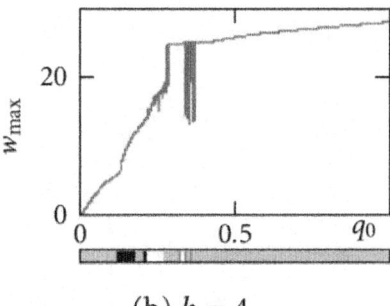

(b) $b = 4$.

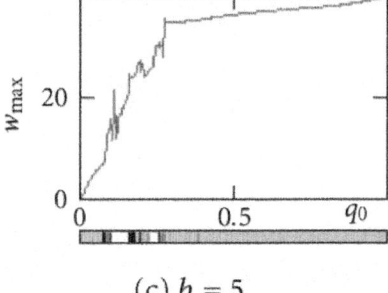

(c) $b = 5$.

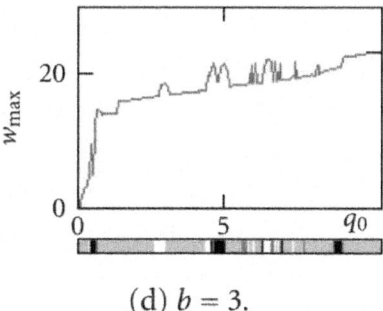

(d) $b = 3$.

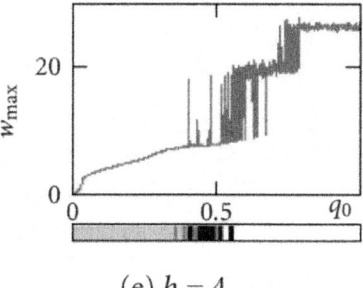

(e) $b = 4$.

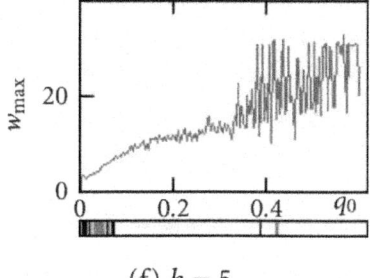

(f) $b = 5$.

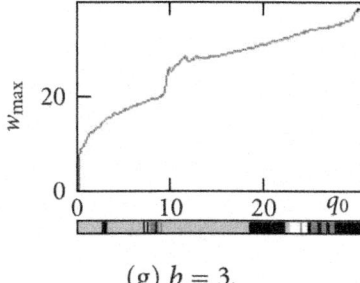

(g) $b = 3$.

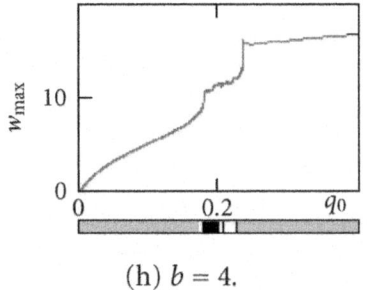

(h) $b = 4$.

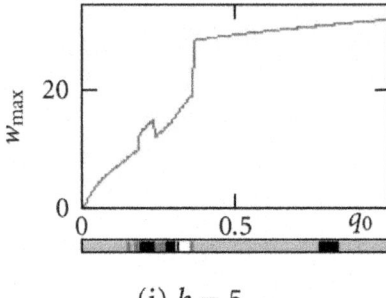

(i) $b = 5$.

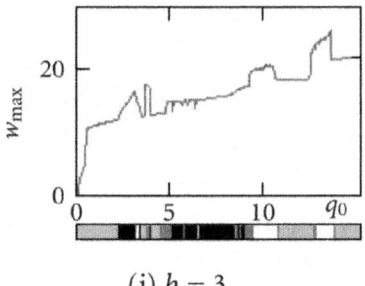

(j) $b = 3$.

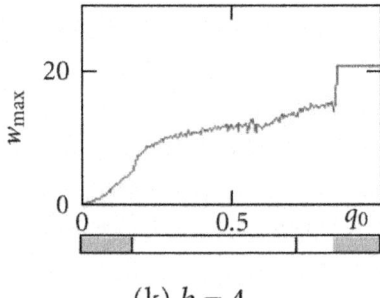

(k) $b = 4$.

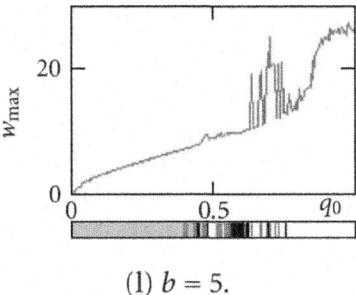

(1) $b = 5$.

Figure 7.2: Dependencies $q_0(w_{max})$. (a)–(c) Ball-type movable resistance contour. (d)–(f) Ball-type unmovable resistance contour. (g)–(i) Sliding clamping. (j)–(l) Stiff clamping.

change of vibrations type is manifested by colors associated with signal types. Owing to the increase of b parameter, chaotic vibrations are observed for the smaller amplitude of the existing load than those for b = 3. The further increase of the sloping parameter up to b = 5 (Figure 7.2(c)) yields the increase of the first-order discontinuities, and of the surface of chaotic zones. The analogical picture is produced also for other boundary conditions.

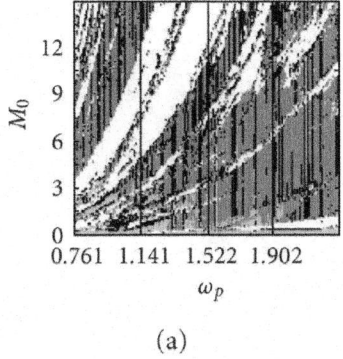

(a)

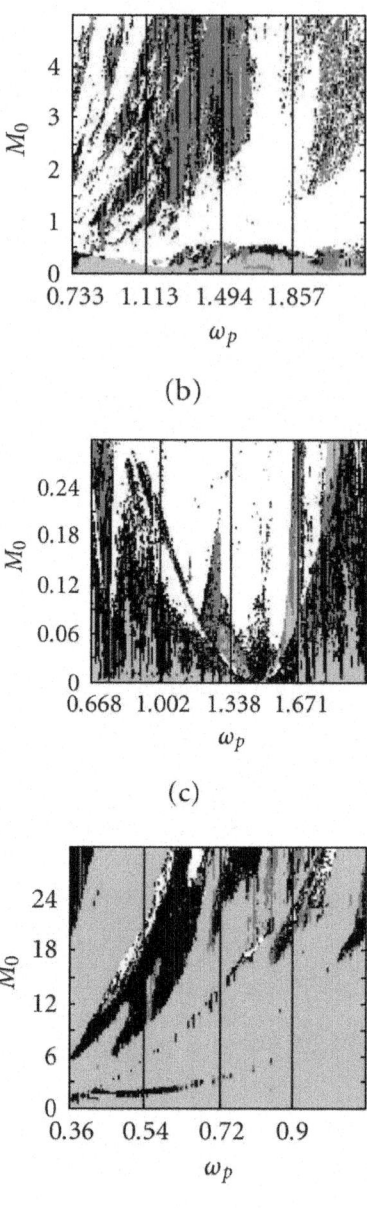

(b)

(c)

(d)

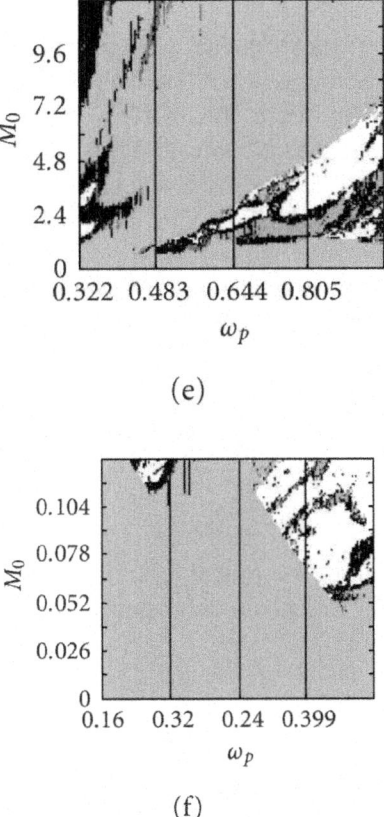

(e)

(f)

Figure 8.1: Charts of control parameters $\{M_0, \omega_p\}$.

COMPLEX VIBRATIONS OF A SHELL SUBJECTED TO HARMONIC MOMENT

In this section, we consider vibrations of the outlined class problems of theory of shells in the case of harmonic resistance moment excitation $M = M_0 \sin(\omega_p t)$. This moment drives both contours of the shell with ball-type support (2.12) for $b = 3$, $b = 4$, $b = 5$, and with unmovable contour (2.14) for $b = 3$, $b = 4$, $b = 5$ for $q = 0$ in (2.1) and (2.10).

For all the considered problems, charts of control parameters $\{M_0, \omega_p\}$ are constructed. For ball-type unmovable contour with $b = 3$, $\varepsilon = 0.1$, see Figure 8.1(a), for $b = 4$, $\varepsilon = 0.1$, see Figure 8.1(b), and for $b = 5$, $\varepsilon = 0.1$, see Figure 8.1(c), as well as for ball-type movable contour $b = 3$, $\varepsilon = 0.1$, see Figure

8.1(d), for b = 5, ε = 0.1, see Figure 8.1(e), for b = 5, ε = 0.1, see Figure 8.1(f).

The analysis of the dependencies {M_0, ω_p} for these types of problems implies that in the case of the shell with ball-type unmovable contour subjected to harmonic moment (Figures 8.1(a), 8.1(b), and 8.1(c)), a transition from harmonic to chaotic vibrations is realized either according to the Feigenbaum [11] or to the modified Ruelle-TakensNewhouse [24] scenarios. In the problem with ball-type movable contour (Figures 8.1(d), 8.1(e), and 8.1(f)), the system is transited into chaos only according to the Feigenbaum scenario. The Feigenbaum model is well verified via numerical experiment applied to simple mathematical models. The same behaviour is also detected in the case of our

Table 8.1

N	1st bifurcation	2nd bifurcation	3rd bifurcation	4th bifurcation
$q_{0,n}$	0137	0.1395	0.140 13	0.140 266 7
d_n	—	3.968 253 968	4.608 632 04	—

analysed shell with a ball-type movable contour and subjected to harmonic moment. The following Feigenbaum constant is computed for b = 4:

$$d_n = \frac{q_{0,n} - q_{0,n-1}}{q_{0,n+1} - q_{0,n}} = 4.608\,632\,04, \quad n = 4 \text{ (see Table 8.1).}$$

(8.1)

Notice that the theoretical value d = 4.669 16224, and hence the difference between theoretical and numerical results is 1.289%.

In two charts of control parameters {M_0, ω_p}, constructed for the shell with a ball-type unmovable contour, large zones of chaotic behaviour are visible. Contrarily, for a balltype movable contour, in the chart of control parameters {M_0, ω_p}, significantly fewer zones with chaotic vibrations appear. It is worth noticing that during the increase of the shell rise, chaotic zones essentially increase. It is clear that the system dynamics essentially depends on loading type for the same boundary and initial conditions, and the shell geometry.

For all problems, the graphs of the dependencies $w_{max}(M_0)$, $F_{max}(M_0)$, $M_{1max}(M_0)$, where $M1 = -(\partial^2 w/\partial r^2) - \nu(\partial w/r \partial r)$ (for $r = \Delta$), are constructed. For a ball-type movable contour with b = 3, ε = 0.1, see Figure 8.2(a), for a ball-type unmovable contour b = 4, ε = 0.1, see Figure 8.2(b), for b = 5, ε = 0.1, see Figure 8.2(c), for a ball-type unmovable contour b = 3, ε = 0.1, see Figure 8.2(d), for b = 4, ε = 0.1, see Figure 8.2(e), and for b = 5, ε = 0.1, see Figure 8.2(f).

During investigations of the shell vibrations, its stress-strain state plays the most important role, and hence its dependence on the amplitude M_0 is studied. One of the key problems is associated with the consideration of the deflection function and stresses versus parameter M0 variations. As it is well known in the theory of flexible shells, the full shell stress consists of the stresses in the mean shell surface and of the bending stresses. The stress function F plays an important role in the stress determination in the shell mean surface, and in the case of bending stresses—it is the moment M_1. Let us construct the characteristics $w_{max}(M_0)$, $F_{max}(M_0)$, $M_{1max}(M_0)$. The analysis of the given characteristics indicates that all of them describe in the same manner the occurrence of dynamical stability loss via stiff bifurcation. It is evidently shown in the characteristic $M_{1max}(M_0)$, since it contains the second derivative of the deflection function with respect to a coordinate. Dependencies $w_{max}(M_0)$, $F_{max}(M_0)$, $M_{1max}(M_0)$, the bifurcation scale, and the Lyapunov exponents well coincide [7]. The analysis of the mentioned characteristics yields the conclusion that the occurrence of timing chaos implies the occurrence of spatial chaos, that is, timing and spatial chaos are mutually coupled. It is worth noticing that chaos of shell's stress-strain state is also observed.

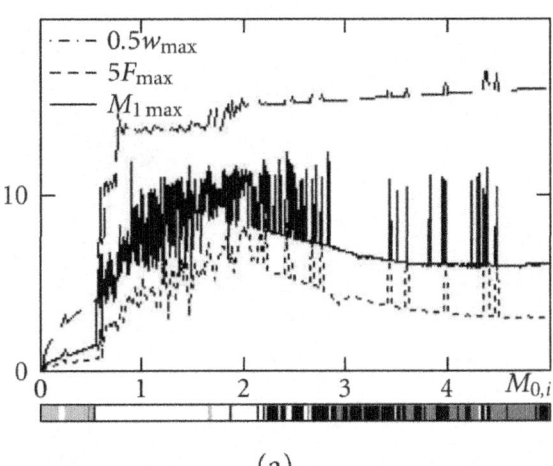

(a)

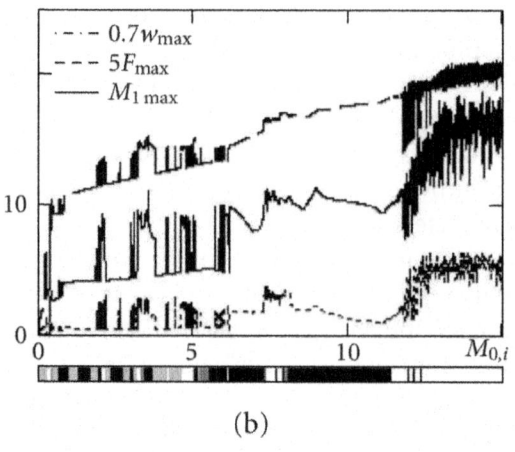

(b)

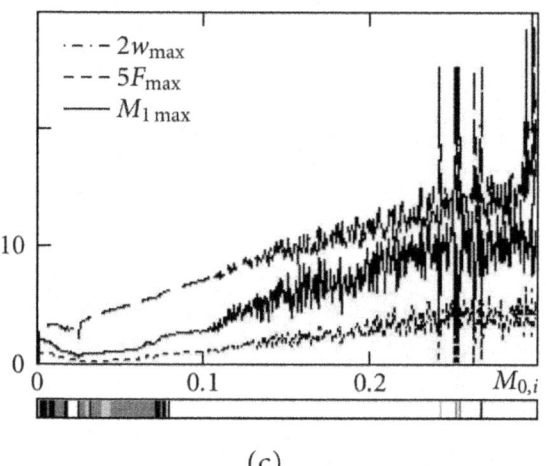

(c)

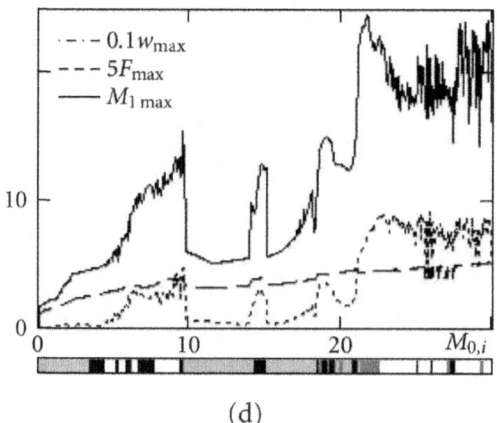

(d)

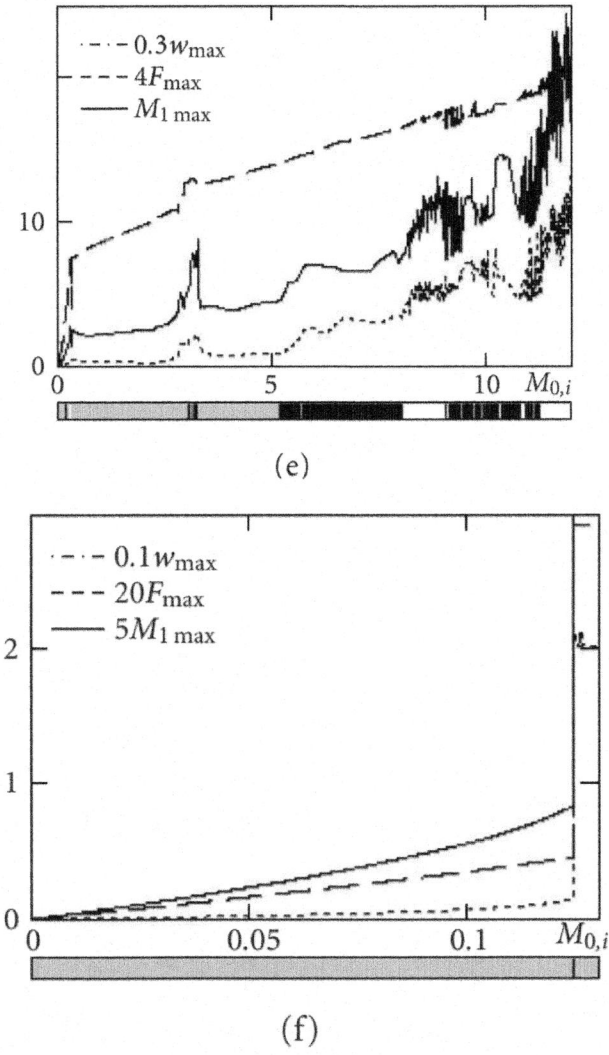

Figure 8.2. Dependencies $w_{max}(M_0)$, $F_{max}(M_0)$, $M_{1max}(M_0)$.

BEHAVIOUR OF SHELLS SUBJECTED TO A LOCAL HARMONIC LOAD

Consider the shell with a ball-type contour subjected to two types of local loads: (i) $q = q_0 \sin(\omega_p t)$ is loaded in five points $8 \leq i \leq 12$, when $0 \leq i \leq n$, i, n $\in$ N, in the neighborhood of the fourth one (in other points $q = 0$); (ii) load $q = q_0 \sin(\omega_p t)$ is applied to five points $0 \leq i \leq 4$, where $0 \leq i \leq n$, i, n $\in$ N, in the neighborhood of the centre (in other points $q = 0$).

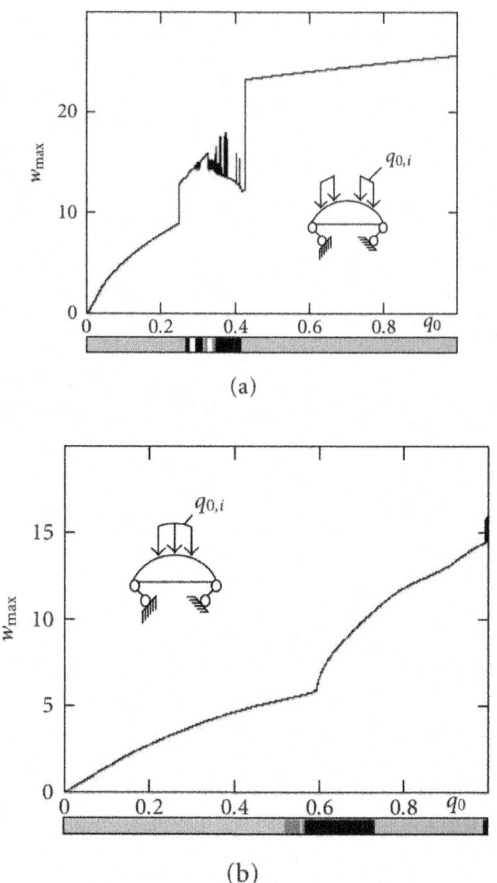

(a)

(b)

Figure 9.1: Dependency $w_{max}(q_0)$.

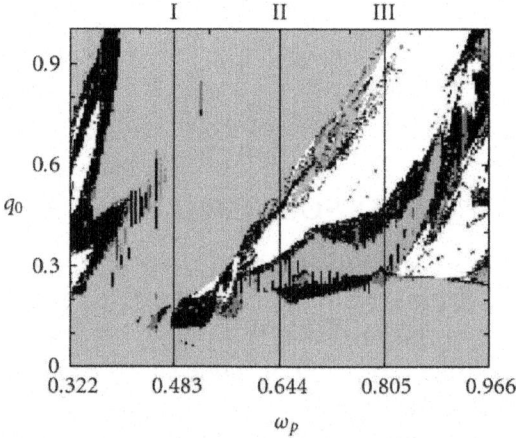

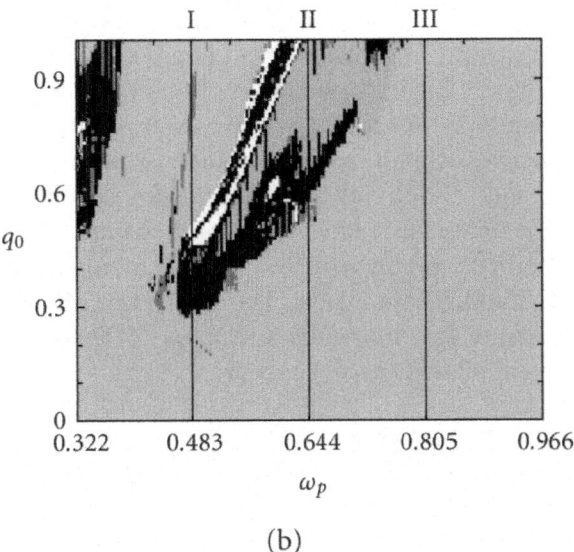

(b)

Figure 9.2: Chart of control parameters $\{q_0, \omega p\}$.

The graphs of the dependency $w_{max}(q_0)$ are given for the first type in Figure 9.1(a), whereas for the second type—in Figure 9.1(b). In the case of first shell loading type, two stiff bifurcations appear. The first stability loss occurs during the transition from harmonic vibrations to first Hopf bifurcation, whereas the second one is associated with the transition from chaotic to harmonic vibrations. The signal-type scale exhibits the existence of the Feigenbaum scenario. The dependency $w_{max}(q_0)$ is more smoothened; there is no first-order discontinuity, as in the first case. In the scale of signal type, a small zone of soft bifurcations appears, and there are no chaotic zones.

Numerical experiment yields five Hopf bifurcations, and the obtained Feigenbaum constant is equal to 4.67784 (the difference in comparison to the theoretical value is 0.168%).

Table 10.1

Problem	Distributed load	Local load	Resistance moment
1	$q = q_0 \cdot \sin(\omega_p \cdot t)$	—	—
2	—	$q = q_0 \cdot \sin(\omega_p \cdot t)$	—
3	—	—	$M = M_0 \cdot \sin(\omega_p \cdot t)$
4	$q = q_0 \cdot \sin(\omega_p \cdot t)$	$q_1 = 0.6 \cdot \sin(0.725 \cdot t)$	—
5	$q = q_0 \cdot \sin(\omega_p \cdot t)$	—	$M_1 = 9.6 \cdot \sin(0.886 \cdot t)$

For both types of local shell loading, the charts of control parameters $\{q_0,\omega_p\}$ are constructed (first loading type is given in Figure 9.2(a), second loading type—Figure 9.2(b)). In the chart of control parameters $\{q_0,\omega_p\}$ and for the first loading type, one may observe a large zone of chaotic vibrations with high frequencies. Small zones of Hopf bifurcations associated with low frequencies and "drops" of independent frequencies as well as their linear combinations are also visible. Vibration character is different now in comparison to the first loading type. In the chart of control parameters, large zones of regular vibrations are visible. The small zone of chaotic vibrations is shifted into the area of low frequencies. "Drops" of independent frequencies and their linear combinations do not appear.

CONTROL OF CHAOTIC VIBRATIONS OF FLEXIBLE SPHERICAL SHELLS

It is well known that a chaotic attractor of a dynamical system consists of the countable set of saddle cycles with different periods and that its representing point (trajectory) visits close neigbourhood of each of them. If in this time instant one tries to stabilize a saddle cycle, then trajectory will remain in its vicinity, and the system exhibits periodic motions. On the other hand, problems of the control of chaos in interacting systems are indirectly associated with problems of synchronization. In what follows, by applying some targetoriented excitations, the defined chaotic subsets corresponding to synchronized motions of identical systems can be transformed to the stable ones with respect to some chosen directions, keeping the rest of them as unstable ones in other directions. One may expect the controlled transition from unsynchronized chaotic vibrations to regime of fully synchronized chaos as a result. In this work, the mentioned control is realized with the help of target-oriented shell excitation. The shell is sinusoidally excited $q = q_0 \sin(\omega pt)$ and two types of periodic excitations are applied:

(i)   additional local transversal harmonic load is applied in five points $8 \leq i \leq 12$, where $0 \leq i \leq n$, $i, n \in Z$

(ii)   additional resistant harmonic moment.

The mentioned excitations are of two types: with a fixed frequency and with synchronized frequencies.

In order to study the behaviour of the shell subjected to two exciting loads with different frequencies (problems 3–5, Table 10.1), the mathematical model of two-frequency vibrating system is realized and charts of vibration character for control parameters $\{q_0,\omega_p\}$ are constructed (see Figure 10.1).

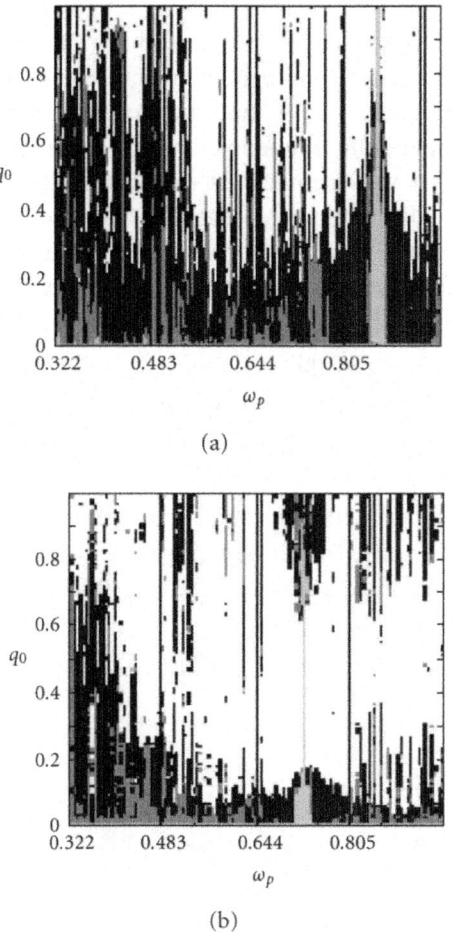

Figure 10.1: Charts of different vibrations in the plane of $\{q_d, \omega_p\}$.

In all charts, large zones of chaotic vibrations are remarkable. The increase of the frequency of local excitation causes the expansion of chaotic zones. The system is transited into harmonic vibrations only in the case when frequencies of two exciting forces are close to each other. In this case, the area of harmonic vibrations (located in vicinity of the frequency of either exciting force or exciting resistance moment (Figure 10.1(b))) on the charts of vibration characters is studied in more detail, with the use of the dependencies $w_{max}(q_0)$. Here, five loading types are reported. Notice that the number of curves shown in Figure 10.2 and the number of problems reported in Table 10.1 are identical. In Figures 10.2(a) and 10.2(b), the graphs related to the problems 1, 2, 4 (1,3,5) are constructed. In curves 1, 2, 3 (Figures 10.2(a) and 10.2(b)), the first-order discontinuities are shown, indicating the existence of stiff stability loss. The occurrence of stiff stability loss is manifested also through signal-type scales, where in the corresponding time instants, a change of vibrations appears. In the scales of signal types 1, 2, 3, one may observe a few zones of chaotic vibrations with independent frequencies and bifurcations. In both problems, the synchronization of two frequencies of exciting loads deletes the zone of chaotic vibrations, bifurcation and that of independent frequencies. In the scales of signal types 4 (Figure 10.2(a)) and 5 (Figure 10.2(b)), only harmonic vibrations are visible. Dependencies $w_{max}(q_0)$4(Figure 10.2(a)) and 5 (Figure 10.2(b)) are smooth, and first-order discontinuities do not appear.

Notice that the values of the frequency ω_p and moment M_0 or ω_p and the amplitude of the exciting force have been chosen according to the experiments carried out. It has been detected that the increase of the added excitation causes the expansion of a zone of harmonic vibrations.

The analysis of the control parameter charts of the problems 4 and 5 (Table 10.1) shows that in the case of frequencies without local and distributed loads or a resistance moment and distributed load $\omega_p = 0.859$, the vertical zone occurs in the charts of control

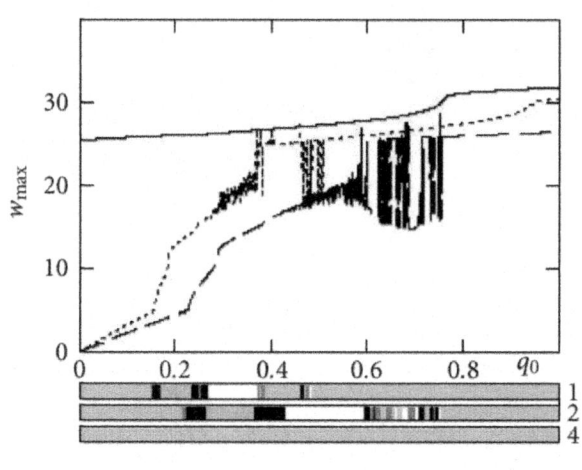

(a)

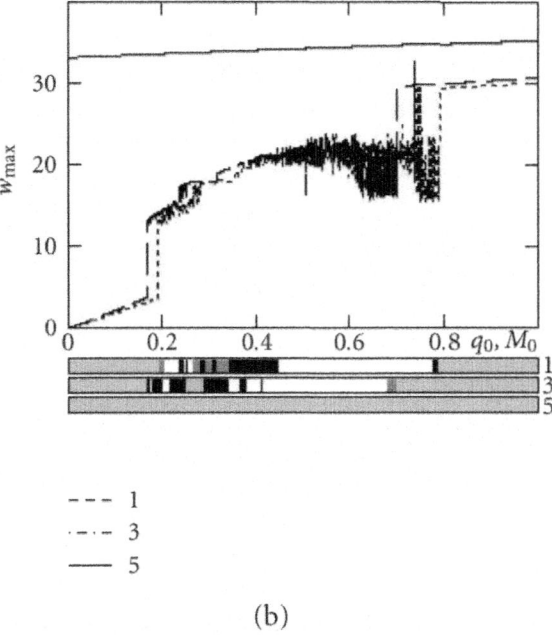

(b)

Figure 10.2: Dependencies $w_{max}(q_0)$.

parameter s, whereas for all $q_0 \in [0,1]$ harmonic vibrations are visible. Therefore, it is required to consider the problem when the shell is subjected to a uniformly distributed load and local load action, or to harmonic resistance moment with synchronized frequencies in both excitations (Table 10.2).

In Figure 10.3, the charts of vibration character for the control parameters $\{q_0,\omega_p\}$, $\{M_0,\omega_p\}$ are reported. The first three charts have been already displayed in Figures 7.1(b), 8.1(e), and 9.2(a). They are repeated here to get more deep insight of the problem. The analysis of the obtained results yields the observation that a forced synchronization of external excitations shifts the analysed mechanical system into another type of vibrations. It is worth noticing that chaotic zones are almost fully cancelled. Recall that they appeared previously in the charts $\{q_0,\omega_p\}$, $\{M_0,\omega_p\}$ (Figures 10.3(a), 10.3(b), 10.3(c), problems 1, 2, 3; Table 10.2) practically in the same places. In the case of fourth loading type (Figure 10.3(d), problem 4; Table 10.2), the exhibited vibrations are mostly harmonic. Only a small zone of chaos and bifurcations with low and high frequencies remain. In the case of fifth type of loading (Figure 10.3(e), problem 5; Table 10.2), there are bifurcation zones, and small zones of chaos only associated with low frequencies are exhibited. Similar considerations have been carried out also for the shell with a ball-type unmovable resistance contour and a sloping parameter b = 4. The results of these considerations are reported in Figure 10.4.

According to earlier obtained results concerning a ball-type moving resistance contour, only some problems of chaos control are addressed. In other words, the methods of chaos control used efficiently before are applied now to the shell with a ball-type unmovable resistance contour. Although charts of control parameters corresponding to the uniformly distributed harmonic load (Figure 10.4(a)) and to harmonic resistance moment

Table 10.2

Problem	Distributed load	Local load	Resistance moment
1	$q = q_0 \cdot \sin(\omega_p \cdot t)$	—	—
2	—	$q = q_0 \cdot \sin(\omega_p \cdot t)$	—
3	—	—	$M = M_0 \cdot \sin(\omega_p \cdot t)$
4	$q = q_0 \cdot \sin(\omega_p \cdot t)$	$q_1 = 0.6 \cdot \sin(\omega_p \cdot t)$	—
5	$q = q_0 \cdot \sin(\omega_p \cdot t)$	—	$M_1 = 9.6 \cdot \sin(\omega_p \cdot t)$

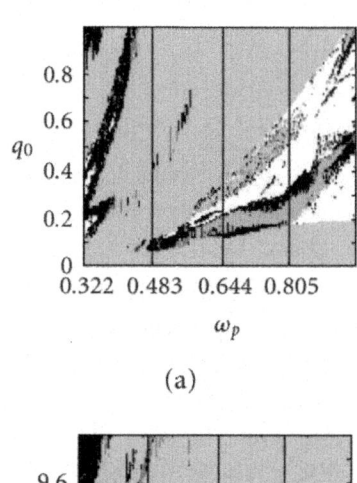

(a)

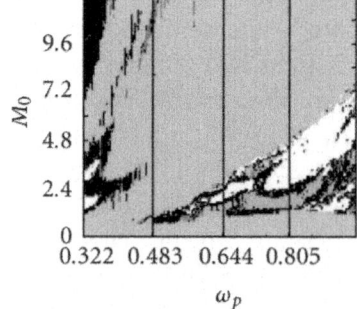

(b)

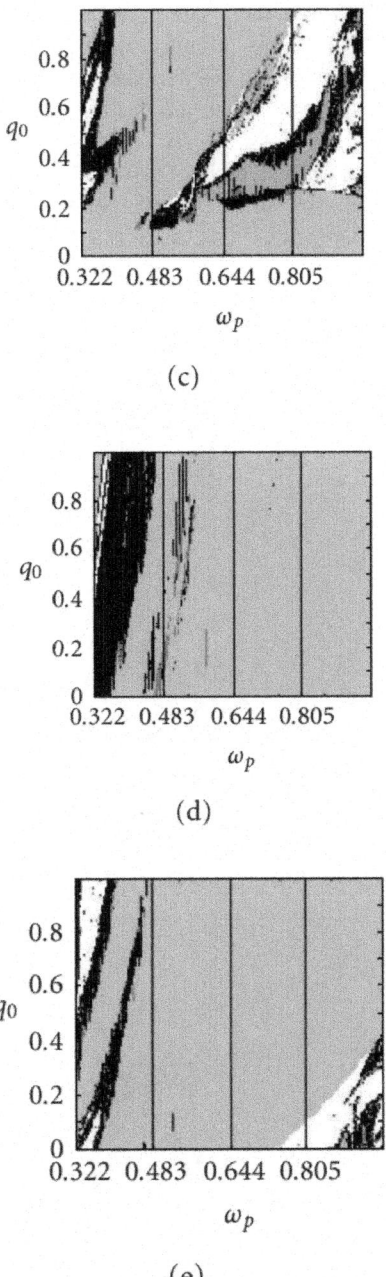

(c)

(d)

(e)

Figure 10.4(b)) have been earlier constructed, they are repeated for the sake of clarity. Besides, charts describing behaviour of the shell subjected to a distributed load and harmonic moment, as well as to local and continuous loading, are constructed. All applied loads are shown in Table 10.3. The phase shift of the local load on amount of π means that the local load works in antiphase with respect to a continuous load. Interaction of both added loads yields the decrease of chaotic vibrations and increase of harmonic vibrations. Notice that the action of the added local load is more efficient from the point of view of energy loss, as well as from the point of view of chaos control. The key result obtained in the course of our study is the following. The methods of control of chaotic vibrations of elastic spherical hells are the same for both ball-type movable and unmovable clampings.

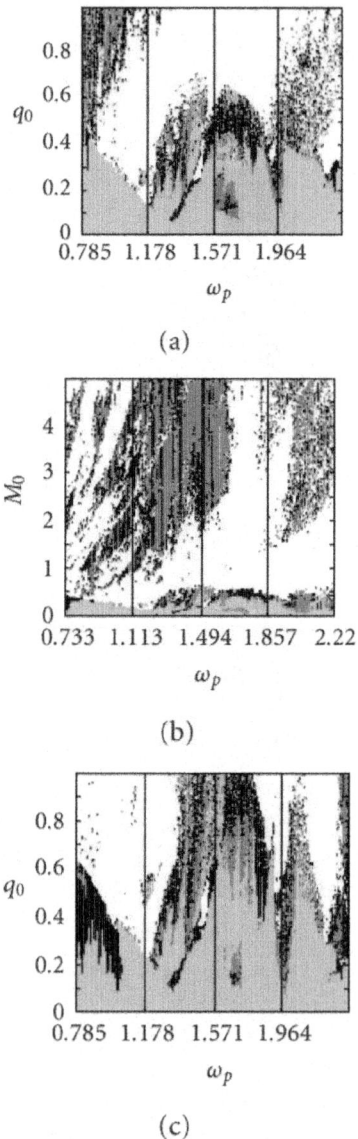

(a)

(b)

(c)

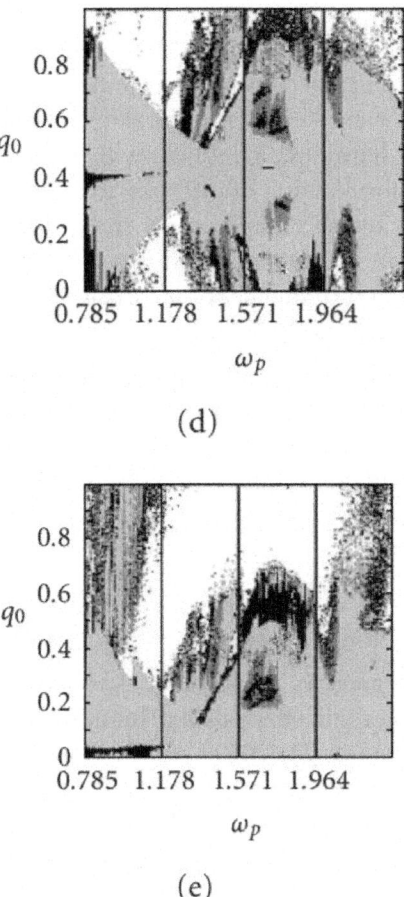

(d)

(e)

Figure 10.4: Charts of various vibrations in the planes $\{q_0, \omega_p\}$, $\{M_0, \omega_p\}$.

Table 10.3

Problem	Distributed load	Local load	Resistance moment
1	$q = q_0 \cdot \sin(\omega_p \cdot t)$	—	—
2	—	$q = q_0 \cdot \sin(\omega_p \cdot t)$	—
3	—	—	$M = M_0 \cdot \sin(\omega_p \cdot t)$
4	$q = q_0 \cdot \sin(\omega_p \cdot t)$	$q_1 = \sin(\omega_p \cdot t - \pi)$	—
5	$q = q_0 \cdot \sin(\omega_p \cdot t)$	—	$M_1 = 0.5 \cdot \sin(\omega_p \cdot t)$

CONCLUSIONS

A brief conclusion follow. Control of chaos is understood as a transformation of chaotic system behaviour into regular or chaotic one but with other properties resulting from target-oriented excitations applied to our shell, that is, harmonic resistance moment or harmonic local load with synchronized frequencies (the so-called forced synchronization of different exciting frequencies). Therefore, by variation of the load, one may control shell vibrations and, for instance, increase a zone of harmonic vibrations. It is worth noticing that the introduced changes of amplitude and frequency of the exciting load allowed to avoid stiff stability loss and to keep only harmonic shell vibrations.

Finally, a package of programs has been developed allowing for the construction of a mathematical model of an elastic spherical shell. Large amounts of numerical experiments have been carried out, yielding new results in the field of stochastic vibrations of flexible shallow spherical shells.

REFERENCES

1.   J. Awrejcewicz and V. A. Krysko, Feigenbaum scenario exhibited by thin plate dynamics, Nonlinear Dynamics 24 (2001), no. 4, 373–398.

2.   J. Awrejcewicz and A. V. Krysko, Analysis of complex parametric vibrations of plates and shells using Bubnov-Galerkin approach, Archive of Applied Mechanics 73 (2003), no. 7, 495–504.

3.   J. Awrejcewicz and V. A. Krysko, Nonclassical Thermoelastic Problems in Nonlinear Dynamics of Shells. Applications of the Bubnov-Galerkin and Finite Difference Numerical Methods, Scientific Computation, Springer, Berlin, 2003.

4.   , Nonlinear coupled problems in dynamics of shells, International Journal of Engineering Science 41 (2003), no. 6, 587–607.

5.   J. Awrejcewicz, V. A. Krysko, and A. V. Krysko, Spatial-temporal chaos and solitons exhibited by von Karm´ an model ´ , International Journal of Bifurcation and Chaos in Applied Sciences and Engineering 12 (2002), no. 7, 1465–1513.

6.   J. Awrejcewicz, V. A. Krysko, and A. V. Krysko, Complex parametric vibrations of flexible rectangular plates, Meccanica. International Journal of the Italian Association of Theoretical and Applied Mechanics 39 (2004), no. 3, 221–244.

7.   J. Awrejcewicz, V. A. Krysko, and G. G. Narkaitis, Bifurcations of a thin plate-strip excited transversally and axially, Nonlinear Dynamics 32 (2003), 187–209.

8.   J. Awrejcewicz, V. A. Krysko, and A. F. Vakakis, Nonlinear Dynamics of Continuous Elastic Systems, Springer, Berlin, 2004.

9.   V. F. Fedos'ev, Application of the step method to analyse stability of compressed rod, Prikladnaya Matematika i Mekhanika 27 (1963), no. 5, 833–841 (Russian).

10. , On the method of solutions of stability of deformable bodies, Prikladnaya Matematika i Mekhanika 27 (1963), no. 2, 265–275 (Russian).

11. M. J. Feigenbaum, The universal metric properties of nonlinear transformations, Journal of Statistical Physics 21 (1979), no. 6, 669–706.

12. A. Hubler and E. L ¨ uscher, ¨ Resonant stimulation and control of nonlinear oscillators, Naturwissenschaften 76 (1989), no. 2, 67–69.

13. E. A. Jackson, The entrainment and migration controls of multiple-attractor systems, Physics Letters. A 151 (1990), no. 9, 478–484.

14. , On the control of complex dynamic systems, Physica D. Nonlinear Phenomena 50 (1991), no. 3, 341–366.

15. V. A. Krysko, J. Awrejcewicz, and V. M. Bruk, On the solution of a coupled thermo-mechanical problem for non-homogeneous Timoshenko-type shells, Journal of Mathematical Analysis and Applications 273 (2002), no. 2, 409–416.

16. V. A. Krysko and I. V. Kravtsova, Stochastic vibrations of flexible flat axisymmetric shells exposed inhomogeneous loading, Proceedings of 7th International Conference on Dynamics of System— Theory and Applications (Łod´ z, 2003), 2003, pp. 189–197. ´

17. , Stochastic vibrations of flexible axially symmetric supported along contour spherical shells, Izviestia VUZ, Maschinostroyeniye 1 (2004), 3–13 (Russian).

18. V. A. Krysko and T. V. Shchekaturova, Chaotic vibrations of shallow shells, Izviesta AN Mekhanika Tviordoga Tela 4 (2004), 140–150 (Russian).

19. L. D. Landau, On the problem of a turbulence, Doklady Akademii Nauk 44 (1944), no. 8, 339–342 (Russian).

20. B. B. Mandelbrot, The Fractal Geometry of Nature, Schriftenreihe fur den Referenten, W. H. ¨ Freeman, California, 1982.

21. P. Manneville and Y. Pomeau, Different ways to turbulence in dissipative dynamical systems, Physica D. Nonlinear Phenomena 1 (1980), no. 2, 219–226.

22. E. Ott, C. Grebogi, and J. A. Yorke, Controlling chaos, Physical Review Letters 64 (1990), no. 11, 1196–1199.

23. V. Petrov, V. Gaspar, J. Massere, and K. Showalter, Controlling chaos in the Belousov-Zhabotinsky reaction, Nature 361 (1993), no. 6409, 240–243.

24. D. Ruelle and F. Takens, On the nature of turbulence, Communications in Mathematical Physics 20 (1971), 167–192.

25. S. J. Schiff, K. Jerger, D. H. Duong, T. Chang, M. L. Spano, and W. L. Ditto, Controlling chaos in the brain, Nature 370 (1994), no. 6491, 615–620.

26. T. Shinbrot, C. Grebogi, J. A. Yorke, and E. Ott, Using small perturbations to control chaos, Nature 363 (1993), no. 6428, 411–417.

27. J. Singer, Y.-Z. Wang, and H. H. Bau, Controlling a chaotic system, Physical Review Letters 66 (1991), no. 9, 1123–1125.

28. S. Smale, Dynamical systems and turbulence, Turbulence Seminar (Univ. Calif., Berkeley, Calif., 1976/1977), Lecture Notes in Math., vol. 615, Springer, Berlin, 1977, pp. 48–70.

29. N. V. Valishvili, Methods of Computation of Rotational Shells, Mashinostroyeniye, Moscow, 1976.

CITATION

CHAPTER 1

Ramon F. Alvarez-Estrada, Classical and Quantum Models in Non-Equilibrium Statistical Mechanics: Moment Methods and Long-Time Approximations, doi:10.3390/e14020291.

CHAPTER 2

F. Argoul, B. Audit and A. Arneodo (2015). Mechanical Sensing of Living Systems — From Statics to Dynamics, Biosensors - Micro and Nanoscale Applications, Dr. Toonika Rinken (Ed.), ISBN: 978-953-51-2173-2, InTech, DOI: 10.5772/60883.

CHAPTER 3

Nikolai A. Magnitskii (2012). Theory of Elementary Particles Based on Newtonian Mechanics, Theoretical Concepts of Quantum Mechanics, Prof. Mohammad Reza Pahlavani (Ed.), ISBN: 978-953-51-0088-1, InTech, DOI: 10.5772/33548.

CHAPTER 4

N. Umakantha, "A New Approach to Classical Statistical Mechanics," Journal of Modern Physics, Vol. 2 No. 11, 2011, pp. 1235-1241. doi: 10.4236/jmp.2011.211153.

CHAPTER 5

M. Flores, G. Urquiza and J. Rodríguez, "A Fatigue Analysis of a Hydraulic Francis Turbine Runner," World Journal of Mechanics, Vol. 2 No. 1, 2012, pp. 28-34. doi: 10.4236/wjm.2012.21004.

CHAPTER 6

M. Toumi, M. Bouazara and M. Richard, "Development of Analytical Model for Modular Tank Vehicle Carrying Liquid Cargo," World Journal of Mechanics, Vol. 3 No. 2, 2013, pp. 122-138. doi: 10.4236/wjm.2013.32010.

CHAPTER 7

G. Ni, S. Chen and J. Xu, "Discrete Symmetry in Relativistic Quantum Mechanics," Journal of Modern Physics, Vol. 4 No. 5, 2013, pp. 651-675. doi: 10.4236/jmp.2013.45094.

CHAPTER 8

Charles A. Osheku, "Mechanics of Static Slip and Energy Dissipation in Sandwich Structures: Case of Homogeneous Elastic Beams in Transverse Magnetic Fields," ISRN Mechanical Engineering, vol. 2012, Article ID 372019, 23 pages, 2012. doi:10.5402/2012/372019.

CHAPTER 9

J. Awrejcewicz, v. A. Krysko, and i. V. Kravtsova, dynamics and statics of flexible axially symmetric shallow shells, doi 10.1155/Mpe/2006/35672

INDEX

A

Angular momentum 132

C

Conventional approach (CA) 100
Cytoskeleton (CSK) 3

D

Differential equations 131
Differential interference contrast (DIC)
15
Digital holographic microscopy (DHM)
15

E

Extracellular matrix (ECM) 3

F

Fluctuation-dissipation theorem (FDT)
7
Focal adhesion (FA) 3

M

Magnetic twisting cytometry (MTC) 5
Maple software 132

Q

Quantum master equation (QME) 15

S

Soft glassy material (SGM) 5
Sprung mass 118, 119, 120, 121, 122,
123, 125, 126, 127, 128, 130, 131,
132, 133, 134, 135

T

Total internal reflection fluorescence
(TIRF) 15
Traction force microscopy (TFM) 9
Two-dimensional (2D) 7

U

Underlying Assumptions 216